DETONATION
Theory and Experiment

Wildon Fickett
and
William C. Davis

DOVER PUBLICATIONS, INC.
Mineola, New York

Copyright

Copyright © 1979 by Wildon Fickett and William C. Davis
All rights reserved.

Bibliographical Note

This Dover edition, first published in 2000, is a slightly corrected, unabridged republication of the work originally published in 1979 by the University of California Press, Berkeley and Los Angeles, under the title *Detonation*. A new Preface by the authors has been specially prepared for this edition.

Library of Congress Cataloging-in-Publication Data

Fickett, Wildon.
   Detonation : theory and experiment / Wildon Fickett and William C. Davis.
      p. cm.
   Originally published: Berkeley : University of California Press, c1979.
   Includes bibliographical references and index.
   ISBN-13: 978-0-486-41456-0 (pbk.)
   ISBN-10: 0-486-41456-6 (pbk.)
   1. Detonation waves. I. Davis, William Chester, 1925–
II. Title.
QC168.85.D46 F53 2000
662'.27—dc21

00-060377

Manufactured in the United States by LSC Communications
41456608   2017
www.doverpublications.com

To Ella Mae and Maggie

Contents

Preface to the Dover Edition xi
Preface xiii
Acknowledgments xiv

INTRODUCTION 1
1A. History 2
1B. Plan of the Book 11

THE SIMPLE THEORY 13
2A. The Simplest Theory 16
    1. Conservation Laws 16
    2. D-Discussion 21
    3. Piston Problem 21
2B. Application of the Simplest Theory; Product
    Equations of State 24
    1. Equations of State Without Explicit Chemistry 25
    2. Equations of State With Explicit Chemistry 31
    3. Kamlet's Short Method 34
    4. Quasistatic Cycle for Detonations 35
    5. Overview 38
2C. The Zeldovich-von Neumann-Doering Model 42
    1. Example 1: Gas 45
    2. Example 2: Solid 46
Appendix 2A. Formulas for Detonation in a Polytropic Gas . . . 52

EXPERIMENTAL TESTS OF THE SIMPLE THEORY 55
3A. Gases 56
    1. Experiment 59
    2. Discussion 65

CONTENTS

3B.	Solids and Liquids	68
	1. Theory	69
	2. Experiment	71

FLOW IN A REACTIVE MEDIUM ... 75

4A.	The Model	76
	1. Chemical Reactions	79
	2. Equation of State and Rate	88
	3. Equations of Motion	89
	4. Other Forms of the Equations of Motion	91
	5. Steady Solutions	97
	6. The Shock-Change Equation	101
4B.*	Material Properties	102
	1. The Polytropic Gas	103
	2. Binary Mixture of Different Polytropic Gases	105
4C.	Representative Flows	108
	1. Flow Without Reaction	109
	2. Reaction Without Flow	109
	3. Sound Waves in a Reactive Mixture	110
	4. Shock Wave in a Reactive Mixture	110
	5. Rarefaction Wave in a Reactive Mixture	114

Appendix 4A. Chemical Reaction Equations ... 114
Appendix 4B. Temperature from Internal Energy ... 121
Appendix 4C. Equations of Motion for Slab, Cylinder, and Sphere Symmetry ... 124
Appendix 4D. Frozen and Equilibrium Sound Speeds and σ ... 126
Appendix 4E. Shock-Change Equations ... 131

STEADY DETONATION ... 133

5A.	One Reaction, $\sigma > 0$	137
	1. Properties at Fixed Composition	137
	2. Properties at Equilibrium Composition	139
	3. Detonation with One Irreversible Reaction	140
	4. Detonation with One Reversible Reaction	143
	*5. Magnitude of the Effects of Reversibility	147
	*6. A Realistic Example: Hydrogen/Oxygen	151
5B.	Two Irreversible Reactions	153
	1. Both Reactions Exothermic	163
	2. Second Reaction Endothermic (Eigenvalue Detonation)	168
5C.*	One Irreversible Reaction with a Mole Decrement (Pathological Detonation)	174
5D.*	Two Reversible Reactions	176
	1. The λ-plane	177
	2. D-Discussion	180

	3. The Piston Problem	183
	4. Examples	189
5E.*	More Than Two Reactions	190
5F.	Inclusion of Transport Effects	191
5G.	Slightly Divergent Flow	199
	1. The Steady-Flow Equations	204
	2. Approximation for the Radial Derivative	208
	3. A Simple Example—Irreversible Reaction in an Ideal Gas	212
	4. Effect of Chemical Equilibrium (Reversible Reaction)	220
	5. The General Case	224
	6. Applications and Results	226

THE NONSTEADY SOLUTION ... 230

6A.	Stability Theory	233
	1. General Theory	233
	2. The Square-Wave Detonation	238
	3. Results	246
	4. Shock Stability	252
6B.*	Approximate Theories	258
	1. Nonlinear Perturbation Theory	259
	2. Geometrical Acoustics	267
	3. One-Dimensional Oscillation in the Square-Wave Detonation	275
6C.	Finite-Difference Calculations	276
	1. One Dimension	278
	*2. Two Dimensions	288

STRUCTURE OF THE FRONT ... 291

7A.	Overview	292
	1. An Intuitive Picture	293
	2. The Triple Point	295
	3. The Simplest Regular Structure	298
	4. Experimental Methods	301
	5. Calculations	303
7B.	Macroscopic Properties	306
	1. Structures	306
	2. Spacing and Acoustic Coupling	312
	3. The Transverse Wave	323
	4. The Sonic Surface	326
7C.*	Details of Structure	327
	1. Marginal Detonation in a Round Tube (Single Spin)	328
	2. Marginal Detonation in Rectangular Tubes	332
	3. Ordinary Detonation	341

CONTENTS

7D.	Comparison of Theory and Experiment	342
	1. Onset of Instability	343
	2. Fast Gallop	343
	3. Cell Size	346
7E.	Liquids and Solids	350
	1. Differences from Gases	351
	2. Light Confinement	353
	3. Heavy Confinement	356
	4. Discussion	360
Appendix 7A. Interpretation of Smear-Camera Photographs		361
Bibliography		364
Index		383

*The starred sections may be skipped with little loss of continuity.

Preface to the Dover Edition

In the years since *Detonation* appeared in 1979, an enormous amount of work concerned with detonation physics has been published by scientists all over the world. Some of the new work was stimulated by this book. We could imagine going through the book chapter by chapter and detailing how knowledge on the subject has expanded in every aspect. And there would be new chapters with new ideas. Such a review would be very valuable, but it is beyond the capabilities of the authors, and would take at least as many pages as there are in the present book. Perhaps it will be undertaken before too long. For the present, the book is reproduced as it was first published, with the errors we are aware of corrected. We are grateful to Dover Publications, Inc. for making it available again, and hope that it will be a useful reference.

Wildon Fickett and William C. Davis
Los Alamos, November 1999

Preface

In the thirty-five years since Zeldovich, von Neumann, and Doering independently formulated treatments of detonation, many workers have contributed to further understanding of the process. The theoretical work has not been widely studied, partly because it is detailed and difficult, and partly because it has not had direct application in engineering. The experimental work has revealed a complex transverse structure in detonation waves, and has shown that the average pressures and densities are appreciably below those predicted by theory. At present it is not at all clear how to relate theory and experiment.

Our subject is the mechanism of propagation of the fully developed, macroscopically steady, plane detonation wave; the important topics of initiation and failure are excluded. Our aim is a reasonably comprehensive, more tutorial than exhaustive, review of the pertinent theory and experiment. To make the theory more accessible, we have worked out a number of simple cases for illustration. The field is still an open one; we hope that we have brought the gaps, both experimental and theoretical, to the attention of readers interested in filling them.

Our treatment of the theory and our discussion of the experiments are limited to detonation that is steady (independent of time), at least macroscopically, and also limited to analysis of the equations, with little use of the computer for solving them. Another book in the Los Alamos Series in the Basic and Applied Sciences, *Numerical Modeling of Detonations,* by Charles L. Mader (1979), extends to many other problems.

Acknowledgments

Both of us have been at the Los Alamos Scientific Laboratory for many years, and throughout this time the Laboratory administrators have encouraged study and research in subjects like detonation theory, subjects having only long-range relevance to the immediate problems they have to solve. The research environment fostered at Los Alamos made our work possible.

Our supervisors, who are our coworkers in the study of detonation, have actively encouraged and supported us through the several years it took us to prepare this book, and we are grateful to them all. Our colleagues at Los Alamos and throughout the world, many of them personal friends, have generously taken the trouble to furnish photographs for this book, and have also allowed us to use their unpublished results; all are acknowledged where they appear. They have also, over the years, furnished us their ideas and insights, contributing greatly to our understanding of detonation, and we acknowledge our debt to them here, as we have not mentioned it elsewhere.

1

INTRODUCTION

The rapid and violent form of combustion called detonation differs from other forms in that all the important energy transfer is by mass flow in strong compression waves, with negligible contributions from other processes like heat conduction which are so important in flames. The leading part of a detonation front is a strong shock wave propagating into the explosive. This shock heats the material by compressing it, thus triggering chemical reaction, and a balance is attained such that the chemical reaction supports the shock. In this process material is consumed 10^3 to 10^6 times faster than in a flame, making detonation easily distinguishable from other combustion processes. The very rapid energy conversion in explosives is the property that makes them useful. For example, a good solid explosive converts energy at a rate of 10^{10} watts per square centimeter of its detonation front. For perspective, this can be compared with the total electric generating capacity of the United States, about 4×10^{11} watts, or the solar energy intercepted by the earth, about 4×10^{16} watts. A 20-m square detonation wave operates at a power level equal to all the power the earth receives from the sun.

The most easily measured characteristic property of a detonation is the velocity at which the front propagates into the explosive. The front of a detonation wave initiated at one end of a large-diameter stick of explosive is found to approach a nearly plane shape and a constant velocity of propagation. Thus it seems reasonable to assume that a limiting velocity exists, and that in the limit the chemical reaction takes place in a steadily propagating zone in the explosive. A mathematically tractable problem related to physical reality in this limit is that of plane, steady detonation. It has been generally assumed that this problem's solution would describe detonation experiments as the results were extrapolated to infinite size. Therefore, experiments designed for comparison with theory have usually been measurements of a set of long cylinders of different diameters initiated at one end, and extrapolation of the results to infinite diameter. Unfortunately no other problem, not even the seemingly simple problem of a spherically expanding detonation, has been properly treated, so there can be no direct comparison (without extrapolation to infinite size) of experiment with theory.

In the absence of a theory for more complicated flow configurations, it has become customary to treat all detonation problems by assuming that the reaction zone differs inappreciably from its plane, steady limit, and to apply the corresponding theory to experiments directly. This assumption seems reasonable when the velocity has come close to the limiting value, which it does very quickly, and when the radius of curvature of the front is much larger than the reaction zone thickness. This approach has proved satisfactory for many engineering applications, particularly if some of the parameters, such as the equation of state, are chosen to fit measurements made on pieces similar to those being calculated. The situation is similar to most applications of theory in engineering. A mathematically tractable but not perfectly applicable treatment, with slight corrections based on previous experience, is often used. One difference is that in other fields the engineer's intuition is sometimes aided by the existence of exact solutions which, although too complicated for routine use, can be used to determine correction factors. Such exact solutions are not available to explosives engineers.

In addition to experimental results presented for comparison with the one-dimensional theory, we discuss some others in detail because they show a whole group of phenomena about which the theory has nothing to say. Although the detonation front is nearly plane and the velocity nearly constant as assumed in theory, close observation reveals a complex, three-dimensional, time-dependent structure. Discrepancies between the one-dimensional theory and experiment are probably largely attributable to the presence of this structure.

1A. HISTORY

In preparing this short history we have of course gone back and read the early papers, but we have not attempted the careful analysis of the professional historian, and have no doubt failed to give proper credit in some cases.

The phenomenon of detonation was first recognized by Berthelot and Vieille (1881, 1882), and by Mallard and Le Chatelier (1881), during studies of flame propagation. The elements of the simplest one-dimensional theory were formulated around the turn of the century by Chapman (1899) and by Jouguet (1905, 1917). Becker (1922) gives a good summary of the state of the subject in its early days.

Because we refer to the results of the Chapman-Jouguet theory throughout the following discussion, we digress now for a definition of terms and a short description of the results. (A more complete discussion is given in Chapter 2.) The entire flow is assumed to be one-dimensional and the front is treated as a discontinuity plane across which the conservation laws for shock waves apply, with the equation of state depending appropriately on the degree of chemical reaction. The chemical reaction is regarded as proceeding to completion within a short distance relative to the size of the charge so that it is, in effect, instantaneous. The state of the explosive products at the point of complete reaction is called the *final state*. As in the theory of shock waves, once the detonation velocity (the propagation velocity of the front) is specified, application of the laws of conservation of mass, momentum, and energy determines the final state behind the front (given the equation of state of the reaction products). The presence of the exothermic chemical reaction does, however, introduce a new and important property. In the nonreactive case, shocks of all strengths are possible, and in the limiting case of a zero-strength shock, the shock velocity is equal to the sound velocity. With an exothermic reaction, the conservation laws have no solution for shocks below a certain minimum velocity. For typical heats of reaction the minimum velocity is well above sound velocity.

Given the detonation velocity D as a parameter, the conservation laws have no final-state solution for D less than this minimum value, one solution for D equal to the minimum value, and two solutions for all larger values of D. The solution for the minimum value of D is called the Chapman-Jouguet or CJ state. For larger values of D, one solution has pressure greater than CJ pressure and is called the *strong* solution; the other has pressure lower than the CJ pressure and is called the *weak* solution. In the coordinate frame moving with the front, the flow is subsonic at the strong point and supersonic at the weak point. Therefore a detonation whose final state is at the strong point may be overtaken by a following rarefaction and reduced in strength, whereas one whose final state is at the weak point runs away from a following rarefaction, leaving an ever-growing, constant-state region between the end of the reaction zone and the head of the rarefaction wave.

The rear boundary condition determines the detonation velocity actually realized. With the usual rear boundary conditions, such as a rigid wall or no confinement at all at the initiating plane, a rarefaction wave must follow the detonation front to reduce the forward material velocity at the final-state point to that required at the rear boundary. The strong solution must be rejected, then, because of its vulnerability to degradation by this following rarefaction wave. The weak solution is also rejected on rather intuitive and arbitrary grounds. The only solution left is that defined by the so-called Chapman-Jouguet hypothesis: that the steady detonation velocity is the minimum velocity consistent with the conservation conditions. At this velocity the flow immediately behind the front is sonic in a coordinate system attached to the front, so that the head of the rarefaction wave moves at precisely the speed of the front.

This simple theory was apparently an immediate success. The earliest workers were able to predict detonation velocities in gases to within one or two percent, even with the crude values of thermodynamic functions available then. Had they been able to measure density or pressure, they would have found appreciable deviations. Even so, the simple theory works remarkably well.

The first indication that real detonations might be more complex than is postulated by the simple theory came with discovery of the spin phenomenon by Campbell and Woodhead (1927). Their smear-camera photographs of certain detonating mixtures showed an undulating front with striations behind it. The most likely explanation of this front is that it contains a region of higher than average temperature and luminosity which rotates around the axis of the tube as the detonation advances. This spin phenomenon was observed in systems near the detonation limit, where the available energy and the rate of reaction are barely sufficient to maintain propagation in the chosen tube diameter. We use the term "marginal" to describe such a system, with no sharp dividing line between marginal and non-marginal systems. With the techniques then available, the appearance of spin appeared to be confined to such marginal systems. Jost (1946) gives an account of the early work in this field.

A fundamental advance that removed the special postulates and plausibility arguments of the simplest theory was made independently by Zeldovich (1940) in Russia, von Neumann (1942) in the United States, and Doering (1943) in Germany, and their treatment has come to be called the ZND model of detonation. It is based on the Euler equations of hydrodynamics, that is, the inviscid flow equations in which transport effects and dissipative processes other than the chemical reaction are neglected. The flow is assumed to be one-dimensional, and the shock is treated as a discontinuity, but now as one in which no chemical reaction occurs. The reaction is assumed to be triggered by the passage of the shock in the material and to proceed at a finite rate thereafter. It is represented by a single forward rate process that proceeds to completion. In the coordinate system attached to the shock, the flow equations have a solution that is steady throughout the zone of chemical reaction. The state at any point inside this zone is related to that of the unreacted material ahead by the same conservation laws that apply to a jump discontinuity, with the equation of state evaluated for the degree of reaction at the point. The state at the end of the reaction zone is, of course, included and thus obeys exactly the same conservation laws as apply to instantaneous reaction, so the reasons for applying the CJ hypothesis are unchanged. The CJ detonation velocity is then independent of the form of the reaction rate law, as in the simple theory, and can be calculated from the algebraic conservation laws and the complete-reaction equation of state. Fortunately, conditions in gas detonations (pressure under 100 atm, temperature under 7000 K) are such that the products are accurately described by the ideal gas equation of state with very small corrections. The standard tables of thermodynamic functions are applicable. Thus the equation of state is known, and a priori calculations can be made exactly.

The work of Berets, Greene, and Kistiakowsky (1950) ushered in an era of extensive comparison of the one-dimensional theory with experimental measurements on gaseous systems. They repeated both the detonation velocity measurements and the calculations made earlier by Lewis and Friauf (1930) on a number of hydrogen-oxygen mixtures, using more recent thermodynamic data. Away from the detonation limits the agreement was good, the calculated velocities being about one percent higher than the experimental ones. In the next few years there were significant improvements in both computational and experimental techniques. Electronic calculating machines permitted extensive and precise calculations including a determination of the chemical equilibrium composition, a very laborious procedure when done by hand. Duff, Knight, and Rink (1958) measured the density by x-ray absorption; Edwards, Jones and Price (1963) measured the pressure using piezoelectric gauges; and Fay and Opel (1958) and Edwards, Jones, and Price (1963) measured the Mach number of the flow by schlieren photography of the Mach lines emanating from disturbances at the tube walls. Careful measurements of the detonation velocity were made by Peek and Thrap (1957) and by Brochet, Manson, Rouze, and Struck (1963). White (1961) made a very comprehensive study of the hydrogen-oxygen system with carbon monoxide and other diluents. In addition to measuring pressure and detonation velocity, he made extensive use of spark interferograms to measure density changes, and visible light photometry to measure a composition product and to detect temperature rises in compression and shock waves generated behind the detonation.

The results of all this work suggested that the effective state point at the end of the reaction zone is in the vicinity of a weak solution of the conservation equations. The measured pressures and densities are ten to fifteen percent below the calculated CJ values, the flow is supersonic with Mach number about ten percent above the calculated CJ values, and the detonation velocities are one half to one percent above, in approximate agreement with the conservation laws. The departures from calculated values became less puzzling when detailed concurrent studies showed that the front was not smooth, but contained a complicated time-dependent fine structure. The measured state values thus represent some sort of average, inherent in the design of the experiment and the apparatus.

Experiments in condensed explosives are much more difficult and costly, and fewer state variables are amenable to measurement. More serious is the lack of knowledge of the product equation of state, precluding a priori calculation with any useful degree of accuracy. (The many published determinations of the constants in approximate theoretical forms amount to little more than complicated exercises in empirical curve fitting.)

Fortunately an indirect but exact method of testing the CJ hypothesis comes to the rescue. It requires no knowledge of the equation of state other than the assumption that it exists. Following earlier work by Jones (1949), Stanyukovich (1955) and Manson (1958) pointed out that, using only very general assumptions, essentially those of the simple ZND theory, with no knowledge of the equation of state, the derivatives $dD/d\rho_0$ and dD/dE_0 determine the detonation pressure.[1] Wood and Fickett (1963) turned the argument around by suggesting that the applicability of the theory could be tested by comparing the pressure calculated from the measured derivatives with that found by the more conventional method of measuring the velocity imparted to a metal plate driven by the explosive. Davis, Craig, and Ramsay (1965) tested nitromethane and liquid and solid TNT and showed that, well outside of the experimental error, the theory does not describe these explosives properly. Unfortunately this indirect method of testing the theory does not indicate the actual state at the end of the reaction zone.

While experiment and theory were being extensively compared, the theory of the one-dimensional, steady reaction zone was extended to include other effects, beginning with the work of Kirkwood and Wood

[1] D = detonation velocity, ρ_0 = initial density, and E_0 = initial internal energy.

(1954). They considered a fluid that obeyed the Euler equations, that is, one in which transport effects are neglected, but in which an arbitrary number of chemical reactions may occur, each proceeding both forward and backward so that a state of chemical equilibrium can be attained. Others considered a fluid described by the Navier-Stokes equations, that is, one in which heat conduction, diffusion, and viscosity are allowed but which has only a single forward chemical reaction. The work of Kirkwood and Wood was extended to include slightly two-dimensional flow caused by boundary layer or edge effects treated in the quasi-one-dimensional (nozzle) approximation.

All of these problems exhibit mathematical complexities not found in the earlier theoretical treatments. In each case, the flow equations are reduced, under the steady state assumption, to a set of autonomous, first-order, ordinary, differential equations with the detonation velocity as a parameter. Without drastic and unwarranted assumptions to simplify the equations, analytical solutions cannot be obtained, and the properties of a solution must be obtained indirectly by studying the topology of the phase plane, with particular attention to the critical points. The qualitative behavior of the integral curve (a solution of the set of differential equations) starting at the initial point (the state immediately behind the shock) must be ascertained for each value of the detonation velocity. The equations have some mathematical similarity to those encountered in the nonlinear mechanics of discrete systems with more than one degree of freedom. An additional complication introduced by the possibility of chemical equilibrium is the appearance of thermodynamic derivatives at both fixed and equilibrium compositions, the most important being the so-called frozen and equilibrium sound speeds. Although this complicates the analysis considerably, the effect on the results is relatively small, particularly for condensed explosives.

The importance of the more elaborate models of steady one-dimensional detonation is that they yield weak solutions. As mentioned above, the laws of conservation of mass, momentum, and energy have no solution for detonation velocities below the CJ value, a unique solution at the CJ value, and two solutions, one weak and one strong, for any velocity above the CJ value. In the simple case of a gas with a single forward reaction in which the number of moles does not decrease (and in which transport effects are neglected), the weak point cannot be reached. The solution always terminates at a strong point if the forward velocity of the rear boundary is large enough, or at the CJ point, which is the lower limit of the set of final states. This is the case treated in the ZND model. The extended theory shows that almost any increase in complexity of the fluid system opens up the possibility of reaching the weak point. In fact, von Neumann (1942) in his first paper on detonation discussed such a case in which the system, with one forward reaction, is such that the number of moles of gas decreases during the reaction. He showed that the state at the end of the steady zone may be a weak point, thus violating the CJ condition. In the language of the extended theory, solutions reaching a strong point encounter only the terminal critical nodal point, which means that this type of termination exists for a continuous range of detonation velocities. Solutions reaching a weak point must first pass through a critical saddle point, so a weak detonation has a unique detonation velocity whose value depends on the properties of the material, including the reaction rate. This situation is described, in the usual language, by the statement that the detonation velocity is an eigenvalue of the set of differential equations.

An important physical insight arising from the work on the extended theory is clarification of the precise way in which the chemical reaction affects the flow. For a nonreactive system, the Euler equations are homogeneous. With properly chosen independent variables the corresponding equations for a reactive system are the same, except for a single source term that makes the system of equations inhomogeneous. For a single reaction this term may be written as σr, where r is the chemical reaction rate and σ is an effective energy release or thermicity coefficient. (For a many-reaction case this product is replaced by a sum of such terms, one for each reaction.) A positive value of σr signifies, roughly, a transfer of energy from chemical bonds to the flow, and a negative value signifies the reverse. The coefficient σ is the sum of two terms, one involving the enthalpy change in the reaction, and the other the volume change. An increase in molar volume (equivalent to an increased number of moles in a gas system) caused by the reaction has an effect on the flow equivalent to some positive heat-release value.

When transport effects are neglected, an oversimplified statement of the results is that attainment of the weak final state requires that the sign of σr be negative in some part of the reaction zone. This condition can be achieved in various ways, such as having a single reaction with positive heat release but negative volume change, or two reactions, one exothermic and one endothermic, or even two exothermic reactions with disparate rates so that one is driven beyond its equilibrium point by the flow in the early part of the reaction zone, with σr thus becoming negative as the composition returns to the equilibrium state. In the quasi-one-dimensional case, the effect of the lateral flow divergence away from the axis enters as an additive term to σr with the sign always the same as that for an endothermic reaction.

When transport effects are included, the weak solution can also be obtained simply as a result of the transport effects, but only by choosing an extremely fast and probably non-physical reaction rate.

The extended theory thus offers several possibilities for reaching weak detonation states like those observed in gases. But this should probably not be taken too seriously, because the one-dimensional theory can apply, at best, only in an average sense because the observed flow is significantly three-dimensional.

Inadequate though they may be for direct application to real detonation, the one-dimensional solutions are important in theoretical development because they are the necessary base for attack on the much more complicated problem of detonation treated in three dimensions. This step is the study of the hydrodynamic stability of the one-dimensional solutions to infinitesimal three-dimensional disturbances. Development of stability theory began at about the time that close observation of the wave front in many detonating systems first revealed that the front is typically not plane but finely wrinkled by small transverse waves traveling back and forth across it.

The first theoretical work on this subject was done by Shchelkin (1959) who offered qualitative arguments based on the "square-wave" model of the reaction zone, in which the entire reaction is supposed to take place instantaneously at a finite distance behind the shock front determined by the induction time. Others continued this work somewhat more quantitatively. The rigorous treatment of the hydrodynamic stability of the general one-dimensional solution of the Eulerian flow equations to infinitesimal three-dimensional perturbations was done by Erpenbeck (1962a). The method is basically the usual one of linearizing the three-dimensional time-dependent equations about the one-dimensional steady solution and inquiring about the growth or decay of solutions of these linearized equations. These solutions are infinitesimal perturbations of the one-dimensional steady solution of the complete equations. Erpenbeck and others obtained a number of results on various aspects of this problem. Strictly speaking, the theory applies only to overdriven (strong) detonations, and specific conclusions have been obtained only for simple ideal gas models of explosive. However, the results suggest that most detonations are unstable to disturbances in some range of transverse wavelength and can thus be expected to have a wrinkled, non-steady front. Although not, strictly speaking, a stability analysis, the theoretical study by Barthel and Strehlow (1966) of the transverse motion of sound waves in the reaction zone provided insight into the amplification mechanism responsible for the instability.

INTRODUCTION   Chap. 1

   Detailed time-dependent numerical finite-difference calculations in
two or three space dimensions of even the simplest flows of this type
have not been performed, although Mader (1967b) reported a
preliminary effort in this direction. However, one special result obtained
by Erpenbeck (1964b) makes a one-dimensional time-dependent
calculation interesting. This result is that, for a reaction rate having a
strong but physically reasonable temperature dependence, cases can be
found for which the steady solution is unstable to *one-dimensional* disturbances. The numerical calculations of Mader (1965) and of Fickett
and Wood (1966) motivated by this result show a detonation pulsating
with large amplitude. Erpenbeck later obtained essentially the same
results from an approximate nonlinear analysis of the flow equations.
   Experimental study of the non-one-dimensional structure of the
detonation front really begins with the early discovery of spin. For a long
time this phenomenon was thought to be isolated, occurring only in
marginal systems. However, careful observations in the late 1950s and
early 1960s revealed that detonation fronts typically contain complicated three-dimensional structure. It was soon realized that spin was
only a special case of a much more extensive phenomenon of transverse
wave motion on the amplitude. Proper adjustment of the initial conditions produces "double-headed" spin, and so on to higher modes. Non-
marginal systems, once thought to be one-dimensional, exhibit many
relatively small transverse waves. So-called "galloping" detonations in
which the principal time-dependence is one-dimensional like the one-
dimensional instability described above have also been observed.
   Front structure in classical or single-headed spin has been quite
thoroughly elucidated by the independent experimental work of Schott
(1965b) and of Voitsekhovskii, Mitrofanov, and Topchian (1963), reported in their book. It is reminiscent of the three-shock configuration
produced when a plane shock wave runs over a wedge of sufficiently
large angle, in which some gas passes through a single shock, the Mach
stem, and the rest is double shocked by an incident and reflected shock
like those occurring in regular shock reflection. In the spinning detonation, a shock extends from a radius of the non-planar front back into the
flow behind the front, and the triple-shock intersection consists of this
shock and the two parts of the front intersecting along this radius. The
entire configuration rotates about the tube axis. Some of the gas passes
through two shocks and some through only one, so that the reaction
products are two interleaved helices separated by slip surfaces.
   Study of the detonation front with many transverse waves begins with
the work of Denisov and Troshin (1959) and of Duff (1961), and continues to the present time with many workers involved. The simple and

powerful technique of soot inscription is extensively used. The walls and end plate of a tube are lined with a soot-coated film. As the detonation passes over or collides with these surfaces, the shock intersections leave their tracks by removing part of the coating. The result provides an overall view of the structure as well as much fine detail. High-speed photography is also important.

A qualitative detonation picture emerges which has a front with a time-dependent cellular structure followed by a three-dimensional, possibly turbulent, flow. The cell boundaries consist of three-shock configurations in transverse motion across the front. In gaseous systems, the cell size is proportional to the reaction time, so it can be increased by lowering the initial pressure or increasing the amount of inert diluent. Less information is available for condensed explosives, but roughly the same structure seems to be present, with edge effects playing an important part.

1B. PLAN OF THE BOOK

As stated in the preface, we limit our subject to the mechanism of propagation of (macroscopically) plane, steady detonation, with only a few passing references to the important omitted topics of initiation, failure, and the small time-dependence in the slow approach to the steady state. Applications, such as rock blasting, are not even mentioned. Recent books which cover some of our topics are those by Strehlow (1968a), Dremin et al. (1970), Gruschka and Wecken (1971), and Johansson and Persson (1970). Aguilar-Bartolome (1972) gives a very detailed introduction and review in Spanish.

Our goal is a fairly comprehensive, but not necessarily exhaustive, tutorial review. To this end we sometimes cite only the most pertinent references, or those which have been most helpful to us. A nearly complete bibliography is, however, contained in the reference lists of the sources we cite.

The reader needs some familiarity with the elements of shock hydrodynamics and supersonic flow, and college mathematics through differential equations. Basic to Chapter 5 are the phase plane, paths in this plane, and the nature of the critical points of the differential equations; we encounter only nodes and saddles.

Each chapter and major section have an introduction, and in some cases a summary at the end. Reading just these will give a brief overview.

Chapter 2 reviews the simple theory of steady, plane detonation, developed between 1880 and 1950, and Chapter 3 compares the simple theory with experiment. The reactive flow equations and some of their more general properties appear in Chapter 4, in preparation for Chapter 5, which covers the variety of steady solutions developed since 1950. Chapter 6 describes the work on hydrodynamic stability of the steady plane solutions, and related calculations of finite-amplitude perturbations. The results agree with experiment in suggesting that steady plane detonations do not exist in nature because they are unstable. Chapter 7 is devoted to experimental observations of the finite structure of the front.

2

THE SIMPLE THEORY

The relatively simple description of steady plane detonation called the ZND model is well-known. This brief review emphasizes features and points of view appropriate to understanding the more detailed theories presented in later chapters. The ideas were developed independently by Zeldovich (1940), von Neumann (1942), and Doering (1943). Evans and Ablow (1961) explain clearly the elements of the simple theory, with a good selection of the applicable portions of compressible flow theory. Books by Taylor (1952) and Zeldovich and Kompaneets (1960) also contain much of the same material.

The ZND description neglects transport processes, and assumes one-dimensional flow. We restrict the term ZND to the case of a single irreversible (forward only) reaction of positive thermicity (conversion of chemical bond energy to macroscopic translational energy); more complicated cases are discussed in Chapter 5. The configuration of the wave is shown in Fig. 2.1. The unreacted explosive ahead of the wave is in metastable equilibrium, with zero reaction rate. The shock at the head of the wave is a jump discontinuity in which no reaction takes place. It heats the material and triggers the chemical reaction. Following the shock is the *reaction zone*. The reaction proceeds in the reaction zone and is complete in the *final state*. The flow in the reaction zone is steady in the coordinate frame attached to the shock; the shock and the reaction zone propagate together at the (constant) *detonation velocity* D. The *following flow* between the final state and the rear boundary is a time-dependent rarefaction wave followed by a constant state in the *unsupported* case, Fig. 2.1a, and a constant state only in the *overdriven* case, Fig. 2.1b.

The velocity and configuration of the complete wave are uniquely determined by the *rear boundary condition*. Throughout, we take for

THE SIMPLE THEORY

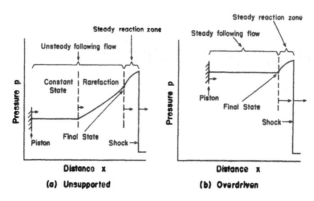

Fig. 2.1. Steady detonations.

this the constant velocity of a hypothetical piston following the wave. The problem is to determine, for a given piston velocity u_p, the detonation velocity D and the complete flow, i.e., both the steady reaction zone and the following flow. Now the following flow is determined by two boundary conditions: the specified piston velocity at the left, and the final state of the reaction zone at the right. But the final state is not known until the reaction zone is determined. The solution procedure is therefore divided into two parts, which we call the D-*discussion* and the *piston problem*. We will follow this same procedure in discussing more complicated steady solutions in Chapter 5.

The D-*discussion* consists of finding all possible steady solutions for the reaction zone, treating D as a parameter. Fixing D determines a unique steady solution. Its all-important final state is the right boundary condition for the following flow. The result of the D-discussion is that steady solutions exist for all values of D above a certain minimum value. This minimum value is the velocity of the *unsupported* wave, Fig. 2.1a. For all higher values of D, we have an *overdriven* wave, Fig. 2.1b. We now have the information we need to proceed to the second step, namely, the dependence of the final-state particle velocity on D. This turns out to be a monotone increasing function u(D).

The *piston problem* consists of finding the single-valued function $D(u_p)$, the detonation velocity as a function of the given piston velocity, and determining the following flow connecting the final state of the reaction zone with the piston. If the piston velocity is greater than the final-state particle velocity for the minimum D, then we have an overdriven detonation. The value of D is that for which the final-state particle velocity is equal to the piston velocity, and the following flow is just a constant state, the same as the final state. If the piston velocity is less than the minimum final-state particle velocity we have an unsupported detonation. The following flow now consists of a time-dependent rarefaction wave which reduces the particle velocity from that at the final state to the piston velocity, followed by a constant state extending back to the piston. (Some authors use the term "unsupported" to mean a free rear boundary, i.e. $p = 0$. Note that in this case the rarefaction terminates in a region of cavitation and the constant state is absent.)

The *simplest theory*, Sec. 2A, is the limit of instantaneous reaction, with the shock and reaction zone collapsed into a single jump discontinuity in which the reaction is complete. The initiation transients also vanish; the detonation runs at constant velocity from the moment of initiation, generating a relatively simple self-similar flow between it and the piston. The final state, now immediately behind the reactive shock, satisfies the usual conservation relations. The Hugoniot curve, though, is different from the usual one for unreactive flow; it is displaced outward from the origin in p-v space by the heat-of-reaction term. It doesn't pass through the initial state point, and there is a minimum value of D which satisfies the conservation relations. For a piston velocity less than the Hugoniot particle velocity for a minimum D, the detonation is *unsupported*; the front propagates at the minimum D and is followed by a centered rarefaction wave, the *Taylor wave*. A very important result is that the detonation Hugoniot curve, and the minimum-D state on it, are determined by the equation of state of the reaction products alone, and the equation of state of the unreacted explosive is not needed.

Practical applications, Sec. 2B, rarely go beyond the simplest theory. Most of the section is devoted to the main problem (for solids and liquids) of determining the products equation of state. All equations of state, because good theory is lacking, require fitting to experimental detonation measurements. They fall naturally into two groups: those in which chemistry does not appear explicitly and whose adjustable parameters relate energy to pressure and volume directly, and those in which chemistry appears and whose parameters are molecular properties, with energy, pressure, and volume obtained from calculation of the equilibrium composition. The first class of course requires far less computation.

ZND theory, Sec. 2C, has finite reaction rate, and some examples of steady (reaction zone) solutions are displayed. A most important point is that the properties of the final state, and thus the D-discussion and the piston problem, are *identical* to those of the simplest theory with its infinite reaction rate and unresolved reaction zone. The final state and the detonation velocity depend only on the equation of state of the reaction products. The reaction-rate law and the equation of state of the partially reacted material affect only the interior of the reaction zone in the steady flow considered here. The initiation transients, which they do affect, are not discussed.

2A. THE SIMPLEST THEORY

The simplest theory assumes the following:

1. The flow is one-dimensional (laminar).

2. The plane detonation front is a jump discontinuity, a shock in which the chemical reaction is assumed to be completed instantaneously. The material emerging from the discontinuity is assumed to be in thermochemical equilibrium, and is thus described by a thermodynamic equation of state.

3. The jump discontinuity is steady (independent of time), so that the state of the material emerging from the front is independent of time. The flow following this point may be time-dependent.

2A1. Conservation Laws

For a detonation wave propagating with velocity D into an explosive initially at rest, and giving the explosive products a velocity u at a point immediately behind the front (the final state), the conservation of mass requires that

$$\rho_0 D = \rho(D - u), \tag{2.1}$$

where ρ_0 is the density of the unreacted explosive (the initial state) and ρ is the density in the final state. Notice the use of the term "final state;" it is the final state of the *steady* part of the flow, and the conservation equations as written apply only to steady flow. The conservation of momentum is expressed by

$$p - p_0 = \rho_0 u D, \tag{2.2}$$

where p_0 and p are the pressures at the initial and final states. When u is eliminated from these two equations, the result defines a line in the p-v plane called the "Rayleigh line" and expressed by

Sec. 2A   THE SIMPLEST THEORY

$$\mathcal{R} = \rho_0^2 D^2 - (p - p_0)/(v_0 - v) = 0, \tag{2.3}$$

where $v = \rho^{-1}$ is the specific volume. Note that density and specific volume are often both used in an expression partly to avoid reciprocals, but mainly because condensed explosives are characterized by initial density, while the p-v plane is most convenient for diagrams of the detonation process. Examples of Rayleigh lines for some arbitrarily chosen values of D are shown in Fig. 2.2. A Rayleigh line passes through the point (p_0, v_0) and has slope $-\rho_0^2 D^2$. The limiting cases are the horizontal, $D = 0$, and the vertical, $D = \infty$. The vertical Rayleigh line, corresponding to an infinite propagation velocity, represents the limiting case of a constant-volume detonation in which all the material reacts at the same instant of time.

If D rather than u is eliminated from Eqs. (2.1) and (2.2), the result is

$$u^2 = (p - p_0)(v_0 - v). \tag{2.4}$$

Thus the curves of constant particle velocity in the p-v plane are hyperbolas, as shown in Fig. 2.3. Given u and D, the final state (p,v) is the intersection of the corresponding particle-velocity hyperbola and Rayleigh line.

The conservation of energy condition is

$$E(p, v, \lambda = 1) + pv + 1/2 (D - u)^2$$
$$= E(p_0, v_0, \lambda = 0) + p_0 v_0 + 1/2 D^2, \tag{2.5}$$

where E is the specific internal energy and λ specifies the degree of chemical reaction, changing from 0 for no reaction to 1 for complete reaction. The extent of transformation of chemical bond energy to heat is thus roughly proportional to λ. The equation for the *Hugoniot curve* in the p-v plane is obtained by eliminating u and D from Eq. (2.5) by using Eqs. (2.1) and (2.2):

$$\mathcal{H} = E(p, v, \lambda = 1) - E(p_0, v_0, \lambda = 0)$$
$$- 1/2 (p + p_0)(v_0 - v) = 0. \tag{2.6}$$

It is determined once the equation of state $E(p, v, \lambda)$ is given. The two equations (2.3) and (2.6) then determine the state (p,v) for a given detonation velocity, as the intersection of the Rayleigh line with the Hugoniot curve.

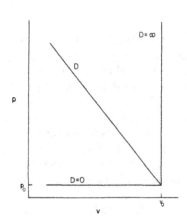
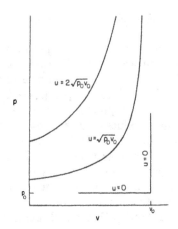

Fig. 2.2. Curves of constant detonation velocity, called Rayleigh lines.

Fig. 2.3. Curves of constant particle velocity.

For a polytropic gas (ideal gas with constant heat capacity) with reaction A → B having constant heat of complete reaction q (= $-\Delta H$) the equation of state is

$$pv = RT \tag{2.7}$$

$$E = C_v T - \lambda q = pv/(\gamma - 1) - \lambda q, \tag{2.8}$$

where C_v is the constant-volume heat capacity. The Hugoniot curve is

$$\left(\frac{p}{p_0} + \mu^2\right)\left(\frac{v}{v_0} - \mu^2\right) = 1 - \mu^4 + 2\mu^2 \frac{q}{p_0 v_0}, \tag{2.9}$$

where $\mu^2 = (\gamma - 1)/(\gamma + 1)$. This is the equation of a rectangular hyperbola in the p/p_0 vs v/v_0 plane, centered at the point $v/v_0 = \mu^2$, $p/p_0 = -\mu^2$, with the minimum distance from this center to the curve a linear function of q. This Hugoniot curve, Fig. 2.4, is the locus of all possible final states (the state as the material emerges from the discontinuity) for any detonation.

The conservation conditions require that the final state point in the p-v plane lie on both this Hugoniot curve and the Rayleigh line, Eq. (2.3). The expression for the Rayleigh line contains the detonation velocity as a parameter, so the whole family of possible Rayleigh lines must be considered. Three of them are also shown in Fig. (2.4), each

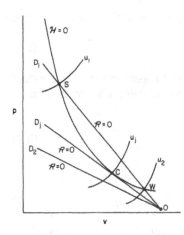

Fig. 2.4. Diagram of the Hugoniot curve, Rayleigh lines, and curves of constant u, for a detonation with instantaneous reaction.

labeled with the value of the detonation velocity. For a sufficiently small value of D, say D_2, the Rayleigh line and the Hugoniot curve have no intersection, so there is no solution that satisfies the assumptions. For a large value, say D_1, there will be two solutions which are marked S and W (for strong and weak) in the figure. For a particular value of D, called the Chapman-Jouguet velocity D_j, there will be a unique solution to the two conditions. It can be shown that the flow at the upper or strong point is subsonic with respect to the front ($u + c > D$), and a disturbance arising behind the front will overtake it. At the lower or weak point the flow is supersonic with respect to the front ($u + c < D$), and a disturbance behind the front will fall farther behind. At the tangent point, called the CJ point, the flow is sonic ($u + c = D$).

Before proceeding, we summarize the properties of the CJ point, some of which are not used until later. The CJ condition is that the Rayleigh line and the Hugoniot curve be tangent, so their slopes are equal at the CJ point. From Eq. (2.3) for the Rayleigh line, we find

$$(dp/dv)_R = - (p - p_0)/ (v_0 - v). \tag{2.10}$$

From Eq. (2.6) for the Hugoniot curve

$$(dp/dv)_{JC} = 2(dE/dv)_{JC}/(v_0 - v) + (p + p_0)/(v_0 - v). \tag{2.11}$$

The CJ condition is that the right hand sides of Eqs. (2.10) and (2.11) are equal, and after some manipulation, we find that

$$(dE/dv)_{JC} = - p. \tag{2.12}$$

However, we also know from $dE = TdS - pdv$ that on an isentrope

$$(\partial E/\partial v)_s = -p, \tag{2.13}$$

and this implies that an isentrope is also tangent to both the Rayleigh line and the Hugoniot curve at the CJ point. Using this result with Eq. (2.10) we obtain

$$-(\partial p/\partial v)_s = (p - p_0)/(v_0 - v), \tag{2.14}$$

or

$$\gamma \equiv -(\partial \ln p/\partial \ln v)_s = (1 - p_0/p)/(v_0/v - 1), \tag{2.15}$$

where γ is defined as the negative logarithmic slope of the isentrope. Elimination of v between (2.15) and the Rayleigh relation (2.3) gives

$$p = \rho_0 D^2/(\gamma + 1 - p_0/p). \tag{2.16}$$

If p_0 can be neglected, as is usually the case in condensed explosives, we have at the CJ point

$$v/v_0 = \gamma/(\gamma + 1), \tag{2.17}$$

$$p = \rho_0 D^2/(\gamma + 1), \tag{2.18}$$

$$u = D/(\gamma + 1), \tag{2.19}$$

and

$$c = D\gamma/(\gamma + 1), \tag{2.20}$$

the last two being obtained by further application of the conservation conditions. These CJ relations (2.10)-(2.20) are not limited to the final Hugoniot curve for complete reaction, but apply equally well to any fixed composition. Equation (2.11) could be obtained by differentiating Eq. (2.6) with λ equal to any arbitrary fixed value instead of $\lambda = 1$. Then the p-v point determined for a particular value of λ is the point where a straight line through (p_0, v_0) and an isentrope for that value of λ are both tangent to the Hugoniot curve. The locus of these points for all values of λ is called the locus of tangents or sonic locus, and is important to the arguments presented in Chapter 5.

2A2. D-Discussion

The D-discussion is concerned with only the steady part of the flow — the steady reaction zone terminating in the "final state." The simplest theory is degenerate in that the reaction is regarded as taking place instantaneously within the shock transition. We are left with just the final state itself, a point on the complete-reaction Hugoniot curve. This state is given by the intersection of the Rayleigh line and the Hugoniot curve, Fig. 2.4. There are three cases to be distinguished:

1. $D < D_j$. There is no steady solution, since the Rayleigh line does not intersect the Hugoniot curve.

2. $D = D_j$. There is one steady solution, the CJ point at which the Rayleigh line is tangent to the Hugoniot curve. The flow at this point is sonic.

3. $D > D_j$. There are two possible solutions, the upper or strong intersection, and the lower or weak intersection. We reject the lower intersection by assumption as inconsistent with the postulated mechanism, leaving only the upper point at which the flow is subsonic.

2A3. Piston Problem

Consider a detonation in a rigid tube, Fig. 2.5, followed by a piston controlled by external forces so that its velocity may be specified. To solve the "piston problem" is to find the complete flow field between the front and a piston of specified constant velocity. We must first find a value of D giving a suitable steady solution, and then find a suitable (in general time-dependent) flow connecting its final state with the piston. In doing this, we will of course have occasion to refer to the constant-u hyperbolas of Fig. 2.4. Again we consider three cases, but reverse the order of the D-discussion.

1. $u_p > u_j$. This case has a very simple solution: a uniform state between the front and the piston, Fig. 2.5a, given by the intersection of the constant-u curve for $u = u_p$ with the Hugoniot curve, Fig. 2.4. This intersection is a strong point (above the CJ point). Since the flow is subsonic, this solution is vulnerable to degradation by rarefaction waves from the rear. With this flow established, a slight decrease in piston velocity will generate a forward-moving rarefaction, Fig. 2.5b, which will overtake the front and eventually produce the uniform steady solution corresponding to the new piston velocity.

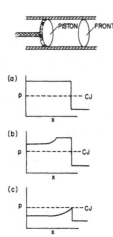

Fig. 2.5. The piston problem.

2. $u_p = u_j$. We still have a uniform state, the CJ state. The flow is sonic so a rarefaction generated at the piston cannot overtake the front.

3. $u_p < u_j$. The front can move no slower than D_j, so we still have the CJ state at the front, but now need a rarefaction wave, Fig. 2.5c, to reduce the velocity at the front to that at the piston. The solution consists of a rarefaction wave followed by a uniform state. The head of the rarefaction remains at the front, since the flow is sonic there. The rarefaction is time-dependent, expanding to fill the space between the front and the uniform region. For zero piston velocity (a rigid rear boundary), the tail of the rarefaction (the slope discontinuity in Fig. 2.5c) is about halfway between the front and the piston (exactly halfway for a polytropic gas, Eq. [2.8], with neglect of p_0).

In the typical application, the piston velocity is less than or equal to zero, so this case ($u_p < u_j$) is the one of greatest interest. In gases we usually have the rigid rear boundary of a closed tube, so $u_p = 0$. In liquids and solids the high pressure pushes the rear confinement (if any) away so $u_p < 0$. This flow configuration is so important that we describe it in some detail.

Going back to the point of initiation $x = t = 0$, we assume that (since the reaction is instantaneous) the front moves with the constant CJ detonation velocity at all times. The flow is then completely determined by the boundary conditions consisting of the CJ state along the line $x = D_j t$ on the right and the prescribed constant piston velocity along the line $x = u_p t$ at the left of the x-t diagram of the flow, shown in Fig. 2.6.

Sec. 2A THE SIMPLEST THEORY

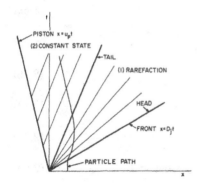

Fig. 2.6. The complete flow in a detonation as given by the simplest (instantaneous reaction) theory.

The fan-shaped region 1 is a centered rarefaction wave (often called the Taylor wave), and region 2 is a constant state.

The light lines are the forward characteristics (see Sec. 4A). Within the rarefaction fan, they emanate from the origin. The state is constant along each forward characteristic, so the flow is self-similar; that is, the state is a function of the single variable x/t only. The entropy is constant throughout at the CJ value. The equations describing the rarefaction wave are

$$p = p(\rho) \tag{2.21}$$

$$c = c(p) = (dp/d\rho)^{1/2} \tag{2.22}$$

$$dp/du = \rho c \tag{2.23}$$

$$u + c = x/t. \tag{2.24}$$

The first equation is the isentropic equation of state at the CJ entropy; the sound speed is its derivative. The last two equations are the characteristic relations. We see from the first two equations that ρ and c are functions of p only; the third equation can then be integrated from the CJ state for p(u). With this solution substituted into the second equation to give c(u) = c(p(u)), the last equation can be solved for u(x/t).

Were the rarefaction running into a constant-state region, its head would be the characteristic (ray) having x/t equal to the value of u + c in that constant state. Here it runs into the CJ state, for which u + c = D, and so coincides with the front. The tail characteristic terminates the rarefaction fan. Its slope is the value of x/t for which u given by Eq. (2.24) is equal to u_p,

$$u_p + c(u_p) = x/t. \tag{2.25}$$

For the polytropic gas

$$p = p_j(\rho/\rho_j)^\gamma \tag{2.26}$$

$$c = c_j(p/p_j)^{(\gamma-1)/2\gamma} \tag{2.27}$$

$$p/p_j = [1 + (\gamma-1)(u - u_j)/2c_j]^{2\gamma/(\gamma-1)} \tag{2.28}$$

$$u - 2c/(\gamma-1) = u_j - 2c_j/(\gamma-1) = \text{const} = L \tag{2.29}$$

$$u = 2(\gamma+1)^{-1}(x/t) + (\gamma-1)(\gamma+1)^{-1}L, \tag{2.30}$$

with $p_0 = 0$, $L = -D/(\gamma - 1)$. The tail value of x/t is obtained from Eq. (2.30) with $u = u_p$. If the rear boundary condition is $p = 0$ (no confinement in vacuum) there is no constant-state region, and at the tail $p = 0$, $c = 0$, and $u = u_e = L$ (the escape speed). If the piston velocity becomes more negative as time proceeds (as from an accelerated rear confining plate), the fan is unchanged with the tail slope determined from the initial piston velocity, but the constant-state region is replaced by a nonself-similar time-dependent flow.

2B. APPLICATION OF THE SIMPLEST THEORY; PRODUCT EQUATIONS OF STATE

The description of the simplest theory completes detonation theory as far as practical calculations are concerned, for almost all of these are based on it, with empirical corrections for other effects in some cases. The only thing required is the equation of state of the reaction products, and that is the topic of this section. Gaseous explosives are well-described by the ideal gas equation of state, and we concentrate our attention on liquids and solids.

In one dimension, the simple flow of Fig. 2.6 describes the propagation of the detonation through the explosive. When it runs into another material, such as a driven plate, the initial reflected and transmitted waves can be obtained by standard shock matching techniques; see, for example, Deal (1957). Beyond that, a computer is required. Representative one-dimensional production codes are SIN (Mader and Gage 1967), KO (Wilkins et al. 1962), and RICSHAW (Lambourn and Hoskin 1970). Their proper use is almost an art, requiring considerable familiarity with pecularities and limitations of the finite-difference scheme. The two-dimensional codes are even harder to use. A representative two-dimensional production code is HEMP (Wilkins 1969). In two dimensions, the front is sometimes propagated by a Huygens construction with constant CJ wave velocity.

The simplest theory has, of course, well-recognized limitations, to which the rest of this book is largely devoted. An obvious one follows from the assumption of instantaneous reaction, because the detonation then has no length or time scale (other than the distance and time it has run), and there can be no edge effects, no initiation transients, no time-dependent build-up of the wave to full steady detonation. Less obvious, but no less serious, is the hydrodynamic instability of the steady solution, discussed in Chapters 6 and 7. Because of the instability, there are transverse waves in the reaction zone, making it importantly three-dimensional, so the one-dimensional conservation laws can apply only in an average sense. Nonetheless, calculations based on the simplest theory are already complicated enough for the available computers, and that theory is the basis for almost everything that has been done.

This section is a review of the most used product equations of state. Other reviews are given by Jacobs (1960) and Skidmore (1967). The equations of state are of two types: those which do not treat the chemistry explicitly, and those which do. The former are simply fits to some experimental data for the particular composition. The latter contain individual equations of state for the component molecules, and a mixture rule for combining them to give an equation of state for any composition. The composition is selected by requiring that the mixture be at chemical equilibrium, and calculating that equilibrium. This can of course be the wrong thing to do when the products have cooled sufficiently in expansion to slow down the reaction rates and effectively freeze the composition.

2B1. Equations of State Without Explicit Chemistry

Solution of the hydrodynamic equations for the flow problems requires only the incomplete equation of state $E = E(p,v)$; temperature is not needed. All the equations of state in this section 2B1 are of this form.

The simplest possibility is the *constant-γ form*

$$E - E_0 = pv/(\gamma - 1) - q, \qquad (2.31)$$

which is Eq. (2.8) with $\lambda = 1$ for products. With p_0 set to zero after the equation is multiplied by it, Eq. (2.9) for the Hugoniot curve becomes

$$p(v/v_0 - \mu^2) = 2\mu^2 q/v_0. \qquad (2.32)$$

Substitution for v and p from the CJ conditions, Eqs. (2.17) and (2.18), yields, for the CJ state,

$$p = 2(\gamma - 1)\rho_0 q \tag{2.33}$$

and

$$D^2 = 2(\gamma^2 - 1)q. \tag{2.34}$$

With $\gamma = 3$, which is a fairly good value for many liquid and solid explosives, the expressions for the following flow are simplified, as can be seen by substituting into Eqs. (2.26) to (2.30). The value of q can be estimated from heats of formation and an assumed product composition. In the infancy of the computer codes, workers used this approach almost exclusively. As calibration data became available, γ and q were treated as adjustable parameters. One defect of this equation of state is that it predicts that detonation velocity is independent of initial density, in disagreement with experiment.

Goranson (1946) suggested that the expansion isentrope and reflected shock Hugoniot curves for an explosive could be measured by allowing a plane detonation wave to drive inert plates of selected thicknesses, and determining the initial free-surface velocity of each plate. By extrapolating the data to zero plate thickness, and using the measured shock velocity through the plates, one could find the pressure at the interface at the instant of first contact. By repeating the experiments using inerts of varying shock impedance, one could map out the curves over any desired pressure region. Duff and Houston (1955) and Deal (1957, 1958) carried out the program for several explosives, and found that the data are fit very well by

$$pv^\gamma = p_j v_j^\gamma. \tag{2.35}$$

This constant-γ isentrope usually leads to the constant-γ form for the energy, Eq. (2.31).

This result presents a problem: how can this experimental result, Eq. (2.35), and its consequence, Eq. (2.31), be used without having the constraint of Eq. (2.34), that predicts (wrongly) that detonation velocity is independent of initial density? Fickett and Wood (1958) proposed that Eqs. (2.35) and (2.31) be assumed to hold only along the principal isentrope, Eq. (2.35), and that this isentrope be used as a reference curve (subscript r). Then the equation of state in the neighborhood of the reference curve is given by the first-order Taylor expansion about it as

Sec. 2B APPLICATION OF THE SIMPLEST THEORY

$$E(p,v) = E_r + \beta v(p - p_r) \tag{2.36}$$

or

$$E(p,v) = E_r + (p/\alpha)(v - v_r), \tag{2.37}$$

where

$$\alpha = p/(\partial E/\partial v)_p \tag{2.38}$$

and

$$\beta = (\partial E/\partial p)_v/v = \Gamma^{-1}, \tag{2.39}$$

where Γ is the Gruneisen coefficient. In the first form, Eq. (2.36), E_r, β, and p_r are functions of v; that is,

$$p_r = p_j(v_j/v)^\gamma, \tag{2.40}$$

$$E_r = p_r v/(\gamma - 1), \tag{2.41}$$

and

$$\beta = \beta(v). \tag{2.42}$$

In the second form, Eq. (2.37), E_r, α, and v_r are functions of p; that is,

$$v_r = v_j(p_j/p)^{1/\gamma}, \tag{2.43}$$

$$E_r = pv_r/(\gamma - 1), \tag{2.44}$$

and

$$\alpha = \alpha(p). \tag{2.45}$$

These forms allow the necessary freedom for fitting the variation of D with ρ_0.

Fickett and Wood (1958) studied the properties of various forms of these equations of state, and compared them with the experimental data available then. Their choice was the *constant-β equation of state*, given by Eqs. (2.36), (2.39), (2.40), and (2.41), with β = constant for all p and v. The expressions for p_j and D_j are complicated, and are not given here

(Fickett and Wood give formulas[1] that can be manipulated to find them). However, some unpublished numerical calculations of Fickett for Composition B are fit approximately, near the reference CJ point, by

$$p \propto \rho_0^{13/6} q^{1/3} \tag{2.46}$$

$$D \propto \rho_0^{2/3} q^{1/6}. \tag{2.47}$$

These expressions can be compared with Eqs. (2.33) and (2.34), which have frequently been used for making small corrections for changes in energy or density, to appreciate that the functional form is sensitive to the choice of equation of state.

The *constant-α equation of state*, proposed by Pike and Woodcock (see Skidmore 1967), is given by Eqs. (2.37), (2.38), (2.43), and (2.44), with α = constant for all p and v. Computer codes usually use $p = p(E,v)$ at each step of the mesh, so the constant-α form is not quite as convenient as constant-β.

Fickett and Wood (1958) investigated and discussed the consequences of other assumptions. One such is the assumption that γ is constant for all p and v, which results in the *general constant-γ equation of state*, with β a function of volume satisfying the thermodynamic restrictions. The equation of state is Eq. (2.36), with Eqs. (2.39), (2.40), (2.41), and (2.42), and with

$$\beta(v) = (1 + Cv^{\gamma-1})/(\gamma - 1) \tag{2.48}$$

C = constant.

This form has nothing special to recommend it, but has been used for computing.

A different reference curve is used in the *JWL (Jones-Wilkins-Lee) equation of state*, Kury et al. (1965), Lee et al. (1968, 1973). The isentrope through the CJ point has the form

$$p = A \exp(-R_1 V) + B \exp(-R_2 V) + CV^{-(\omega+1)}, \tag{2.49}$$

where A, B, C, R_1, R_2, and ω are constants, and $V = v/v_0$. The energy along this isentrope is obtained by integrating the isentropic equation

---

[1] The awkward asymmetric definition of Eqs. (2.38) and (2.39) follows Jones (1949). Fickett and Wood defined α as the inverse of Jones' α, and β as here. Note also that β^{-1} is the Gruneisen coefficient Γ.

Sec. 2B APPLICATION OF THE SIMPLEST THEORY

$dE/dv = -p$. The equation of state can be written, using the energy and pressure on the isentrope as reference values in Eq. (2.36), as

$$e = \rho_0 E = (A/R_1)\exp(-R_1 V) + (B/R_2)\exp(-R_2 V) + (C/\omega)V^{-\omega}$$
$$+ (V/\omega)\{p - [A \exp(-R_1 V) + B \exp(-R_2 V) + CV^{-(\omega+1)}]\}, \qquad (2.50)$$

where e is the internal energy/unit volume. The JWL is a constant-β equation of state, with $\omega = \beta^{-1} = \Gamma$, the Gruneisen coefficient. Notice that the terms containing C in E_r and p_r cancel, see Eq. (2.36); the equation is usually written without them. Notice also that for large V the exponential terms in Eq. (2.50) become small, and $e = pV/\omega$, similar to Eq. (2.31) but with a different zero energy point. The JWL equation of state has been calibrated down to about 0.1 GPa, and is widely used for practical calculations. Tables of values for the parameters are given by Dobratz (1972) and Lee et al. (1973), for many different explosives.

Although these equations of state are all expressed as $E = E(p,v)$, and we have emphasized that temperature is neither needed nor known, still it is possible to examine the thermal properties of the forms for the equation of state. For example, on any isentrope one can obtain T/T^* and $(C_p/C_v)/(C_p/C_v)^*$, where the asterisks indicate the value at some reference point on the isentrope, as functions of p and v (see Appendix B of Chapter 4). Even though little is known about what the thermal properties should be, it turns out that study of these expressions provides a very sensitive test of an equation of state, because we do know what is ridiculous or impossible.

Fickett and Wood (1958) examined the simple constant-γ, the constant-β, and the general constant-γ equations of state. They noted, of course, that the simple constant-γ form does not give the observed variation of detonation velocity with density, and that the general constant-γ form leads to impossibly large values of β at large volume (see Eq. 2.48). They also found that the temperature varies too much to be realistic along the isentrope of the simple constant-γ form, and too little along the isentrope of the general constant-γ form. Along the isentrope of the constant-β form, the variation is nearly what one might guess, perhaps a little too small. The specific heat ratio is invariant in the constant-γ form, and varies the wrong way in the general constant-γ form. In the constant-β form, it varies the right way and about the right amount over the region of usual interest, but goes to infinity at low pressure. Thus, of these three forms, the constant-β equation of state has the most reasonable properties in the region of interest. Other forms apparently have not been studied from this point of view.

For calibrating an equation of state, values for α, β, and γ at the CJ point can be found from a few relatively simple measurements. If the detonation velocity and the CJ pressure are known, Eq. (2.18) yields the value of γ at the CJ point. Measurement of the dependence of detonation velocity on initial density gives α and β in the following way. Differentiation of the Hugoniot curve Eq. (2.6) and the Rayleigh line Eq. (2.3) with respect to ρ_0, along with the CJ conditions Eqs. (2.17) and (2.18) and the thermodynamic identity (see Eq. (3.3), p. 69)

$$\alpha\beta\gamma = 1 + \alpha \tag{2.51}$$

and some manipulation, provides the relation

$$d \ln D/d \ln \rho_0 = (\gamma - 1 - \alpha)/(2 + \alpha). \tag{2.52}$$

Thus α at the CJ point can be found from $D(\rho_0)$, and β is then given by Eq. (2.51). Jones (1949) first obtained Eq. (2.52) and used it to estimate CJ pressures before they had been measured. He measured $D(\rho_0)$ and guessed the value of α, then obtained γ from Eq. (2.51), and the CJ pressure from Eq. (2.18). Since $\alpha \ll 2$, his pressure value is insensitive to reasonable errors in his guess.

There are no simple rules to help one decide which equation of state to use, and what experimental results to use to determine the values of the parameters. The choice is always influenced by the particular application, the accuracy required, and the computer code available. The calibration experiments should be as much like the application as possible, or should at least sample the pressure-volume region where the main energy transfer occurs in the application. It is unrealistic to expect the simple forms discussed here to apply to an extended region of p-v space, derived as they are from the first-order expansion about a reference curve. Therefore a set of parameters which gave excellent results when used for calculations for a particular application may not be the optimum set for a new problem.

These equations of state have proven remarkably useful and successful for many problems. The reasons that their limitations cause as little difficulty as they do are as follows:

1. The equations of state are always calibrated to experimental results from systems as much like the ones to be calculated as the experimenters can make them.

2. The additional entropy produced in the products by shock waves in the following flow is usually small, so the deviation from the reference isentrope is small.

3. In many applications, most of the energy transfer from the explosive products to the things they are driving takes place over a relatively small pressure range.

2B2. Equations of State with Explicit Chemistry

The equations of state described in the preceding section give $E(p,v)$ for a particular explosive at a particular density, and must be calibrated for each explosive. Using the assumptions that the detonation products can be described as a mixture of molecules, and that the CJ state is a state of chemical equilibrium, one can proceed in a more fundamental way to find an equation of state, the detonation properties, and the products composition, with only the atomic composition, heat of formation, and initial density as input.

The idea is to develop somehow an equation of state for each possible molecular species which might be present in the products. Then, with a mixing rule, and with the constraint that the number of atoms of each element is fixed, one obtains an equation of state for all possible compositions of the products. Finally, by calculating the free energy of the mixtures, and finding the composition which has minimum free energy, an equation of state for the products at chemical equilibrium is found.

The starting point for these calculations is a table of thermodynamic functions for the species, at standard pressure (say one atmosphere) and for a range of temperatures. Although an equation of state in the form

$$f(p,v,T) = 0$$

is insufficient for calculating all the thermodynamic functions when used by itself, it is enough, when used with the table, for calculating them at any pressure and temperature. Once such an equation of state is chosen, and the calculations performed, the equation of state for the explosive products at equilibrium is known.

Although the procedure for obtaining the equation of state is easy to outline and simple in concept, it is complicated and tedious to execute. We present no detail here. A careful exposition is given by van Zeggeren and Storey (1970). Calculations for gas detonations are discussed by Strehlow (1968a), and for solid explosive detonations by Taylor (1952).

An introduction to the subject, with examples, is available in Taylor and Tankin (1958). Beattie (1955) discusses the thermodynamics of real gases, and presents many details for using an equation of state to find properties at some state when they are known at a standard state. Hirschfelder, Curtis, and Bird (1964) treat the whole subject at some length.

The choice of an equation of state to use for calculating the thermodynamic functions away from the tabulated values is a crucial step, but no one knows how best to do it. Little is known about the properties of gases in the neighborhood of the CJ point of liquid and solid explosives. Several approaches have been used; three of them are described here.

The Kistiakowsky-Wilson equation of state

$$pv/RT = 1 + xe^{bx}, \qquad (2.53)$$

where

$$x = k/v(T + \theta)^a,$$

and

$$k = \kappa \Sigma_i x_i k_i,$$

with a, b, κ, θ, and k_i empirical constants, is the most extensively calibrated and widely used. The constants k_i, one for each molecular species, are the covolumes, and they form the equations of state for each species. For the mixture, each k_i is multiplied by x_i, the mole fraction of species i, and summed, to find an effective covolume. Equation (2.53) has an unphysical minimum in p vs. T (at constant v and x_i), and θ is added to T to move the minimum outside the region of interest. Cowan and Fickett (1956) calibrated this equation of state to detonation measurements, treating a, b, κ, and all the k_i's as adjustable for matching the measurements. Mader (1967a) recalibrated it, and extended it to other species. There are three computer codes in common use today which employ this equation of state. They are BKW (Mader 1967a), RUBY (Levine and Sharples 1962), and TIGER (Cowperthwaite and Zwisler 1973). These codes give temperature and composition, in addition to internal energy. Within the range of calibration, they are especially useful for predicting the effects of changes in the formulation of explosives.

Fickett (1962, 1963) made an ambitious attempt to apply a physically based theory which offered the hope of a reasonably good a priori calculation, independent of any calibration to detonation experiments. The equation of state was the Lennard-Jones-Devonshire free-volume theory. This single-species equation of state was extended to mixtures; several different mixture theories were tried and compared. As usual, the equilibrium composition was calculated. The primary inputs were the parameters describing the intermolecular pair potentials of the different product species — radius, well-depth, and repulsive exponent.

For the a priori calculation, the potential parameters were chosen from experiments such as atomic scattering and second-virial-coefficient measurements, not detonation measurements. The result was only moderately successful. For a group of common explosives, errors in detonation velocity ranged from -2 to $+7\%$, and errors in CJ pressure ranged from -5 to $+16\%$. The effects of varying most of the available parameters were investigated, but a systematic calibration of the individual parameters of the potential to detonation experiments was not carried out. Simple changes, such as scaling all the molecular sizes by the same factor gave somewhat better agreement with experiment. In general, with changes chosen to give the best over-all agreement with measured detonation velocities, calculated CJ pressures remained somewhat too high.

The JCZ (Jacobs-Cowperthwaite-Zwisler) equations of state are based on a semi-empirical fit to Lennard-Jones-Devonshire (LJD) and Monte Carlo results for various pair potentials. The LJD free-volume model works best at high densities, and less well as the density is reduced. The Monte Carlo method provides results in the intermediate region where there is no order and the LJD (or similar approaches) doesn't work. At very low density, of course, models for small corrections for gas imperfection give good approximations. Jacobs (1969) found a way to fit the best results over the whole range, and these fits provide the necessary single-species equations of state. Cowperthwaite and Zwisler (1976) provided mixture rules, and incorporated the complete equation of state into TIGER for calculations.

Finger et al. (1976) did a large number of experiments with a variety of explosives, and compared their results with a variety of calculations. Their explosives were four composed of the elements C, H, N, and O, three which contained only H, N, and O, one with C, N, and O, one with C, H, N, and F, and one with C, H, N, O, and F. This wide range of elemental compositions helps to separate the equations of state for the various single species. The initial densities of their explosives range from

1130 kg/m³ to 1890 kg/m³, and the CJ points for these vary widely. They used the results to recalibrate the BKW equation of state, and to calibrate a JCZ equation of state. They compared calculations for an old BKW calibration, their new BKW calibration, and their JCZ calibration, both among themselves, and with fits to performance measurements which sample the whole expansion isentrope. They conclude that the JCZ, or another similar equation of state, based on intermolecular potentials, holds the most promise for predicting detonation properties, but that much recalibration, and perhaps some modification, still remain to be done.

2B3. Kamlet's Short Method

Kamlet and Jacobs (1968) found an empirical fit to a large set of BKW calculations for many explosives. Their form for the CJ state is

$$p_J = k\rho_0^2 \phi \qquad (2.54)$$

$$D = A\phi^{1/2}(1 + B\rho_0) \qquad (2.55)$$

$$\phi = NM^{1/2}q^{1/2}, \qquad (2.56)$$

with k = 0.762, A = 22.3, B = 0.0013 (SI units).[2] The heat of reaction per kg of explosive q, the number of moles of gas per kg of explosive N, and the average molecular weight of the gas M are found from the chemical reaction equations with an assumed equilibrium composition. Although N, M, and q are all strongly affected by just what composition of the products is assumed, Kamlet and Ablard (1968) showed that ϕ, the important parameter, is reasonably insensitive to the assumption. Kamlet and Dickenson (1968) and Kamlet and Hurwitz (1968) compared the results of the fit with experimental measurements, and concluded that the fit was as good as could be expected, considering the uncertainties in the data.

If the explosive products contain no solid species, then the number of moles of gas per kg of explosive N and the average molecular weight of the gaseous products M in kg/mole are reciprocals, $N = M^{-1}$. Then the behavior of the explosive is determined by $\phi = (Nq)^{1/2}$. In this case

$$p_J = k\rho_0^2(Nq)^{1/2} \qquad (2.57)$$

[2]For RDX the input numbers are q = 6.20 MJ/kg, N = 33.8 mole/kg, and M = 0.0272 kg/mole, which give ϕ = 13,880. For initial density 1802 kg/m³, one finds D = 8782 m/s, and p_J = 34.3 Gpa.

and

$$D = A(1 + B\rho_0)(Nq)^{1/4} \qquad (2.58)$$

We can compare these expressions with Eqs. (2.33) and (2.34), and with Eqs. (2.46) and (2.47), and find that Nq replaces q, and that we have yet another set of exponents.

Hurwitz and Kamlet (1969) also obtained fits to the BKW isentropes, which are given by

$$p = 0.090\phi\rho_0^{-0.8}v^{-b},$$

where

$$b = 2.90 - 0.33v$$

for

$$v_J \leq v \leq 1.75 \text{ m}^3/\text{Mg},$$

and, in the low pressure range

$$p = 3.72\phi\rho_0^{-4/3}v^{-2}$$

for

$$v > 1.75 \text{ m}^3/\text{Mg}.$$

In these expressions, v is in m^3/Mg (or cm^3/g), so the volumes are numbers from 0.4 to 5 for the calibration range. The high pressure approximation doesn't fit the CJ point and the slope there as well as one could wish and the high and low pressure curves don't match together very well.

It seems that Kamlet's short method is useful for getting quick approximate values, and perhaps for furnishing some insight into what parameters are important.

2B4. Quasistatic Cycle for Detonations

Despite the fact that thermodynamics is a study of equilibrium states, it is frequently used to infer results for processes. This is done by considering a sequence of equilibrium states through which the process passes. A property commonly inferred is the maximum work which

THE SIMPLE THEORY
Chap. 2

could be obtained from an ideal engine, as, for example, using the Carnot cycle. Unlikely as it seems for so irreversible a process, a quasistatic cycle exists for detonation, and can be used to compute the maximum work that the explosive could perform.

To define the cycle, we need only specify the sequence of equilibrium states. However, some ideas about how we might conceive of an actual engine make it easier to understand and remember. The concept was described by Jacobs (1956). To construct the engine, confine the explosive in an upright cylinder of unit length and cross section, closed at the top by a rigid cap, and at the bottom by a movable piston driven by a reversible work source. Assume that all confining materials, including the piston, are rigid, massless, nonconductors of heat. Move the piston into the cylinder with constant velocity u_1 greater than or equal to the CJ particle velocity. As the piston starts to move, instantaneously initiate the detonation at the piston surface. The detonation front moves upward with wave velocity D_1 determined by u_1. The detonation products are in a uniform state with pressure p_1 and velocity u_1. When the detonation wave reaches the upper end of the cylinder, attach the piston to the cylinder at its position at that instant, and remove the driving force from the piston. Allow the cylinder to move upward under the deceleration of gravity until its velocity is reduced to zero. Extract work reversibly by first slowly lowering it to its original position, and then by releasing the piston and allowing the products to expand adiabatically and reversibly to a final pressure equal to the initial pressure. Extract heat by cooling the products at constant pressure to the initial temperature. Add heat, equal to the enthalpy change of the chemical reaction, to react the products, at constant pressure and temperature, back to the original explosive. A diagram of the complete cycle is shown in Fig. 2.7.

During the detonation part of the cycle, the piston moves a distance u_1/D_1 with force p_1, doing work on ρ_0 kg of material, so the work, per unit mass, done by the piston is

$$W_p = p_1 u_1 / \rho_0 D_1. \tag{2.59}$$

Using the conservation relation $p_1 = \rho_0 u_1 D_1$ (neglecting p_0), we find

$$W_p = u_1^2. \tag{2.60}$$

The kinetic energy, which is converted to gravitational potential energy, and then reversibly to work returned to the reversible work source, is, for unit mass of explosive,

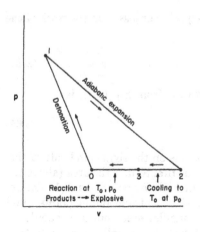

Fig. 2.7. Diagram of the quasistatic cycle for detonation.

$$W_g = -1/2\, u_1^2, \qquad (2.61)$$

and it is negative because it is work done by the material. The work done by expansion of the reaction products is

$$W_x = -\int_{v_1}^{v_2} p_i(v)\,dv, \qquad (2.62)$$

where $p_i(v)$ is the pressure on the isentrope through state 1, and v_2 is the specific volume on that isentrope at the initial pressure p_0. The net work exchanged between the explosive and the reversible work source is $W_p + W_g + W_x$; it is negative because the explosive does work on the source. The cooling from state 2 to state 3 exchanges heat

$$Q_c = \int_{T_2}^{T_0} C_p\,dT, \qquad (2.63)$$

which is also a negative quantity. The final step, from state 3 to the initial state, requires that a reversible heat source furnish the enthalpy change of the chemical reaction q. The sum around the cycle is zero.

In most applications of explosives, the heat rejected, Eq. (2.63), is usually lost. It does no useful work, and it is not rejected to a reversible heat source. (In underwater explosions, it may contribute by heating water to steam.) The useful work is equal to or less than the work of the quasistatic cycle,

$$W_u \leq - (W_p + W_g + W_x). \tag{2.64}$$

An alternate form, found by substituting from Eq. (2.4), is

$$W_u \leq - 1/2\ (p - p_0)\ (v_0 - v_1) - W_x. \tag{2.65}$$

The geometrical significance of the terms on the right-hand side of this equation may be seen from Fig. 2.7. The first term is the area (above the baseline $p = p_0$) under the Rayleigh line joining points 0 and 1. The second term is the area under the expansion isentrope joining points 1 and 2. Thus the right-hand side is the triangular area shown. It should be noted that the figure as drawn is not correct in detail: for a CJ detonation, the isentrope is tangent to the Rayleigh line at point 1, and for an overdriven detonation the isentrope is slightly steeper than the Rayleigh line at point 1, so that the upper part of the isentrope lies below the Rayleigh line. Even in the ideal case of the quasistatic cycle, the area is less than the enthalpy of formation of the explosive by the cooling heat exchange Q_c. Thus an explosive with higher heat of formation than another does not necessarily do more work.

Table 2.1 lists some values computed with an LJD equation of state for an ordinary CHON explosive, RDX, and for an aluminized explosive with very high heat of formation, both having the same initial density. The heat of formation of the aluminized explosive is 55% greater than that of RDX, but the maximum available energy is only 18% greater. The high temperature of the fully expanded products represents energy not available for useful work.

2B5. Overview

The explosives engineer regularly faces the demand that he characterize a new explosive. Characterization means, of course, information about sensitivity and hazard, mechanical properties, cost and availability, and, the only important one for our present purpose, performance. The characterization of performance requires an equation of state, so the behavior in the system of interest can be calculated.

The first step, which should precede the final choice of composition, is to get an equation of state using the methods of Sec. 2B2, with a BKW, LJD, or JCZ method. (Kamlet's Short Method, Sec. 2B3, might be useful as a preliminary approach.) One must remember that these are

Table 2.1 ENERGY STEPS IN THE DETONATION CYCLE

Process	Energy Change	RDX[a]	2Al/NH$_4$NO$_3$[a]
Detonate and extract kinetic energy	$W_p + W_g$	2.014 MJ/kg	1.923 MJ/kg
Adiabatic expansion	W_x	−8.205	−9.251
Cooling at p_o	Q_c	−0.130	−2.467
Reverse reaction	q	6.321	9.795
Useful work	$-(W_p + W_g + W_x)$	6.191	7.328
Efficiency	$-(W_p + W_g + W_x)/q$	98%	75%
Temperatures	T_1	4123 K	6357 K
	T_2	374	1694
	T_o	298	298

[a] Initial density 1800 kg/m³.

calibrated differently, and the results should be compared only with other results from the same code. The maximum available work can be found using the calculated isentrope with the quasistatic cycle, Sec. 2B4. Especially if the products composition, and therefore the shape of the isentrope, is different from explosives used previously, a realistic calculation of the performance of the explosive (as described by the calculated equation of state) in the particular application should be made and compared with the maximum work, and with other explosives in the same calculation. These results, taken together with considerations not related to performance, narrow the choices of compositions.

The next step is to measure the detonation velocity for comparison with the calculated value, to see whether the calculation is valid or must be reconsidered. At the same time, the determination of failure diameter indicates whether, in the size of interest for the application, one can expect the assumption of instantaneous reaction, central to the theory, to hold satisfactorily.

The final step begins with a study of the realistic calculations of the explosive performance in the device being considered. The idea is to identify the region of the p-v plane where the important energy transfer from the explosive products to the driven material occurs. Then one must devise and perform experiments which sample that region with adequate precision and detail. Finally, the data from the experiments are used to calibrate an equation of state of the sort found in Sec. 2B1, say a constant-β or JWL equation of state, for use in the detailed design of the device.

For the purpose of designing experiments to sample the p-v plane, it is useful to think qualitatively about what regions are accessible. For this end, a distorted diagram of the p-u plane is shown in Fig. 2.8, and its image in the p-v plane in Fig. 2.9. Only the initial match of the products at the CJ point can be considered in these diagrams, and the the effects of the Taylor wave, or any more complicated flow, must be neglected. Consider, in Fig. 2.9, a plane-wave CJ detonation, represented by point J, where the CJ Rayleigh line is tangent to the detonation Hugoniot curve. Since usually in a detonation system there are no processes which reduce the entropy, the region below the isentrope through J (curve 3) is unaccessible. The reflected shock Hugoniot curve centered at J (curve 4) is the locus of points accessible in a reflected shock. The shock which

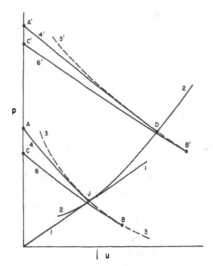

Fig. 2.8. Diagram in the p-u plane, distorted for clarity, showing accessible regions. Curve 1 is the CJ Rayleigh line, curve 2 is the detonation Hugoniot curve, and curve 3 is the principal isentrope. See the text for a description of the other curves.

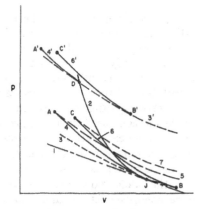

Fig. 2.9. Diagram in the p-v plane, distorted for clarity, showing accessible regions. The numbering of the curves is the same as in Fig. 2.8.

brings the products to rest (at point A), achieved when the detonation runs head-on into another detonation wave, is the strongest reflected shock in most systems. Expansion from point A is down the isentrope (curve 5) through that point. Expansion after any weaker shock will be along an isentrope between curves 3 and 5. If the products, starting at J, first expand down the isentrope (curve 3) to point B, and are then subjected to a reflected shock, the locus of shock states is the Hugoniot curve centered at point B (curve 6), and the strongest shock is to point C. (The highest possible pressure for point C would represent the case of two head-on detonation waves separated by a vacuum gap.) Isentropic expansion from point C is to states along the isentrope through C (curve 7), or, more likely, its reflection into the region of negative u. The images of these curves in the p-u plane, with the same numbers for the curves and letters for the points, are shown in Fig. 2.8.

Overdriven detonations, represented by point D in Figs. 2.8 and 2.9, are achieved in practice in convergent detonation waves, and, more commonly but less well recognized, when two detonation waves intersect at a small angle to produce a Mach-stem detonation wave. Overdriven states are also produced by flying plates impacting on explosive, and by driving with a more powerful explosive. All the possibilities described in the last paragraph with J as a center can also exist with D as a center. The corresponding curves are shown in the figures with the numbers and letters having primes on them.

It may be possible to use semiquantitative analysis like this to get an adequate idea of what regions of the p-v plane are important in the device under consideration, and also to design experiments to sample that region satisfactorily. A more accurate analysis of the correspondence between the test experiments and the device is a difficult and tedious process. Apparently design engineers have developed satisfactory procedures without making much effort to understand or test the correspondence.

To conclude this section on equations of state, we offer some warnings which reflect the unsatisfactory present state of calculations for design of explosives systems. First, the simple theory (see Sec. 3B for discussion) is not really applicable to real explosives, even to the extrapolation to infinite size. Second, infinite size is not of interest to a designer, and the finite reaction time, which leads to the failure to reach steady state in the distances used, and to complex edge effects at boundaries, is not usually taken into account explicitly. Third, the hydrodynamic codes, used both for analysis of the experiments and for the design itself, have errors in their results which may be several percent in the two-dimensional codes. It is customary to modify the equation of state so that the calculations agree with experiment, at least in the interpretation of the experiments used to develop the equation of state. Thus an "equation of state for explosive products" is not what its name implies it is, but is an engineering approximation for simulating, in a particular code, the behavior of explosive pieces of a certain shape and size, initiated in some chosen way. Failure to recognize this fact may lead one to use an "equation of state" under conditions where it doesn't work properly. Appreciable error can result.

2C. THE ZELDOVICH-VON NEUMANN-DOERING MODEL

If the simplest theory is extended to include a finite reaction rate, the treatment becomes that usually called the Zeldovich-von Neumann-Doering or ZND model. The shock wave is assumed to be much thinner than the zone of chemical reaction,[3] so the shock can be considered a discontinuous jump. This assumption is physically reasonable, because a very few collisions in the material will bring about mechanical equilibrium behind the shock, but many will usually be required for one to occur with enough energy to initiate the chemical reaction. The detonation front is is pictured as a shock wave that initiates a zone of chemical reaction behind it. The explicit assumptions are as follows:

1. The flow is one dimensional,

2. The shock is a jump discontinuity, because transport effects (heat conduction, radiation, diffusion, viscosity) are neglected,

3. The reaction rate is zero ahead of the shock and finite behind, and the reaction is irreversible (proceeds in the forward direction only),

4. All thermodynamic variables other than the chemical composition are in local thermodynamic equilibrium everywhere.

Under these assumptions the solutions of the flow equations are steady (in a coordinate frame attached to the shock) throughout the region of space in which the reaction takes place. Here the entire steady configuration of the shock and the chemical reaction zone replaces the jump discontinuity of the simpler model, and a flow must be found to join it to the driving piston. The state at the end of the reaction zone must be considered the final state. Demonstration of the existence of the type of solutions assumed here does not give any information about their

---

[3]In the present discussion, we include any relaxation of internal degrees of freedom in the term "chemical reaction."

Sec. 2C ZELDOVICH-VON NEUMANN-DOERING MODEL

hydrodynamic stability, nor does it preclude the existence of nonsteady solutions under the same set of rear boundary conditions. In this sense, the discussion is incomplete.

Using the steady assumption, it is easily shown that the flow equations have integrals in the form of the conservation conditions, Eqs. (2.1) to (2.4), which hold between any point in the assumed uniform constant state ahead of the shock and any point in the interior of the steady reaction zone. The mass and momentum equations are the same as those for the jump discontinuity, Eqs. (2.1) and (2.2). The energy or Hugoniot equation now depends on the extent of chemical reaction λ which varies continuously from 0 to 1, giving as the obvious generalization of Eq. (2.6)

$$\mathcal{H} = E(p,v,\lambda) - E(p_0,v_0,\lambda=0) - 1/2\,(p + p_0)(v_0 - v) = 0. \quad (2.66)$$

For an ideal gas of constant heat capacity this becomes, as the generalization of Eq. (2.9),

$$\left(\frac{p}{p_0} + \mu^2\right)\left(\frac{v}{v_0} - \mu^2\right) = 1 - \mu^4 + 2\mu^2\,\frac{\lambda q}{p_0 v_0}, \quad (2.67)$$

where $\mu^2 = (\gamma - 1)/(\gamma + 1)$.

The p-v diagram is sketched in Fig. 2.10. The Hugoniot curve with $\lambda = 1$ is the complete-reaction curve of Fig. 2.4. Because all terminal states must be on this curve, the D-discussion and piston problem are exactly the same as for the simplest (instantaneous reaction) treatment, Sec. 2A. The flow between the final state and the piston will have the same properties as found before: for piston velocities greater than u_j the detonation velocity is determined by the piston velocity.

The detonation has been assumed steady through the reaction zone, so there is just one value of D for the wave. Therefore the single variable λ, the degree of reaction, defines the state completely as the state point moves down the Rayleigh line, as diagrammed in Fig. 2.10. Immediately behind the shock the state point is N on the Hugoniot curve for $\lambda = 0$. As the reaction proceeds the state point moves down the Rayleigh line until reaction is complete at point S. At each point on the Rayleigh line between N and S there is a unique value of λ determined from the Hugoniot relation. The corresponding values of p and v can be obtained from the conservation relations, the Hugoniot curve, Eq. (2.67), and the Rayleigh line, Eq. (2.3), and then u is given by Eq. (2.4). Thus the state is completely specified at every point between N and S.

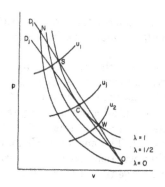

Fig. 2.10. The partial-reaction Hugoniot curves and the curves of Fig. 2.4 for a detonation with a reaction of finite rate.

The dependence of λ on distance x from the shock, or on particle time t since the fluid element passed through the shock, still has to be determined. When a complete equation of state is known, and a chemical reaction rate law

$$d\lambda/dt = r(p,T,\lambda) \tag{2.68}$$

is given, the rate law can be integrated to find $\lambda(t)$. The equation of motion of a particle with respect to the shock

$$dx/dt = D - u \tag{2.69}$$

can then be used to find $\lambda(x)$. Unfortunately, interesting forms of r for Eq. (2.68) cannot be integrated in closed form, and the solutions must be obtained numerically. Two examples, one for a gas and one for a solid, are given below. They are relatively simple, but qualitatively representative. A realistic hydrogen/oxygen example is given in Sec. 5A6.

The ZND solution for the reaction zone is based on simple and reasonable assumptions about how to model the real physical phenomenon. In this model the final state of the reaction zone can be a strong point S or the CJ point. There is no path to a weak point W. This result justifies the unsatisfactory intuitive rejection of the weak final state used for the simplest model, Sec. 2A2. Slightly more complicated assumptions, still physically reasonable, are discussed in Chapter 5, where it is shown that there can be paths to the weak point, and solutions which terminate there.

It is obvious from Fig. 2.10 that there could also be shockless steady solutions with λ increasing up the Rayleigh line from the initial point O to a weak point W. In the present context these are rejected as nonphysical, because the reaction rate would have to be finite in the initial

Sec. 2C   **ZELDOVICH-VON NEUMANN-DOERING MODEL**

state, without a shock to start it. If some sort of external igniter wave were provided, these solutions would be of interest. They are sometimes useful for thought experiments, especially in the limit of infinite D, which gives a constant-volume detonation without mass motion.

2C1. Example 1: Gas

We take the simplest possible reaction with no mole change

$$A \to B, \tag{2.70}$$

and assume constant heat capacity, the same for A and B, and so use the polytropic gas equation of state

$$pv = RT, \tag{2.71}$$

$$E = pv/(\gamma - 1) - \lambda q, \tag{2.72}$$

with $q = 50\, RT_0$ and $\gamma = 1.2$. We take the simple first-order Arrhenius form for the reaction rate,

$$d\lambda/dt = k(1 - \lambda)\exp(-E^\dagger/RT), \tag{2.73}$$

with the activation energy $E^\dagger = 50\, RT_0$. The value of the rate multiplier k serves only to set the time scale. We rewrite the rate equation as

$$d\lambda/d(t/t^*) = (1 - \lambda)\exp(-E^\dagger/RT), \tag{2.74}$$

$$t^* = k^{-1},$$

choose the value of k (after the integration is complete) so $\lambda = 1/2$ at $t/t^* = 1$, and present the results as functions of t/t^*. To integrate the rate equation, we need T as a function of λ; this is obtained by solving the partial-reaction Hugoniot equations at the given value of D, here D_j. These equations are collected in Appendix 2A of this chapter.

The partial-reaction Hugoniot curves are shown in both the p-v and p-u planes in Fig. 2.11. The steady solution for the unsupported (CJ) detonation is shown in Fig. 2.12, with both time and distance from the shock as the independent variable. (Where time is the independent variable, we are following the history of a fluid particle through the steady solution, taken to pass through the shock at $t = 0$.) The maximum in the temperature (and therefore also the sound speed) near the end of the reaction zone can be explained by the geometry of the

THE SIMPLE THEORY   Chap. 2

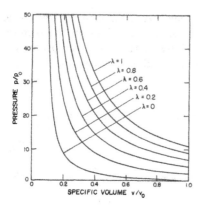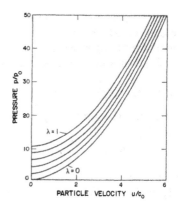

Fig. 2.11. Partial-reaction Hugoniot curves in the (a) p-v, and (b) p-u planes for the polytropic gas explosive $E = pv/(\gamma - 1) - \lambda q$, with $\gamma = 1.2$, $q = 50\,RT_0$. The axis labels for this and the following figures (2.12) are pressure, p/p_0, specific volume, v/v_0, particle velocity, u/c_0, rate, r, temperature, T/T_0, extent of reaction, λ, density, ρ/ρ_0, acoustic impedance, $\rho c/\rho_0 c_0$, sound speed, c/c_0, time, t/t^*, distance, $x/c_0 t^*$. Here t^* is the half-reaction time (see text); the value of the rate multiplier k is 0.69315.

isotherms and isentropes near the CJ point, sketched in Fig. 2.13. The Rayleigh line is tangent to the isentrope at the CJ point. The isotherms are less steep but concave upward, so one of them will be tangent to the Rayleigh line somewhere above the CJ point. If this point of tangency lies below point N, the steady solution will have a temperature maximum.

2C2. Example 2: Solid

We take the fairly typical hypothetical solid explosive used by Fickett and Rivard (1974),

$\rho_0 = 1600$ kg/m³

$D_J = 8500$ m/s

$p_J = 28.9$ GPa,

described by the constant-γ equation of state

$E = pv/(\gamma - 1) - \lambda q$

$\gamma = 3$, $q = 4.5156$ MJ/kg.

Sec. 2C ZELDOVICH-VON NEUMANN-DOERING MODEL

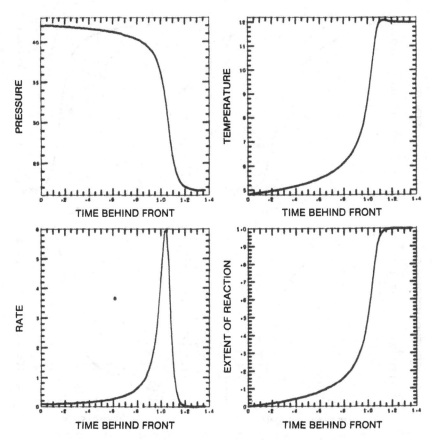

Fig. 2.12. The steady solution for the system of Fig. 2.11 with the rate function $r = k(1 - \lambda)e^{-E^\dagger/RT}$; $E^\dagger = 50\ RT_0$, at $D = D_J = 6.22\ c_0$. (a) Particle histories, (b) snapshots.

All these quantities are expressed in SI units. Since temperatures are known only with considerable uncertainty, and there is virtually no information on reaction rates, we will not try to calculate the temperature, and will use the simple rate function chosen by Fickett and Rivard (1974)

$$d\lambda/dt = r = 2(1 - \lambda)^{1/2},$$

which integrates to

$$(1 - \lambda) = (1 - t)^2.$$

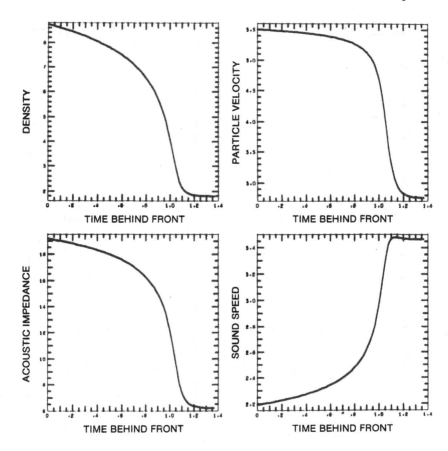

Fig. 2.12a. (cont)

The steady solution with λ as the independent variable is the same as that for the gas of the preceding section, but is simplified by setting p_o to zero, a very good approximation for condensed explosives. The results are given in Appendix 2A, Eqs. (2A-16) to (2A-25). The relation between x and t (distance and time behind the front for a given particle) is

$$x = D[(\gamma - 1)t + 1/2\ t^2]/(\gamma + 1).$$

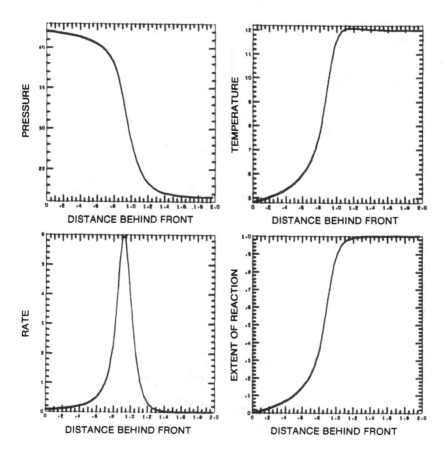

Fig. 2.12b.

Fig. 2.14 shows the partial-reaction Hugoniot curves. The one for $\lambda = 0$ is vertical, a consequence of taking $p_0 = 0$ (taking $p_0 = 1$ atm would hardly change it). A typical realistic $\lambda = 0$ Hugoniot curve is shown as a dashed line; the constant-γ result is not too bad for the von Neumann point N, but of course becomes unrealistic at lower pressures.

Table 2.2 gives the steady CJ solution. With this simple rate function p and u are linear functions of t, and ρ is nearly linear.

THE SIMPLE THEORY Chap. 2

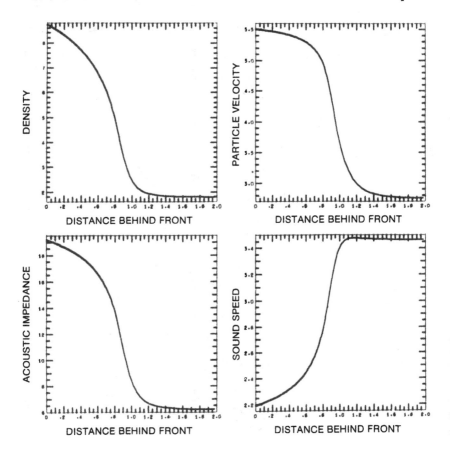

Fig. 2.12b. (cont)

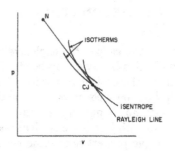

Fig. 2.13. Explanation of the temperature maximum in the steady solution.

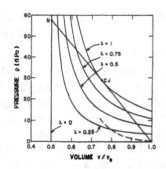

Fig. 2.14. The partial-reaction Hugoniot curves for a hypothetical solid explosive, example 2. The dashed curve is a more realistic $\lambda = 0$ Hugoniot curve.

Table 2.2 STEADY CJ SOLUTION FOR EXAMPLE 2

t μs	x mm	p GPa	u m/s	ρ kg/m³	c m/s	λ
0.0	0.000	57.80	4250	3200	7361	0.000
0.1	0.436	54.91	4038	3048	7352	0.190
0.2	0.893	52.02	3825	2909	7324	0.360
0.3	1.371	49.13	3613	2783	7278	0.510
0.4	1.870	46.24	3400	2667	7212	0.640
0.5	2.391	43.35	3188	2560	7127	0.750
0.6	2.933	40.46	2975	2462	7022	0.840
0.7	3.496	37.57	2763	2370	6896	0.910
0.8	4.080	34.68	2550	2286	6747	0.960
0.9	4.686	31.79	2338	2207	6574	0.990
1.0	5.313	28.90	2125	2133	6375	1.000

THE SIMPLE THEORY Chap. 2

APPENDIX 2A. FORMULAS FOR DETONATION IN A POLYTROPIC GAS

For discussion of detonation the simplest models of the explosive are very useful, because the important qualitative behavior can be studied without unnecessary complication of the mathematical expressions. One useful model is the polytropic gas with γ not only constant but also the same in both reactants and products, with no change in the number of moles of gas as the reaction proceeds, and with just one chemical reaction. This model represents the macroscopic properties of some actual gas systems in a restricted region of the p-v plane, with A and B each regarded as a composite representation of a mixture of several species.

The reaction is

$$A \to B, \qquad (2A\text{-}1)$$

and the equation of state is

$$E = pv/(\gamma - 1) - \lambda q, \qquad (2A\text{-}2)$$

where γ is the isentropic derivative

$$\gamma = -(\partial \ln p/\partial \ln v)_s, \qquad (2A\text{-}3)$$

a constant independent of the degree of reaction. Dimensionless variables, defined as

$$Q = q/RT_0$$

$$V = v/v_0$$

$$U = u/c_0$$

$$M = D/c_0$$

$$P = p/p_0 \qquad (2A\text{-}4)$$

$$C = c/c_0,$$

are convenient. The velocity u is measured in the laboratory reference frame, with the unreacted material at rest.

App. 2A DETONATION IN A POLYTROPIC GAS

Given q and γ, the variables at the CJ state point are

$$M_j - M_j^{-1} = [2Q(\gamma^2-1)/\gamma]^{1/2}, \tag{2A-5}$$

$$P_j = (1 + \gamma M_j^2)/(1 + \gamma), \tag{2A-6}$$

and

$$V_j = (1 + \gamma M_j^2)/[M_j^2(1 + \gamma)]. \tag{2A-7}$$

For any given D the strong and weak intersections with the partial reaction Hugoniot curves are given by

$$V = \frac{\gamma + M^{-2}}{\gamma + 1} \pm x \tag{2A-8}$$

$$P = \frac{\gamma M^2 + 1}{\gamma + 1} \pm \gamma M^2 x \tag{2A-9}$$

and

$$U = \frac{M - M^{-1}}{\gamma + 1} \pm Mx \tag{2A-10}$$

where

$$x = \left[\frac{(M - M^{-1})^2 - 2(\gamma^2 - 1)(\lambda Q/\gamma)}{(\gamma + 1)^2 M^2}\right]^{1/2} \tag{2A-11}$$

and on the sonic locus, including the CJ state where $\lambda = 1$,

$$M_{sonic}^2 = (\gamma^2 - 1)(\lambda Q/\gamma) + 1 + \{[(\gamma^2 - 1)(\lambda Q/\gamma) + 1]^2 - 1\}^{1/2} \tag{2A-12}$$

The sonic locus, which is the locus of points at which a Rayleigh line is tangent to some partial reaction Hugoniot curve, is given by

$$P = (1 + \gamma M^2)/(1 + \gamma), \tag{2A-13}$$

$$V = (1 + \gamma M^2)/[M^2(1 + \gamma)]. \tag{2A-14}$$

These combine to give the sonic locus in p-v as

$$P = V/[(1 + \gamma)V - \gamma]. \tag{2A-15}$$

Slightly different forms are given by Pukhnachev (1963) and by Erpenbeck (1964b).

With neglect of p_0, appropriate for a liquid or solid, the partial-reaction Hugoniot equations are

$$p = fp_j(1 \pm g) \tag{2A-16}$$

$$v = v_j(1 \mp g/\gamma) \tag{2A-17}$$

$$u = (1 - v/v_0)D \tag{2A-18}$$

$$f = (D/D_j)^2 \tag{2A-19}$$

$$g = (1 - \lambda/f)^{1/2}. \tag{2A-20}$$

The CJ state is given by

$$D_j^2 = 2(\gamma^2 - 1)q \tag{2A-21}$$

$$p_j = \rho_0 D_j^2/(\gamma + 1) \tag{2A-22}$$

$$v_j/v_0 = \gamma/(\gamma + 1) \tag{2A-23}$$

$$u_j = D_j/(\gamma + 1) \tag{2A-24}$$

$$c_j = \gamma D_j/(\gamma + 1). \tag{2A-25}$$

Except for (2A-21), these CJ relations hold for any equation of state with γ evaluated at the CJ state.

3

EXPERIMENTAL TESTS OF THE SIMPLE THEORY

The simple theory has been extensively tested by comparing its predictions with experiment. The situation is so different for gases and for liquids and solids that we consider the two cases separately.

In gases, Sec. 3A, we are in the fortunate position of having an equation of state which is exactly known: the ideal gas relation $pv = RT$ (with very small corrections for gas imperfection) plus the standard tabulations of the thermodynamic functions for the internal degrees of freedom. Thus the calculations can be done essentially exactly. The difficulty comes in the nature of the real system, which has, in most if not all cases, the inherent transverse structure described in Chapter 7, and is thus one-dimensional only in some transverse-average sense. Most of the measurements, such as density by x-ray absorption, are also transverse averages. The results are that the one-dimensional conservation conditions are found to apply with an error of at most a few percent, so that the state point lies approximately on the detonation Hugoniot curve. Thus overdriven detonations are calculated quite accurately. The CJ condition, however, is another matter. In unsupported detonations the state point still lies approximately on the calculated Hugoniot curve, but well down on the weak branch, with pressure and density 10% to 15% below the CJ value and with the flow supersonic with a Mach number of 1.10 to 1.15.

Liquids and solids, Sec. 3B, are much harder to study. The equation of state is so poorly known that a direct comparison of an a priori calculation with experiment is useless. Clean-cut experiments are more difficult, because no rigid confinement is available. A steady solution can be assured and the effects of lateral expansion removed by working with long charges of several large diameters, and extrapolating to infinite charge diameter. This is too tedious and expensive for measurement of any quantity other than detonation velocity. Measurements of pressure and particle velocity are usually done in large-diameter systems with relatively short lengths of run (of the order of 100 mm), initiated with a plane-wave lens, so there is a central portion of the front still unaffected by side rarefactions. However, this brings up the question of whether a steady state has been reached. Run lengths of this order occur in many practical applications, and this question is currently under study, particularly for solids, with conflicting results.

The front pressure is inferred from measurements of the free-surface velocity of plates driven by the explosive, extrapolated to zero plate thickness. There are some difficulties in interpreting the results. Particle velocities in the interior can be measured at the cost of introducing a perturbation, by inserting thin metal foils and following their motion by radiography or electromagnetic methods. Finally, transverse structure is known to be present in many systems, although little is known about its detailed form.

Fortunately, there is one method of testing the simple theory which does not require a known equation of state. Assuming only that an equilibrium products equation of state *exists*, and that the simple theory (one-dimensional conservation conditions plus the CJ condition) applies, it can be shown that measurements of the variation of detonation velocity D with initial density and energy determine, with no approximation, the CJ pressure. The test consists of comparing the pressure obtained in this way with that more directly measured by the plate method. This test has been carried out for the liquid explosive nitromethane with the result that pressure inferred from the variation of D is 15% less than that directly measured. This comparison of course gives no other information, such as whether the state point lies on the Hugoniot curve.

3A. GASES

A concise summary of the experimental tests of the simple theory may be found in the short review of the current status of gas-detonation research by Schott (1965a); see also Strehlow (1968b) and White (1961). The book by Greene and Toennies (1964) is a good general compendium of shock-tube studies.

The most commonly investigated systems consist of mixtures of carbon- or hydrogen-containing fuels with oxygen. If sufficient oxygen is present, the reaction products are all gases whose thermodynamic properties are well-known. The final detonation pressure in most experiments is well under 100 atmospheres, so the equation of state is the

Sec. 3A GASES

ideal-gas equation of state with very minor corrections for gas imperfection effects. It is assumed that the products are in thermodynamic equilibrium. The Hugoniot equations can be solved numerically to obtain with very little uncertainty the Hugoniot curve on which the state at the end of the steady zone must lie if the flow is one-dimensional and laminar. The Chapman-Jouguet condition is then used to determine the propagation velocity and the state on the Hugoniot curve corresponding to the unsupported detonation. The test of the simple theory consists of comparison of the calculated and experimental results.

There are two effects which make the comparison only approximate. The first is the possibility of vibrational non-equilibrium in some systems. The second is that the flow is not strictly one-dimensional. The relatively trivial non-one-dimensionality due to boundary layers at the walls can be satisfactorily accounted for and effectively removed from the results by making measurements at several diameters and extrapolating the results to infinite diameter. More serious is the inherent non-one-dimensional structure due to the self-sustaining transverse disturbances on the front described in Chapter 7. At the time most of the experiments considered here were done, little was known about this fine structure. Single spin was recognized, but generally regarded as an anomalous phenomenon found near the detonation limits, and it was avoided by doing experiments well away from the limits. As a fortunate consequence, the wavelength of the transverse disturbances was small compared to the tube diameter, and their amplitude considerably smaller than in extreme cases such as spin. When the tube diameter is only a few times the wavelength, the deviations from the simple theory are much larger.

The fine structure is powered by the energy of the chemical reaction, and tends to decay as a fluid particle falls behind the front. The pressure perturbations decay rapidly, but the density and entropy perturbations do so only by heat conduction, so on the time scale of interest they remain essentially forever. Observations show a decay zone, following the reaction zone and several times as long, over which the mean pressure and density decay to an essentially constant value (the detonation having been allowed to run long enough that the effect of the following rarefaction wave is negligible). The comparisons between one-dimensional theory and experiment are made at the end of this decay zone, where the transverse perturbations are as small as they are going to get. The one-dimensional theory does not apply exactly, of course, but only in the transverse-average approximation.

The quantities available for measurement are shown in Fig. 3.1, reproduced from Schott's review. The table in the figure gives calculated

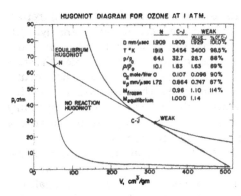

Fig. 3.1. Comparison of the calculated state at the CJ point and an arbitrarily chosen point on the weak branch of the detonation Hugoniot curve for a detonation in pure O_3. M is the Mach number $(D - u)/c$ at the end of the reaction zone. From Schott (1965a).

states for a detonation in pure ozone at the CJ point, together with an arbitrarily chosen point on the weak branch of the Hugoniot curve. This point, chosen so the detonation velocity is 1% above the CJ value, is typical of those observed. It is seen that the detonation velocity and the temperature are least sensitive to the displacement, and that the changes in all the other quantities listed are much larger and comparable to each other.

The "equilibrium Hugoniot" shown in the figure is the Hugoniot curve calculated by assuming that the reaction-product composition at all points is such that the system is in chemical equilibrium. The CJ point shown is the "equilibrium CJ point" at which the Rayleigh line is tangent to this equilibrium Hugoniot. The corresponding sound speed entering into the CJ condition is the "equilibrium sound speed" involving an adiabatic displacement in which the composition shifts so as to maintain chemical equilibrium. Early in this period of experimentation, it was thought that the equilibrium Hugoniot condition should be used, but that the appropriate CJ condition should use the "frozen sound speed" in which the composition remains fixed during the adiabatic displacement. The frozen CJ point typically lies below the equilibrium CJ point, the displacement being about the same as that of the arbitrarily chosen weak point of Fig. 3.1. With better theoretical understanding of the problem, the equilibrium CJ condition came to be regarded as the correct one to use. A reader perusing the experimental literature should be aware of this point, which is discussed in Sec. 5A.

The results of the comparison of observation and the simple theory may be summarized as follows:

1. The state point for both unsupported and overdriven detonations lies approximately on the calculated equilibrium Hugoniot curve.

2. For overdriven detonations, the simple theory is approximately correct, with measured densities, for example, approximately equal to those predicted from the calculated equilibrium Hugoniot curve at the measured detonation velocity.

3. For unsupported detonations, the observed state point lies on the weak branch of the calculated Hugoniot curve, about as shown in Fig. 3.1, and is consistent with the measured detonation velocity, which does not determine the point well because of the near tangency.

These conclusions are of course not airtight, but are fairly well established. There are discrepancies up to a few percent for the overdriven case. In the unsupported case, the location of the state point on the weak branch implies supersonic flow with the Mach number M about 1.1. This property has been directly verified by observations of the Mach angle, and of the failure of following waves to overtake the front.

3A1. Experiment

Detonation velocities have been measured and compared with the calculated values in a number of systems. Typical studies are those of Peek and Thrap (1957), Brochet, Manson, Rouze, and Struck (1963), Getzinger, Bowen, Oppenheim, and Boudart (1965), and Strauss and Scott (1972). The tube must be long enough to ensure a steady velocity in the measurement section. The measurements should be done in tubes of several different diameters, and the results extrapolated to infinite diameter to eliminate wall effects. The reported measurement uncertainties in the infinite-diameter velocity are as small as a few tenths of a percent. The calculations are also very good, with uncertainty (mostly caused by uncertainty in the heats of formation) of a few tenths of a percent. The corrections for gas imperfection are very small, as are the differences between frozen and equilibrium detonation velocity.

The scatter of the available data about the calculated values is more than might be expected if the uncertainties are really as small as a few tenths of a percent. The measurements are within ±2% of the calculated values, with perhaps a slight preponderance of experimental values higher than the calculated ones.

Because the detonation velocity is relatively insensitive to a displacement of the state point away from the CJ point, these comparisons might best be regarded as indicating that the displacements are not very large. Their magnitude is better determined by measuring some other quantity.

By means of x-ray densitometry, Duff, Knight, and Rink (1958) measured the density in mixtures of C_2H_2, O_2, and Kr, and in mixtures of C_2N_2, O_2, and Kr, the Kr being included to enhance the x-ray absorption. Tubes 25 and 75 mm in diameter and 2 to 5 m long were used. The initial pressures ranged from 2.5 to 11 kPa (1 atm = 101.325 kPa). The standard deviation of the measurements is 0.5% or less, and the absolute error is believed to be about the same. The experiments were compared with calculations made with the equilibrium CJ condition and with gas imperfection effects included. Calculations with the frozen CJ condition were also made for comparison. In overdriven detonations with the final density about 50% above the CJ density, the measured densities are 0.5% to 1.5% above the calculated value, which is within the experimental error. Most of the difference is probably attributable to finite-tube effects. In unsupported detonations agreement is not found. For one mixture the densities were measured at two diameters to permit extrapolation to infinite diameter. For other mixtures the measurements were made in the larger tube, with an approximate correction. The measured densities are 5% to 9% below the calculated values.

Vasiliev, Gavrilenko, and Topchian (1973b) made extensive pressure measurements in $2H_2 + O_2$, $2H_2 + O_2 + 3Ar$, and $C_2H_2 + 2.5O_2$ in a long tube of 80-mm diameter, for initial pressures ranging from 5.1 to 51 kPa. They used pressure transducers as small as 1-mm diameter with a time resolution of 0.05 μs. For the lower pressure mixtures the cell sizes (see Chapter 7) were much larger than the gauges, and the reaction zone was well resolved. The measured von Neumann spike (front shock) pressures were in good agreement with the calculated values. The measured "CJ" pressures were taken as averages over some distance behind an experimentally determined sonic plane, taken to be the plane at which the bow shock separated from a thin plate passing through the reaction zone, as described in the next paragraph. For these mixtures the calculated frozen CJ pressures lie 5 to 7% below the equilibrium CJ pressure. The measurements have a 10% scatter, with most of the mean values lying 4 to 7% below the calculated equilibrium CJ value. Similar results were obtained earlier by Edwards, Jones, and Price (1963).

Vasiliev, Gavrilenko, and Topchian (1973a) made a unique measurement of the rear-facing characteristic velocity $c - u$. This quantity changes more rapidly than any other along the detonation Hugoniot curve; for a 0.1% change in D it changes by 20% while the density changes by 4%. They used stoichiometric mixtures of hydrogen/oxygen, acetylene/oxygen, and methane/oxygen, with the first two diluted with argon. Tube diameters were 21 and 80 mm, and initial pressures ranged from 5.1 to 300 kPa. The characteristic velocity was measured by mounting a thin plate perpendicular to the front, with its edge along a tube diameter. As the detonation passes over the plate, a bow shock is formed on its edge. When the sonic point in the reaction zone reaches the edge of the plate, the bow shock detaches itself and moves downstream with velocity $c - u$. Photographic observation of the bow shock motion then gives the desired velocity. The results for all of the mixtures were found to lie on a single curve in a plot of $(c - u)/D$ vs. d/Z, with d the tube diameter and Z the cell size defined in Chapter 7. Z is approximately inversely proportional to the initial pressure. The measurements range from $d/Z = 6$ to $d/Z = 350$. As d/Z increases from 6, the value of $(c - u)/D$ first falls sharply from a value of about 0.3, and then falls much more slowly, passing through 0.14 at $d/Z = 350$. Typically, the experimental curve passes through the value of $(c - u)/D$ calculated with the frozen sound speed at a value of d/Z near the upper end of its range. The calculated equilibrium value is much lower; for acetylene/oxygen, for example, the calculated frozen and equilibrium values of $(c - u)/D$ are 0.141 and 0.076, respectively. Thus these measurements, like the pressure measurements, place the observed state near the frozen CJ point.

The Mach number of the flow behind the front can be measured directly if the Mach angle of a weak disturbance propagating into it from the tube wall can be observed. Such disturbances have been seen in schlieren photographs by Fay and Opel (1958) and by Edwards, Jones, and Price (1963). Their origin is not clear; they appear to be caused by some effect in the boundary layer near the end of the reaction zone. Figure 3.2 is a smear-camera photograph, reproduced from Edwards,

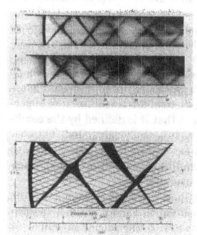

Fig. 3.2. Schlieren smear-camera photographs of detonation waves in a 38.1 mm (1.5 in.) square-section tube for (a) $2H_2 + O_2$ and (b) $H_2 + O_2$ at an initial pressure of 101 kPa. (c) Diagram of the main features of (b). From Edwards, Jones, and Price (1963).

Jones, and Price, of a hydrogen/oxygen detonation in a tube of square cross section. Other observations confirm that the pattern is stationary with respect to the front. Fay and Opel present a similar photograph for a 40% C_2H_2/60% O_2 mixture in a tube 50 mm square. Both show supersonic flow with Mach number about 1.15 for the flow with respect to the detonation front.

White (1961) has made a systematic and thorough study of the system $2H_2 + O_2 + 2CO$. His measurements were made in a relatively large tube of 82.5-mm square cross section with a 2-m-long driver section and 8 m of run followed by the measurement section. The results were not extrapolated to infinite diameter. Initial pressures ranged from 1 to 10 kPa for overdriven detonations and from 7 to 140 kPa for unsupported detonations. Detonation velocities were measured in the usual way with ionization gauges.

The precision of the velocity measurements seems to be somewhat less than those discussed above, perhaps because the mixture composition was not controlled quite so closely. Pressures were measured by means of quartz transducer gauges with an error of about 2%. Spark interferograms covering the full tube width were taken; some examples are shown in Fig. 3.3. These show the transverse structure, and, with the use of the specific optical refractivities of the product species, yield densities with an accuracy of about 3% in regions where the transverse perturbation is not too severe.

Finally, the product of the concentrations of atomic oxygen and carbon monoxide [CO] [O] were obtained from photomultiplier measurements of luminosity at two different wavelengths, λ = 412 nm and λ = 546 nm. (The constant of proportionality between the luminosity and the concentration product was obtained by comparing calculated concentration products and observed luminosity over a wide range of overdriven detonations.) The maximum radical concentration in the reaction zone is expected to occur when the relatively fast bimolecular reactions have equilibrated among themselves, but the slower three-body recombination reactions have not yet proceeded to an appreciable extent. Schott (1960) discusses this hypothetical intermediate state of partial equilibrium in detail, and points out that it is defined by the condition that the net mole change be zero, so it is easily calculated. The concentration measurements described next appear to be consistent with this description of the reaction path, and they constitute a rough observation of some of the reaction-zone structure.

In rather strongly overdriven detonations, with pressures ranging from about 1.5 to 4 times the CJ value, measurements made within a few centimeters of the front agree approximately with the calculated values. At

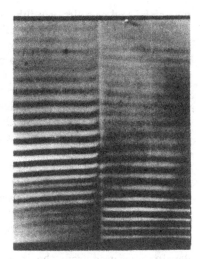

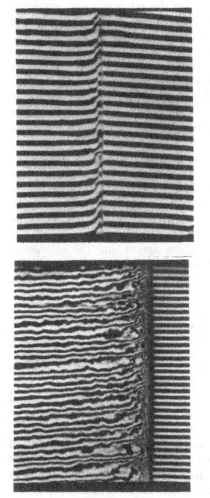

Fig. 3.3. Spark interferograms of detonations in $2H_2 + O_2 + 2CO$. (a) Overdriven, $D = 2700$ m/s, $p_0 = 2.7$ kPa. (b) Overdriven, $D = 3100$ m/s, $p_0 = 2.7$ kPa. (c) Unsupported, $D = 2200$ m/s, $p_0 = 3.0$ kPa. From White (1961).

the lower end of the overdrive range within the reaction zone, the calculated partial equilibrium value of the [CO] [O] concentration product is about eight times as great as the complete equilibrium value. As the pressure increases, the two approach each other, becoming equal at the upper end of the overdrive range. The photomultiplier measurements show an abrupt rise to a peak luminosity followed by a decrease to a relatively steady value. The measured peak, presumably corresponding to the partial equilibrium state, corresponds to a concentration product that lies about halfway between the values calculated for partial and complete equilibrium over the entire range of overdrive, with the experimental and both calculated values coming together at the upper end of the range. Experimental limitations such as collimation slit width and shock-front curvature are expected to reduce the measured peak, and may be responsible for all the difference. The densities measured by Duff, Knight, and Rink (1958) in overdriven detonations tend to lie slightly above the calculated values, and White's measured densities are also above the calculated values by as much as 5% in some cases. Uncertainties in the specific refractivities may account for some of this discrepancy.

The full range of measurements was made on unsupported detonations with quite different results. Interferograms, of which Fig. 3.3c is typical, show strong transverse perturbations. The structure decays over a distance of a few centimeters, and as it does so the density and luminosity fall. This transition region is followed by an extended region of the flow in which the density and luminosity change very slowly. The density measured from the interferogram agrees with the calculated value at the immediate rear of the reaction zone, but falls about 15% below that value at the end of the transition region. Because of the rise time of the gauge, about 25 μsec, details of the pressure change in this region cannot be obtained, but measured values corresponding to a distance of several centimeters behind the front are 15% to 20% below those calculated. The [CO] [O] concentration product, from the luminosity measurement, also decreases in the decay region. The ratio of its observed value at the end of the transition zone to the calculated CJ value is approximately the square of the ratio of observed and calculated pressures at this point. This result is as expected because, for small changes in pressure or temperature, the concentration product should be proportional to the square of the density. Finally, measured detonation velocities range from about 1.5% below the calculated ones at an initial pressure of 10 kPa to slightly above the calculated ones at 100 kPa. Note that no extrapolation to infinite diameter was made, and that the diameter effect (i.e., boundary-layer effect) is larger at low pressure.

Of great significance for the interpretation of the results is White's observation that a third class of "intermediate" detonations can be distinguished experimentally. When one tries to overdrive an unsupported detonation by increasing the driver pressure above the CJ value, the detonation velocity remains nearly constant until the pressure measured at a point about 10 cm behind the front has increased to 30% or 40% above the value for the unsupported wave. White made a number of detailed observations of the flow profile behind the front over a distance of about 50 cm. Separate experiments had shown that over the range of interest the luminosity is approximately proportional to the square of the pressure, so the pressure profile could be inferred from the luminosity profile obtained from the photomultiplier, which has a fast time response. White used two photomultipliers, one twice as far from the diaphragm as the other, to record the entire profile at two times. A clear picture of the nature of the flow emerges from these observations. In the intermediate range, between the unsupported detonation and the overdriven one, the detonation is followed by a shock wave (or a compression wave, depending on initial and driving pressures) that moves more slowly than the front, and therefore lags farther and farther behind. In the region between the front and the following shock, the flow is essentially the same as in the unsupported detonation. As the driver pressure is increased, the strength and speed of the following shock increases until, in the region of true overdrive, it has merged with the front, and the flow has a single-wave structure.

3A2. Discussion

The experimental evidence seems to agree, in general, with the hypothesis that the state point at the end of the reaction zone is on the weak branch of the detonation Hugoniot curve. Most measurements of detonation velocity, pressure, density, and final Mach number are consistent with the hypothesis. White's measurements of the pressure profile in detonations intermediate between unsupported and overdriven are especially convincing in showing that the following flow is supersonic, as expected for a weak detonation. In this section we discuss some alternative hypotheses. It seems to us that the weak-detonation hypothesis is the only tenable one.

Because all experiments are done in finite-sized tubes, the flow is neither exactly steady (independent of time) nor exactly one-dimensional, so the state point does not lie exactly on the Hugoniot curve and the Rayleigh line because these are the one-dimensional conservation conditions. It has not been shown by direct measurement of two quantities (say density and pressure) with sufficient accuracy that the final state point lies on the Hugoniot curve. The small divergence of the flow, shown by the fact that the wave front is very nearly plane, indicates that the departure from one-dimensionality is very small.

In any test like this, the calculations made for comparison must correctly represent the theory. There is no reason to believe that there are large errors in any of the calculations or in their input data.

Another possible source of error arises from the fact that the measurements are made in tubes of finite size, and are not made at the center of the tube where the flow is most nearly described by the simple theory.

The pressure is measured at the wall and is thus affected by boundary layer flow, and the density is an integral measurement made by absorption or refraction of radiation so that it represents a transverse average across the tube. Changing the size of the tube changes the relative thickness of the boundary layer, and changing the initial pressure changes the absolute thickness of the boundary layer. The experiments show that the measured values must be approximately correct, and that the errors are not large.

White's measurements of the behavior of following waves, and the measurements of the propagation velocity of weak disturbances by Fay and Opel and by Edwards, Jones, and Price are interpreted to mean that the flow behind the reaction zone is supersonic, and the supersonic flow means that the detonation final state is on the weak branch of the detonation Hugoniot curve. In the long tubes used in the experiments, the expansion in the rarefaction wave following the detonation is very slow, and one-dimensional theory predicts a correspondingly slow increase of the Mach number from unity. However, finite tube diameter introduces another possibility. The boundary layer is formed by the drag of the tube wall on the moving gas behind the front, so the material near the wall moves more slowly than that in the center, and therefore falls behind the front faster. In addition, the heat loss to the wall cools the gas in the boundary layer so that its density is higher. The combination of high density and low velocity removes a considerable amount of material from the gas that is following the front at the CJ mass velocity. The effect on the main flow is as if some fluid were leaking through the wall of the tube. The rarefaction behind the front is thus stronger than expected in any treatment that neglects the boundary layer, and the flow Mach number increases more than expected.

Fay (1959) offers a qualitative treatment of this effect in the quasi-one-dimensional (nozzle) approximation. He notes that the strictly one-dimensional theory of Zeldovich (1940), in which the wall forces are considered to be distributed uniformly across the tube, is inadequate. With average wall forces determined from pipe flow measurements, this theory gives velocity deficits a thousand times smaller than those measured experimentally. Fay's treatment of the problem begins with an estimate of the total mass flux in the boundary layer and from this he obtains the equivalent rate of expansion of a nozzle that has no boundary layer. The estimate is based on experimental measurements of boundary layer growth in shock tubes. From this point on, the flow problem is that of a quasi-one-dimensional flow in a slowly enlarging channel. We discuss the same problem in Sec. 5G, except that there the rate of area increase is related to the shock-front curvature, whereas Fay assumes a plane front and uses his boundary-layer estimate of equivalent rate of area increase. Although the set of ordinary differential equations defining the channel flow could be integrated directly, it is simpler to formulate the problem in terms of the usual one-dimensional equations with a small but unknown correction term containing an integral of the true pressure over the length of the reaction. Fay estimates the reaction-zone length by imposing the condition that the flow be frozen sonic at the end of the reaction zone, and assuming a single reaction and a simple form for the dependence of the progress of reaction on distance near equilibrium. Using this treatment, he is able to predict the velocity deficits of several different gaseous systems with an accuracy of 10% to 30%.

This quasi-one-dimensional theory gives the Mach number in the supersonic flow behind the reaction zone as a function of distance, and Fay and Opel (1958) suggest that the observed increase in Mach number and decrease in density might be accounted for in this way. Duff and Knight (1958) challenge this interpretation and present arguments against it. More recent detailed comparison of the predictions of the quasi-one-dimensional theory with experiment in large-diameter tubes by Edwards, Jones, and Price (1963) show that it does not agree with the observed results. Somewhat more convincing is White's (1961) remark that the ordinary one-dimensional treatment seems adequate to describe overdriven detonations, even though the flow parameters, the front curvature, and the boundary layer are not much different from those of the unsupported case. In conclusion, it seems likely that the effect of the inherent transverse structure of the unsupported detonation wave is more important than the effect of finite tube diameter, at least in the large tubes used for the experiments described here.

One of Fay's predictions is that a rapid change in density immediately behind the front is expected. A principal argument of Duff and Knight (1958) is that the density change is not observed, and the experimental results are presented by Duff, Knight, and Rink (1958), who found the density to be constant (within their precision of about 1/2%) over a distance of 20 mm. Their experiment was done on a mixture of $0.3C_2H_2$ + $0.3O_2$ + 0.4Kr at an initial pressure of 37 kPa, and although they do not say so explicitly, presumably in their 75-mm tube. On the other hand, White (1961) found that the density decreased by about 15% over a 30-mm distance behind the front in a mixture of $2H_2 + O_2 + 2CO$ at initial pressure of 15 to 30 kPa in his 82.5-mm square tube, when the detonation was unsupported. The overdriven detonation, however, had essentially constant density behind the wave front, showing that Fay's boundary layer explanation was not the correct one. White believes that the decay of transverse structure in the unsupported detonation causes the density drop. The difference between White's observations and those of Duff, Knight, and Rink for the unsupported detonation is probably due to the difference in spatial scale of the disturbances on the front in the two systems. The calculated CJ temperature for the $C_2H_2/O_2/Kr$ system is about 5800 K compared to 3300 K for the $H_2/O_2/CO$ system. Observations suggest that the higher the temperature, and thus the shorter the reaction time, the smaller the disturbances on the front will be. It may well be that the density decrease White observed is also present in the other system, but is too rapid to be seen.

3B. SOLIDS AND LIQUIDS

As mentioned in the introduction, the lack of a sufficiently accurate products equation of state precludes the straightforward comparison of theory and experiment used for gases. But there is an alternative: comparison of the CJ pressure, measured conventionally, with the CJ pressure inferred (rigorously through the simple theory) from measured changes in the detonation velocity D under variations in the initial state.

The measurement of D is done in the conventional way, using long rate sticks (cylinders), and removing the effects of finite charge diameter by extrapolation to infinite diameter. The conventional pressure measurement (see, for example, Deal 1957) requires measuring the initial free-surface velocity of a series of plates of different thickness driven by the explosive, and then extrapolating the results to zero plate thickness. There are some difficulties in interpreting the results; an example is mentioned in Sec. 3B1. Applying the method to the geometry used for the measurement of D involves a double extrapolation in plate thickness and charge diameter and is quite expensive, though it has been done. The more common method is to work in a one-dimensional system: the central portion of the end of a short (50 to 100 mm of run), large-diameter (150 to 200 mm) piece of explosive initiated with a plane-wave lens. With the length/diameter ratio thus restricted, a central one-dimensional region, not yet affected by edge rarefactions, is available at the end of the charge. Working in this geometry of course requires an explosive with short reaction time so the initiation transients will be over quickly and a steady reaction zone will be achieved in the relatively short length of run available. The liquid explosive nitromethane chosen for this experiment satisfies this requirement. It also has the advantage over any pressed or cast solid of being perfectly homogeneous.

For the solid explosives used in most applications, the question of the validity of the simple theory for the steady reaction zone may be academic. The steady state may not be reached in the available length of run, typically 100 mm or less. Although much of the current research suggests that the steady state is not achieved, the answer is not yet definite. See Rivard et al. (1970), Davis and Venable (1970), Mader and Craig (1975), Bdzil and Davis (1975), and Davis (1976). These papers also throw some light on the interpretation of the conventional CJ-pressure measurements.

3B1. Theory

The first work on the general problem of the effect of initial-state variations on detonation velocity was done by Jones (1949), who considered the information about the product equation of state which could be obtained from measurements of detonation velocity as a function of loading density. The theory presented here is due originally to Stanyukovich (1955, Chapter 7, Sec. 40) and Manson (1958). Wood and Fickett (1963) reviewed the theory and proposed its application to an experimental test of the Chapman-Jouguet theory.

The assumptions are essentially those of the simple theory described in Chapter 2. The flow must be steady and one-dimensional (laminar), and the reaction products must be in equilibrium at the end of the reaction zone, so that, at least on the time scale of the experiment, they are described by an equation of state of the form $E = E(p, v)$. Then, given the Chapman-Jouguet hypothesis that the Rayleigh line and the Hugoniot curve are tangent for an unsupported detonation, one can show that the CJ detonation velocity varies in a simple way with the energy and density of the initial state.

The equations used to obtain the desired relations are the Hugoniot and Rayleigh equations (2.3) and (2.6),

$$E - E_0 = 1/2 \, p(v_0 - v) \tag{3.1}$$

$$\rho_0^2 D^2 = p/(v_0 - v) , \tag{3.2}$$

and the thermodynamic identity Eq. (2.51)

$$\alpha \beta \gamma = 1 + \alpha , \tag{3.3}$$

where

$$\gamma = -(\partial \ln p/\partial \ln v)_s , \tag{3.4}$$

$$\alpha = p(\partial v/\partial E)_p, \ \beta = (\partial E/\partial p)_v/v, \tag{3.5}$$

and the relations which apply at the CJ point

$$p = \rho_0 D^2/(\gamma + 1) \tag{3.6}$$

$$v = v_0\gamma/(\gamma + 1). \tag{3.7}$$

The initial pressure p_0 and its variations are neglected here because the application is to condensed-phase explosives, and the variations that can be produced are very small. The quantity α is the same as that used by Jones (1949). Differentiation of the Hugoniot equations gives

$$(\partial E/\partial p)_v dp + (\partial E/\partial v)_p dv - dE_0 = 1/2\, p(dv_0 - dv)$$
$$+ 1/2\, (v_0 - v)dp, \tag{3.8}$$

which becomes, after elimination of $(\partial E/\partial p)_v$ by means of the thermodynamic identity (3.3) and substitution of v/γ for $(v_0 - v)$ from Eq. (3.7)

$$dp/p + \gamma dv/v = [(\gamma + 1)dv_0/v_0 + 2\gamma dE_0/pv]/[1 + 2\alpha^{-1}]. \tag{3.9}$$

Differentiation of the Rayleigh equation (3.2) gives with the same substitution for $(v_0 - v)$

$$dp/p + \gamma dv/v = 2dD/D + (\gamma - 1)dv_0/v_0. \tag{3.10}$$

The same combination of dp and dv occurs in both equations by virtue of the tangency of the two curves at the CJ state. Thus these differentials can both be eliminated. The variables p and v can also be eliminated by using the CJ relations, Eqs. (3.6) and (3.7). The result for dD/D is

$$dD/D = -A dv_0/v_0 + B dE_0/D^2, \tag{3.11}$$

where

$$A = (\gamma - 1 - \alpha)/(\alpha + 2), \quad B = \alpha(\gamma + 1)^2/(\alpha + 2).$$

The coefficients A and B are to be evaluated at the CJ state, and they contain only two unknowns, the derivatives α and γ at the CJ state. There is no assumption that α and γ are constant.

Now if some state variable X of the initial state of the explosive (such as T_0 for a liquid explosive) is varied, the change in detonation velocity can be measured at several values of X to determine the derivative dD/dX. The procedure can be repeated with another state variable Y, and the two equations can be written

$$D^{-1}dD/dX = -(A/v_0)dv_0/dX + (B/D^2)dE_0/dX \qquad (3.12)$$

$$D^{-1}dD/dY = -(A/v_0)dv_0/dY + (B/D^2)dE_0/dY. \qquad (3.13)$$

The derivatives dv_0/dX, dv_0/dY, dE_0/dX, and dE_0/dY can all be determined from measurements of the thermodynamic properties of the material. The two equations thus can be solved for the unknowns A and B, and also for α and γ, provided linearly independent variations are chosen (i.e., the ratio dv_0/dE_0 is not the same in the two sets of experiments). Substitution of γ back into the equation for the pressure at the CJ point, (3.6), gives the pressure determined by derivatives, which is to be compared with the pressure measured directly.

An interesting by-product is the first-order expansion of the equation of state in the form E(p, v) about the CJ point,

$$E(p, v) = (\partial E/\partial p)_v (p - p_j) + (\partial E/\partial v)_p (v - v_j) + \dots . \qquad (3.14)$$

One partial derivative of the energy is p/α, and the other is βv, Eq. (3.5).

3B2. Experiment

Wood and Fickett (1963) describe a number of ways of actually carrying out the desired initial-state variations discussed in the previous section, including those used in the experimental test made by Davis, Craig, and Ramsay (1965). The useful ones, of course, are those such that one variation is mainly a variation in initial density and the other mainly a variation in initial energy.

Wecken (1959) and Manson (1958) used the variation method to determine detonation pressure, but they did not compare the result with pressure measured in another fashion to test the theory. A clear-cut experimental test of the Chapman-Jouguet theory as defined above has been made by Davis, Craig, and Ramsay (1965), using nitromethane (CH_3NO_2) as the test explosive. A change in initial temperature causes a change in both density and initial energy, but the effect of the density change is large compared to that of the energy change, so the initial temperature was used as one initial-state variable. The energy-change derivative was obtained by mixing the nitromethane with another explosive with the same atomic composition but different heat of formation. Because the atomic composition is the same, the assumption that chemical equilibrium obtains at the end of the reaction zone ensures that the product composition at a given state point and, therefore, the product equation of state are the same. The explosive used is called Acenina, and is a mixture of equimolar amounts of acetonitrile (CH_3CN), nitric acid (HNO_3), and water. The densities of nitromethane and Acenina are nearly the same, 1159 kg/m^3 (1.159 g/cm^3) and 1139 kg/m^3 respectively at 4°C, but their initial energy densities differ by 1.410 MJ/kg (337 cal/g), with nitromethane having the higher energy. An approximate calculation of the CJ state of nitromethane by Fickett (1962) gives 5.84 MJ/kg as the heat release of nitromethane, so the difference is about 25%. The two materials, nitromethane and Acenina, are sufficiently miscible that measurements can be made at several different internal energies, and the derivative can be evaluated for pure nitromethane.

The two variations of the initial state are carried out from a reference state, pure nitromethane at 4°C.
They are

1. Variation of (mainly) the initial energy by changing the composition of a mixture of nitromethane and Acenina at constant temperature.

2. Variation of (mainly) the density by changing the initial temperature of pure nitromethane.

The changes in both variables in each variation were of course carefully determined. The detonation velocity was measured over a temperature range from −26°C to +34°C, and over a composition range from pure nitromethane to pure Acenina. The dependence of D on T_0 and the dependence of D^2 on composition are both linear. Least-squares fits to these linear functions were used to evaluate the required differential changes from the reference state. The pressure was then calculated from these results, as described in the previous section.

The conventional method of obtaining pressure consists of using the explosive to drive a material of known equation of state and measuring the initial free-surface velocity for a sequence of plate thicknesses. The extrapolation to zero plate thickness is made, and the pressure in the inert corresponding to that free-surface velocity obtained. The pressure in the explosive is obtained from an impedance-match solution. The results given for nitromethane at 4°C are

$p = (12.65 \pm .54)$ GPa (126.5 kbar) from the initial state variation,

$p = (14.81 \pm .1)$ GPa from the conventional method.

New values for the heat of formation of nitromethane have been given by Lebedev et al. (1970) and Wagman (1968), that change the pressure from the initial state variation. Also, Davis and Venable (1970) reviewed the conventional measurements, including some later ones, and corrected an error made when reducing free-surface velocity to pressure. The new results for nitromethane at 4°C are

$p = (12.2 \pm .6)$ GPa from the initial state variation,

$p = (14.2 \pm .4)$ GPa from the conventional method.

The $\pm$ terms are the estimated 95% confidence limits. The error of measuring the change of detonation velocity with temperature is responsible for most of the uncertainty in the first value. The conventional pressure measurements show a structure in the pressure profile that is not predicted by theory, and therefore present some difficulty in interpretation, as discussed by Craig (1965) and Davis and Venable (1970). There has been some criticism of the measurements and the interpretation given them.[1] We believe that the conventional pressure measurement is correct, that the difference between the two values is significant, and that it indicates a failure of one or more of the assumptions of the simple theory.

Davis, Craig, and Ramsay (1965) also present a number of other experimental results that cast doubt upon the simple theory. These are not reviewed here, except for one rather simple result. The initial-state variation experiments were also carried out for TNT, using melting of the material as the means for changing the initial energy. The density change was easily made in the solid by controlling loading density, and in the liquid by varying the initial temperature as was done for nitromethane. After correcting for the slight change in initial energy due to heating of the liquid, the values of $(\partial D/\partial \rho_0)_T$ for solid and liquid TNT both at 1450.2 kg/m³ (1.4502 g/cm³) density are

3.187 $\pm$ .027 m/s per kg/m³ for the solid at 25°C,

4.267 $\pm$ .5 m/s per kg/m³ for the liquid at 93°C,

---

[1] Petrone (1968) presents an alternate interpretation of the free-surface velocity measurements which would lower the implied detonation pressure to approximately the value obtained from the initial-state variation. His interpretation is based on a calculated reaction-zone length approximately three orders of magnitude larger than that estimated by Craig (1965). It appears to us that this result should be questioned, for it is based on an equation of state with an unphysically large heat capacity and, hence, low shock temperature, and appears to be inconsistent with some of Craig's planewave results. Duff (1965) has also discussed the interpretation of these results.

where the ± terms are 95% confidence limits. Because the detonation products are expected to be in almost the same thermodynamic states (the initial energy difference is only 0.19 MJ/kg [45 cal/g]), the two slopes are expected to be nearly the same. The large difference suggests immediately that the simple theory does not apply to at least one of the materials. This result has the advantage that there is no question of misinterpretation of pressure measurements. Comparison of the pressure obtained from the initial-state variation with that obtained from the conventional method shows a difference about the same as that for nitromethane. The differences in the velocity derivatives and in the pressures indicate that, for TNT as well as nitromethane, one or more of the assumptions of the simple theory fail.

4

FLOW IN A REACTIVE MEDIUM

This chapter contains an introductory treatment of flow in a reactive medium, which forms the basis for considering steady detonations in Chapter 5 and their hydrodynamic stability in Chapter 6.

The object is to understand how, for the same initial and boundary conditions, the behavior of a fluid with chemical reaction differs from that of one without reaction. The differences can be striking, for the presence of an exothermic chemical reaction makes self-sustaining detonation waves possible. In steady flow, the effect of chemical reaction is essentially the same as that of heat addition, for it transforms chemical bond energy into that of heat and bulk motion.

Except for a few very special cases, there are no explicit solutions of the equations. They are studied by theoretical analysis of their general properties and by numerical calculation of particular cases.

Because they are relatively tractable, the steady detonation solutions discussed in the next chapter play an important part in this study. In this case the partial differential equations reduce to ordinary ones. Much can be learned about the nature of their possible solutions by theoretical study of their phase space, particularly the nature and location of the critical points. The analysis is similar to that of nonlinear mechanical systems. With this knowledge of the phase space as a necessary guide to proper procedure in the neighborhood of the critical points, quite accurate numerical calculations of particular cases are possible.

In this chapter we give a more general discussion. The first step is to write the equations in different forms, including the characteristic form, which emphasize different physical relationships. The properties of the different forms are discussed, and the quantities which appear are related to more familiar ones. The next step is the consideration of the types of wave motions to be expected. Here useful results can be obtained by qualitative discussion, and, in some cases, by linearization.

Section 4A presents the equations, the assumptions on which they are based, and the general analytic study of their properties, including specialization to the case of steady flow. Section 4B collects the properties of the simple fluid used in most of the discussions of the following two chapters—a binary mixture of polytropic gases. Section 4C discusses possible wave motions, principally the propagation of shock and rarefaction waves moving into a medium initially in chemical equilibrium.

The thermodynamic properties of reactive flow were first discussed comprehensively by Campbell (1951). The pioneering work in the formulation of the equations of motion with an arbitrary number of chemical reactions and their application to detonations is that of Kirkwood and Wood (1954), and also Wood and Kirkwood (1957). We follow their notation. Some general references are Vincenti and Kruger (1965), Strehlow (1968a), Wegener (1969), Gruschka and Wecken (1971), and Clarke and McChesney (1976).

4A. THE MODEL

The equations of motion are the Euler equations with the addition of chemical reaction, i.e., the equations of compressible flow in which transport properties (viscosity, heat conduction, diffusion, radiation) are neglected. For the most part these are specialized to time-dependent laminar flow in one space dimension with plane (slab) geometry.

The necessary mathematical assumptions will be written down as needed. Here we discuss the physical ideas on which they are based. The fluid is imagined to be subdivided into elements which are sufficiently small to be idealized as mathematical points but sufficiently large and homogeneous to be characterized by values of the state variables, which thus become smooth functions of position and time. It is assumed that local (i.e., within each element) partial thermodynamic equilibrium exists with respect to all degrees of freedom except chemical reaction. An equivalent statement is that at any given time each fluid element has, given its current chemical composition, a Boltzmann distribution of energy states. The state defined by this distribution is not one of chemical equilibrium, which is approached at a finite rate. All other rate processes (e.g., vibrational relaxation) are treated as infinitely fast so that they are always in equilibrium. Of course these other processes can be and often are assigned finite rates and included by a formalism similar to that for chemical reaction, but for simplicity we omit them here.

Sec. 4A   THE MODEL

This description of the state of affairs is seen to be a realistic one if we consider the collision processes which produce it. Since only a small fraction of the collisions produce chemical change, while nearly all are effective in contributing to the equilibration of other degrees of freedom, chemical equilibration is relatively slow. It is also clear that in regions where the flow gradients are appreciable in a distance comparable to a mean free path, such as in a shock transition, even an efficient collision process may be unable to maintain equilibrium in the face of a rapidly changing state, so that the description will not be realistic within such regions.

Since chemical reaction is the only process treated as not in equilibrium, it is the only mechanism for entropy production other than the shock transition, which, with the neglect of transport effects, appears as a jump discontinuity. In contrast to many force-flux problems, the chemical reaction rates are assumed to be functions of the local thermodynamic state and *not* of the flow gradients.

In Sec. 4A1 we give a detailed account, with numerous examples, of how to put the chemistry into the equations of motion. The main task is finding an economical formulation of the law of definite proportions. This constraint — that all composition changes take place through chemical reaction — allows both the composition and the reaction kinetics to be reduced to the specification of a small set of formally constructed *stoichiometric reactions*. The *progress variables* λ_i of these reactions replace the composition (mole fractions of all the species). A relatively straightforward problem is transforming from chemist's units (moles) to physicist's units (ordinary mass units).

In Sec. 4A2 we set down the functional forms and general properties of the equation-of-state and reaction-rate functions. We also impose some conditions on the first and second derivatives of the equation of state which qualify it as "well-behaved."

In Sec. 4A3 we write the equations of motion and specialize them to one dimension. The constitutive relations now consist of both the equation of state and the reaction rate

$$E(p,v,\lambda)$$

$$r(p,v,\lambda),$$

and to the equations of motion is added the equation for the rate of reaction

$$\dot\lambda = r(p,v,\lambda).$$

We have here specialized to a single reaction. The dot denotes a Lagrangian (particle) time derivative.

In Sec. 4A4, we examine the equations of motion and discuss some of their general properties. It seems to us conceptually important to note that the equations can be written down and solved without any mention of entropy and without any reference to thermodynamics, other than the statement that the equation of state as defined above exists. However, thermodynamics is of some use in clarifying the physical meaning of some of the terms, and in relating them to more familiar quantities.

What one might call the master equation — the relation between p and ρ in an adiabatic displacement — is obtained immediately by expressing the differential of E in the energy equation $\dot{E} = -p\dot{v}$ in terms of those of its arguments. It is (for one reaction)

$$\dot{p} = c^2\dot{\rho} + \rho c^2 \sigma \dot{\lambda}$$

with

$$\rho^2 c^2 = (p + E_v)/E_p$$

$$\rho c^2 \sigma = -E_\lambda/E_p.$$

At this point, c and σ are defined by these relations, but are not further identified. We do see immediately, however, that in the absence of reaction ($\dot{\lambda} = 0$), c^2 is just the usual sound speed, and that the product $\sigma\dot{\lambda}$ or σr, which we call the *thermicity*, or *thermicity product*, gives the pressure change due to reaction. Its sign depends on the sign of the dimensionless *thermicity coefficient* σ, as well as on the direction of reaction (sign of $\dot{\lambda}$).

The characteristic analysis gives u + c as the characteristic speed, the propagation velocity of infinitesimal disturbances or derivative discontinuities, thus identifying c as a sound speed. Application of thermodynamics shows its to be the *frozen sound speed*, the adiabatic derivative at fixed composition, which is the acoustic speed in the high-frequency limit. The *equilibrium sound speed*, the adiabatic derivative at equilibrium composition, which is the acoustic speed in the low-frequency limit, does not appear explicitly in the analysis at this point. The complete significance of these two velocities is not apparent until a global analysis of some simple flow problems is made in Sec. 4C.

The significance of the thermicity coefficient σ is brought out by applying thermodynamics to express it in other forms. It can be expressed as the sum of two terms: the volume increase due to reaction at constant pressure, and the heat release of the reaction at constant pressure. It

Sec. 4A   THE MODEL

thus represents the transformation of energy, obtained from breaking chemical bonds and from the change in the number of particles or molecular sizes by chemical reaction, into energy of heat and motion. The importance of the volume term has often been overlooked; it can be quite important in practical applications.

In Sec. 4A5 we specialize the equations to one-dimensional steady flow. These are formally equivalent to those for heat addition in a constant-area channel (except that the rate of heat addition is a function of the local state, instead of being controlled independently), and they have the same striking dependence on the sign of the local sonic parameter $1 - M^2$ (M being Mach number), with sonic points being of special interest. The steady equations have as integrals the shock conservation relations for mass, momentum, and energy, which relate any two points in the steady flow. The energy equation, however, now has a heat-of-reaction term, and gives the family of partial-reaction Hugoniot curves with parameter λ described in Sec. 2C.

In Sec. 4A6 we present a useful relation which we call the shock-change equation. It relates the rate of change of shock strength to the reaction rate and the pressure- or velocity gradient immediately behind the shock. We do not make much use of it in this book, but it does offer some additional practical insight.

4A1. Chemical Reactions

A single progress variable λ specifies completely the transformation of chemical species in the reaction $A \to B$. We spent very little effort in Chapter 2 to describe its meaning. The real chemistry of explosives is more complicated than $A \to B$, and it is not immediately obvious how to translate the notation and language of chemists into that of fluid dynamicists. Two problems, how to describe chemical reactions with a small number of variables, and how to transform the units, are the subject of this section.

The set of mole numbers per unit mass of system

$$\mathbf{n} = (n_1, n_2, ..., n_s)$$

defines the composition of a chemical system composed of s species. The equation of state

$$E = E(p, v, \mathbf{n})$$

depends on the composition. The chemical reaction equations, because they are constrained by the requirement that the number of each kind of atom remains constant, provide relationships among the components of **n**. We can replace **n** by a smaller number of independent variables λ, and write

$$E = E(p,v,\lambda).$$

The additional requirement that all the components of **n** be nonnegative means that the **n**-space, and therefore the λ-space, has stoichiometric boundaries. The λ-space is used for many discussions of the equations of motion. Fortunately, all, or almost all, the forms of detonation are illustrated when λ has only two components, and our discussions are limited to that case. A reader interested in the physics of detonation, but who does not intend to make detailed chemical calculations, might skip most of the details in the rest of this section.

The system for this discussion is a fluid element or particle which is by our assumption a closed adiabatic system, i.e., one in which no heat or matter enters or leaves, although of course external forces acting on the boundaries can do work on (or have work done on them by) the element. Our system is also by assumption a spatially uniform one; it is the usual fluid element which is small enough to be considered uniform, but large enough to be considered continuous without any statistical effects of the individual atoms. The properties of our fluid element change with time, and the time derivative is denoted by a dot over the appropriate symbol.

Changes in chemical composition take place through chemical reaction. Let there be s species and t reactions. Let n_i, $i = 1, ..., s$, be the number of moles of species i per unit mass. The n_i are not all independent because they are connected through the chemical reaction equations. It is convenient to replace the n_i by a set of progress variables Λ_j, $j = 1, ..., t$, one for each reaction. These also do not in general form an independent set since there are usually more reactions of kinetic interest than the number of Λ's needed to describe the composition. For our theoretical discussion we need an independent set to describe the composition, and this set, fewer in number than the Λ_j, can be obtained from linear combinations of them. The independent set is not unique, and in addition to the requirement that it provide a complete description of the composition we can require that it be convenient to use. Reducing the number of progress variables used is equivalent to reducing the number of equations describing the chemical transformation, and care must be taken to include the rates of all reactions correctly. Much of Appendix A is devoted to the methods used to effect the reduction.

Sec. 4A THE MODEL

It is obvious that we will be concerned with the s species, the t reactions, and some number of independent progress variables, and that there will be numbers of moles per unit mass, mass fractions, stoichiometric coefficients, rates, molecular weights, etc., to go with them. To simplify the expressions we use the usual vector notation in which the symbol written in boldface indicates the collection of all the individual subscripted members with that symbol, such as $\boldsymbol{\Lambda}$ to indicate the collection of all t of the Λ_j; $\boldsymbol{\Lambda} = (\Lambda_1, \Lambda_2, ..., \Lambda_t)$.

A chemical reaction is written as

$$\Sigma_i a_i X_i \leftrightharpoons \Sigma_i b_i X_i, \quad i = 1, ..., s \; ; \; \text{e.g.}, \; 2H + O \leftrightharpoons OH + H,$$

in which X_i represents a molar mass or a molecule of species i, and the dimensionless stoichiometric coefficients a_i and b_i (some of which may be zero) give the minimum number of molecules of each type participating in the reaction. We require that the chemical reaction equation represent an actual kinetic (collisional) mechanism, and we call it an *elementary* or *actual* reaction, distinguished, for example, from $2H_2 + O_2 \leftrightharpoons 2H_2O$, which simply indicates that hydrogen and oxygen combine by some mechanism to form water.

We can imagine a bimolecular reaction involving a two-body collision,

$$H + OH \leftrightharpoons H_2 + O,$$

and a corresponding termolecular reaction involving a three-body collision,

$$H + OH + O_2 \leftrightharpoons H_2 + O + O_2$$

with the O_2 contributing or carrying off some kinetic energy and momentum but otherwise unchanged in the collision. In each of these reactions the same chemical change takes place, and they are said to be *stoichiometrically* equivalent, but the kinetic mechanisms are different and they have different reaction rates. A set of elementary reaction equations may be linearly combined into a smaller set of stoichiometric reaction equations describing the chemical transformations. This is what is done in selecting the linear combination of Λ's. The effective reaction rates for the stoichiometric reactions are linear combinations of those for the elementary reactions. The procedure is described, with an example, in Appendix 4A.

In writing the equations of motion we need a mathematical expression of the conservation-of-mass constraint (the law of definite proportions) on the chemical transformation. This is obtained by introducing a slightly different set of net stoichiometric coefficients

$$\tilde{\nu}_i = b_i - a_i$$

and writing the reaction equation as

$$\Sigma_i \, \tilde{\nu}_i \, X_i = 0.$$

A coefficient $\tilde{\nu}_i$ is negative for reactants and positive for products, and is the net value if the same species appears as both reactant and product. For example, the elementary reaction

$$2H + O \rightleftharpoons OH + H$$

is written as

$$H + O \rightleftharpoons OH.$$

We also need for the equations of motion an expression for the rate of change of the concentration of species. As the reaction proceeds, the rates of change of the concentrations of the species involved are related by the law of definite proportions

$$\dot{n}_1/\tilde{\nu}_1 = \dot{n}_2/\tilde{\nu}_2 = \ldots \dot{n}_i/\tilde{\nu}_i = \ldots = \dot{n}_s/\tilde{\nu}_s.$$

A natural definition of the progress variable for the reaction is

$$\dot{\Lambda} = \dot{n}_i/\tilde{\nu}_i.$$

A reaction rate $\tilde{r}$ is defined for each reaction in terms of the rate of change of Λ for that reaction. It is usually given in units of the change of the number of moles per unit volume per unit time, so

$$\tilde{r} = \rho \dot{\Lambda}.$$

The most commonly used form of reaction rate is the dilute-gas or mass-action rate

Sec. 4A THE MODEL

$$\tilde{r} = k_f \Pi_i c_i^{a_i} - k_b \Pi_i c_i^{b_i}$$

$$c_i = \rho n_i, \qquad (4.1)$$

where c_i is the molar concentration in moles per unit volume, and k_f and k_b are the rate multipliers for the forward and back reactions. The multipliers depend on the state variables other than the composition. The Arrhenius form

$$k_f = k e^{-E^\dagger/RT}$$

with constant multiplier k and activation energy $E^\dagger$ is often used. The form for k_b is the same, of course, but k and $E^\dagger$ have different values.

Ordinary mass units, rather than molar units, are used in writing the equations of motion, and thus to add chemical reaction equations to the set of equations we need to express the reaction formalism in these units. We replace specific mole numbers n_i with mass fractions m_i, stoichiometric coefficients $\tilde{\nu}_i$ with new ones ν_i which make the equations represent the transformation of a unit mass of material, the progress variable Λ with the dimensionless progress variable λ, and reaction rate $\tilde{r}$ with r, defined as follows:

$$m_i = n_i M_i$$

$$\nu_i = \tilde{\nu}_i M_i / M \text{ with } M = 1/2 \, \Sigma_i |\tilde{\nu}_i| M_i$$

$$\dot{\lambda} = M \dot{\Lambda}$$

$$r = \dot{\lambda},$$

where M_i is the molecular weight of species i. The quantity M is the mass transformed in the reaction; it could as well have been defined as a sum over the net reactants or the net products, eliminating the 1/2 and the absolute value sign. Combining these definitions with the earlier ones, we find the new and old rates related by

$$r = (M/\rho)\tilde{r}$$

and the rate of change of mass fraction given by

$\dot m_i = \nu_i \dot\lambda.$

This equation can be integrated to give

$m_i - m_i^o = \nu_i(\lambda - \lambda^o);$

we usually take $\lambda^o = 0$, and choose the m_i^o to describe the initial composition, so that

$$m_i = m_i^o + \nu_i \lambda. \tag{4.2}$$

The sum of the mass fractions must be one, $\Sigma_i m_i = 1$, independent of λ. This condition is satisfied because our definition of ν_i is such that

$\Sigma_i \nu_i = 0.$

The chemical reaction equations are written in these new units as

$\Sigma_i \nu_i \{X_i\} = 0,$

where $\{X_i\}$ represents unit mass of species i. The equation we wrote in elementary form as

$2H + O \leftrightarrows OH + H,$

and in net value form as

$H + O \leftrightarrows OH,$

becomes in the new units (with $M = 17$)

$-1/17\ \{H\} - 16/17\ \{O\} + 17/17\ \{OH\} = 0.$

Thus 1/17 unit mass of reactant H (the negative sign identifies a reactant) combines with 16/17 unit mass of reactant O to form unit mass of product OH. The rate of the reaction can be written (in the mass-action approximation) as

$$r = M[\rho^{a-1} k_f \Pi_i (m_i/M_i)^{a_i} - \rho^{b-1} k_b \Pi_i (m_i/M_i)^{b_i}], \tag{4.3}$$

Sec. 4A   THE MODEL

where

$$a = \Sigma_i a_i$$

and

$$b = \Sigma_i b_i.$$

In this rate expression we have factored out ρ to display the density dependence of the rate. The rate expression, complicated as it is, with $k_f = k_f(\rho, T)$ and $m_i = m_i(\lambda)$, is already a gross simplification; however for modeling purposes we often choose the extreme simplification

$$A \to B \quad \text{(forward reaction only, } M = 1\text{)}$$

$$r = k(1 - \lambda)$$

with k = constant. Fortunately most of the interesting phenomena can be illustrated without using a realistic rate function.

Generalization to a set of t chemical reactions is straightforward, and we write the set of reaction equations as

$$\Sigma_{i=1}^q a_{ji} X_i \rightleftharpoons \Sigma_{i=1}^q b_{ji} X_i, \quad j = 1, \ldots, t$$

and define

$$\tilde{\nu}_{ji} = b_{ji} - a_{ji}$$

$$\nu_{ji} = \tilde{\nu}_{ji} M_i / M^j \text{ with } M^j = 1/2 \, \Sigma_i |\tilde{\nu}_{ji}| M_i$$

$$\dot{\lambda}_j = M^j \dot{\Lambda}_j$$

$$r_j = \dot{\lambda}_j.$$

The mass fraction is now given by

$$m_i = m_i^0 + \Sigma_j \nu_{ji} \lambda_j$$

taking $\lambda_j^0 = 0$ for all j. Once again the condition $\Sigma_i m_i = 1$ independent of λ is satisfied, since $\Sigma_i \Sigma_j \nu_{ji} = 0$ because, as discussed before for the single reaction, $\Sigma_i \nu_{ji} = 0$ for every reaction. The rate is related to the usual chemist's rate (the one usually found in tables) by

$$r_j = (M^j/\rho)\tilde{r}_j \qquad (4.4)$$

and in the mass action approximation is

$$r_j = M^j \left[\rho^{(a_j-1)} k_f \Pi_i (m_i/M_i)^{a_{ji}} - \rho^{(b_j-1)} k_b \Pi_i (m_i/M_i)^{b_{ji}} \right],$$

where $a_j = \Sigma_i a_{ji}$ and $b_j = \Sigma_i b_{ji}$.

As stated earlier, the set of λ's arising from these manipulations does not in general form a set of independent variables for the composition. There is a λ for each elementary reaction, and the number of elementary reactions may be quite large. The number considered depends on how closely one wishes to approximate the actual kinetic mechanisms, of which there are a great many. It is possible, as described in Appendix A, to reduce the number of λ's to a smaller number which does form an independent set of variables defining the composition, by expressing some as linear combinations of the others and thus eliminating them from the set. In all subsequent discussion we assume that the reduction to an independent set has been performed. Note, however, that in computational work the simplest algorithm results from integrating the rate equations for the elementary reaction set, since the reduction is avoided. It of course takes more computation time and storage space, a consideration which has become less important as computer power has increased.

Suppose there are n independent reactions. In later discussions we make much use of n-dimensional $\boldsymbol{\lambda}$-space, the space of their independent λ's. In this space the physically accessible region is the convex polytope defined by the condition that all mass fractions be nonnegative:

$$m_i \geq 0, \quad i = 1, ..., s.$$

Each face of the polytope is an $(n-1)$-dimensional hyperplane given by one of the equations $m_i = 0$. A few examples are given here and in Appendix A.

Example 1. Reaction $A \leftrightharpoons B$; initial conditions $m_A^0 = 1$, $m_B^0 = 0$. The stoichiometric coefficients are $\nu_A = -1$ and $\nu_B = 1$, or we

Sec. 4A THE MODEL

may write $\mathbf{v} = (-1,1)$; $m_A = 1 - \lambda$, and $m_B = \lambda$, and the two equations $m_i = 0$ give the region of accessible λ-space as $0 \leq \lambda \leq 1$. Here the polytope reduces to a line and the hyperplanes to the points $\lambda = 0$ and $\lambda = 1$, which are the pure-species points.

Example 2. Parallel reactions $A \rightleftarrows B$ and $A \rightleftarrows C$; initial conditions $m_A^0 = 1$, $m_B^0 = m_C^0 = 0$. The stoichiometric coefficients are $\mathbf{v}_1 = (-1,1,0)$ and $\mathbf{v}_2 = (-1,0,1)$, so the mass fractions are $m_A = 1 - \lambda_1 - \lambda_2$, $m_B = \lambda_1$, and $m_C = \lambda_2$. The $m_i = 0$ equations give the boundaries of the accessible space; the polytope is a triangle and the hyperplanes its sides, as shown in Fig. 4.1a. The corners of the triangle are pure-species points.

Example 3. Consecutive reactions $A \rightleftarrows B$, and $B \rightleftarrows C$; initial conditions $m_A^0 = 1$, $m_B^0 = m_C^0 = 0$. Thus $\mathbf{v}_1 = (-1,1,0)$ and $\mathbf{v}_2 = (0,-1,1)$, so $m_A = 1 - \lambda_1$, $m_B = \lambda_1 - \lambda_2$, $m_C = \lambda_2$. The polytope is again a triangle, but located differently from that of Example 2, as shown in Fig. 4.1b; note that λ_1 fixes the amount of A transformed and that the maximum λ_2 at given λ_1 lies on the line $m_B = 0$. The corners are pure-species points.

Example 4. Same as Example 1, but with 50% inert diluent. Reaction $A + I \rightleftarrows B + I$; initial conditions $m_A^0 = 1/2$, $m_B^0 = 0$, $m_I^0 = 1/2$. Thus $\mathbf{v} = (-1,1,0)$ and $m_A = 1/2 - \lambda$, $m_B = \lambda$, $m_I = 1/2$. This time the $m_i = 0$ equations limit the accessible region to $0 \leq \lambda \leq 1/2$.

Example 5. Reactions $A + B \rightleftarrows C$ and $A \rightleftarrows D$; initial conditions $m_A^0 = 2M_A/(2M_A + M_B)$, $m_B^0 = M_B/(2M_A + M_B)$, $m_C^0 = m_D^0 = 0$, so we start with two molecules of A for each molecule of B. In the previous examples all the M_i were equal, and the ν_{ji} either 0 or 1; here $\mathbf{v}_1 = [-M_A/(M_A + M_B), -M_B/(M_A + M_B), 1, 0]$ and $\mathbf{v}_2 = (-1,0,0,1)$, and $m_A = 2M_A/(2M_A + M_B) - [M_A/(M_A + M_B)] \lambda_1 - \lambda_2$, $m_B = M_B/(2M_A + M_B) - [M_B/(M_A + M_B)] \lambda_1$, $m_C = \lambda_1$, and $m_D = \lambda_2$. The accessible region is a trapezoid, and the corners are points where only two species coexist, as shown in Fig. 4.1c.

87

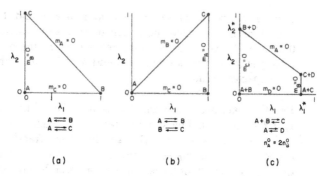

Fig. 4.1. Accessible regions of λ-space for some examples. In (c), the intercepts are $\lambda_1^* = 2M_A/(2M_A + M_B)$, $\lambda_2^* = (M_A + M_B)/(2M_A + M_B)$.

4A2. Equation of State and Rate

In discussing the equations of motion we choose to proceed as far as possible with a minimum set of assumptions before bringing in classical thermodynamics and the entropy. To this end we write the equation of state and rate in terms of the variables which appear in the equations of motion as they are derived from first principles. We thus write them in the form

$$E = E(p, \rho, \lambda)$$

$$r = r(p, \rho, \lambda).$$

We note again that the rate is assumed to depend only on the local state, and not on the flow gradients.

This form of the equation of state is "incomplete" in that it does not specify the temperature, which must be calculated by the integration procedure described in Appendix 4B.

As we develop the equations of motion, without using thermodynamics, we find that sound waves propagate with velocity

$$c = v[(p + E_v)/E_p]^{1/2},$$

where

$$E_v = (\partial E/\partial v)_{p,\lambda}$$

$$E_p = (\partial E/\partial p)_{v,\lambda}.$$

Sec. 4A THE MODEL

To assure that the sound speed be real, a necessary condition for our equations to make sense, it is sufficient to assume

$E_v > -p$

$E_p > 0.$

We make these assumptions throughout.

This is all that needs to be said before proceeding to the discussion of the equations of motion, but we digress here to relate E_v and E_p to quantities which may be more familiar, using classical thermodynamics.

With entropy $S = S(p,v,\lambda)$ we have, from the first law,

$E_p = TS_p$

$E_v = TS_v - p.$

The sound speed c is given by

$\rho^2 c^2 = -(\partial p/\partial v)_{s,\lambda} = S_v/S_p = (p + E_v)/E_p.$

Thus we see that the definition of c in terms of E_v and E_p agrees with the usual one in terms of the isentropic derivative, and also that the requirement $E_p > 0$ is equivalent to the requirement $S_p > 0$, which is commonly used.

As we go on to the discussion of shock and detonation waves, it simplifies the discussion to make an additional assumption to assure that the Rayleigh line intersects the Hugoniot curve in no more than two points, and we often call the equation of state satisfying this assumption "well-behaved." A simpler sufficient condition for most of the discussion is

$(\partial^2 p/\partial v^2)_{s,\lambda} > 0,$

which assures that the isentropes are concave upward. For a discussion of possible equation-of-state restrictions and their consequences see Courant and Friedrichs (1948), p. 141, and references given there, and Cowan (1958).

4A3. Equations of Motion

The equations of motion express the conservation of mass, momentum, and energy for a fluid. We use the inviscid equations, often called

FLOW IN A REACTIVE MEDIUM Chap. 4

the Euler equations, plus those describing the progress of the chemical reactions:[1]

$$\dot{\rho} + \rho \, \text{div} \, \mathbf{u} = 0 \qquad (4.5a)$$

$$\dot{\mathbf{u}} + v \, \mathbf{grad} \, p = 0 \qquad (4.5b)$$

$$\dot{E} + p\dot{v} = 0 \qquad (4.5c)$$

$$\dot{\boldsymbol{\lambda}} = \mathbf{r}. \qquad (4.5d)$$

For n reactions there are n + 5 equations. The dependent variables are p, ρ, $\boldsymbol{\lambda}$, and $\mathbf{u}$ (we use ρ and v interchangeably throughout), and the independent variables are x, y, z, and t (position and time). The dot over a symbol denotes its time derivative along a particle path, i.e.,

$$f = f(x,y,z,t)$$

$$\dot{f} = f_t + (\mathbf{u} \cdot \mathbf{grad})f,$$

where f is any function of the independent variables, e.g., E in Eq. (4.5c).

Specialized to one dimension (slab geometry), and a single reaction, these equations become

$$\dot{\rho} + \rho u_x = 0 \qquad (4.6a)$$

$$\dot{u} + v p_x = 0 \qquad (4.6b)$$

$$\dot{E} + p\dot{v} = 0 \qquad (4.6c)$$

$$\dot{\lambda} = r. \qquad (4.6d)$$

This special case is sufficient for most of our discussions. Other special cases are given in Appendix 4C.

To simplify the notation, most of the following discussion is for the case of a single reaction. The extension to more than one reaction is obvious in most cases. For some examples, see Appendix 4D.

---

[1] Notice that $\mathbf{u}$ and the operator **grad** are ordinary vectors in three-dimensional space, while $\boldsymbol{\lambda}$ and $\mathbf{r}$ are vectors in the reaction hyperspace.

4A4. Other Forms of the Equations of Motion

We display here other forms of the equations of motion, and discuss the significance of the various quantities which appear and the influence of the chemical reaction on the motion.

We begin by obtaining the relation between the changes in p, ρ and λ in a fluid element, an important relation which is fundamental to the rest of the discussion. We first eliminate $\dot{E}$ from the energy equation of motion $\dot{E} + p\dot{v} = 0$ by using the differential relation for the equation of state $E(p,\rho,\lambda)$,

$$dE = E_p dp + E_v dv + E_\lambda d\lambda,$$

and then solving for $\dot{p}$ to obtain the desired result:

$$\dot{p} = -\rho^2 c^2 \dot{v} + \rho c^2 \sigma \dot{\lambda} = c^2 \dot{\rho} + \rho c^2 \sigma \dot{\lambda} = c^2 \dot{\rho} + \rho c^2 \sigma r \qquad (4.7a)$$

$$c^2 = v^2(p + E_v)/E_p \qquad (4.7b)$$

$$\sigma = (\partial p/\partial \lambda)_{E,v}/\rho c^2. \qquad (4.7c)$$

The precise physical significance of the velocity c and of the dimensionless coefficient σ remain to be determined. We see that for the case of no reaction, $\dot{\lambda} = 0$, we have just $\dot{p} = c^2 \dot{\rho}$; since the change is adiabatic, c is the sound speed. With reaction, c turns out to be the characteristic speed, which is also the frozen sound speed defined below. The last term $\rho c^2 \sigma r$ gives the deviation in p-ρ space from the corresponding displacement without reaction. If σr is positive, the pressure change for a given density change is larger than it would be without reaction. The product σr is, roughly speaking, proportional to the rate at which chemical bond energy is transferred to the flow, that is, the rate at which it is transformed into thermal energy and motion.

Our equations of motion turn out to be of hyperbolic type. The important physical property of equations of this type is that they exhibit wave motion — the propagation of disturbances which are in some sense sharp-fronted. There are two different but essentially equivalent methods for the analytical study of such equations: the method of characteristics and the study of acceleration waves.

The method of characteristics (Courant and Friedrichs, 1948, Chapter 2; Thompson, 1972, Chapter 8; Chou and Hopkins, 1972, Chapter 6) is more familiar to workers in gas dynamics; we follow it here.

Although we will make no use of the acceleration-wave method here, it is sufficiently important and relevant that we comment on it briefly before proceeding. An acceleration wave is defined as a derivative discontinuity in p and u such as that at the head of a rarefaction wave. Its strength is defined as the jump in the value of the derivative. The method consists of studying the propagation and variations in strength of such waves. The relation between the two methods is apparent: acceleration waves propagate along characteristics. Although very old, this acceleration-wave approach has been developed extensively only in recent times, beginning with the work of Truesdell and his students. See, for example Truesdell (1969), and Chou and Hopkins (1972), Chapter 5.

The first step in putting the equations into characteristic form is to eliminate $\dot{\rho}$ from the mass conservation equation so that it contains derivatives of p and u only, as does the momentum equation. Using the adiabatic relation just derived to perform this elimination, we obtain the set

$$\dot{p} + \rho c^2 u_x = \rho c^2 \sigma r$$

$$\dot{u} + v p_x = 0$$

$$\dot{p} - c^2 \dot{\rho} = \rho c^2 \sigma r$$

$$\dot{\lambda} = r,$$

where we have chosen to write, as the third equation, the adiabatic relation, Eq. (4.7a), in place of the one for energy conservation. The first and third equations are now inhomogeneous, with the thermicity term $\rho c^2 \sigma r$ appearing explicitly on the right. Although the internal energy has been eliminated, the equation of state is still needed to calculate c.

The characteristic analysis finds certain directions, the characteristic directions, along which the partial derivatives combine into total derivatives. It results in a set of equations in the so-called characteristic form, which define the characteristic directions and define the differential relationships which must be satisfied along them. Our equations in characteristic form are

$$(dx/dt)_+ = u + c \tag{4.8a}$$

$$(dx/dt)_- = u - c \tag{4.8b}$$

$$(dx/dt)_0 = u \tag{4.8c}$$

$$(dp/dt)_+ + \rho c(du/dt)_+ = \rho c^2 \sigma r \tag{4.8d}$$

$$(dp/dt)_- - \rho c(du/dt)_- = \rho c^2 \sigma r \tag{4.8e}$$

$$\dot{p} - c^2 \dot{\rho} = \rho c^2 \sigma r \tag{4.8f}$$

$$\dot{\lambda} = r. \tag{4.8g}$$

The first three equations define the characteristic directions. The last of these states that the direction of the particle paths is characteristic. The last two equations (4.8f,g) are thus unchanged by the transformation since they contain only total derivatives in this direction. The rest of the transformation is straightforward: The equations (4.8d,e) relating p and u along the other characteristic directions are obtained from the first two of the original set by writing out $\dot{p}$ and $\dot{u}$ in terms of their partials with respect to x and t, multiplying the second equation by ρc, adding it to or subtracting it from the first, and collecting terms to form the total derivatives in the characteristic directions.

In problems involving the propagation of disturbances there must be a wavefront which separates the undisturbed region from the disturbed region. The importance of the ± characteristics, besides the simplification of the equations, is that, for weak disturbances, one of them is the wavefront. Their slopes $u \pm c$ are just the expression of the idea that sound waves (weak disturbances) propagate at velocity c with respect to the fluid particles moving at velocity u. The waves corresponding to the + characteristics, Eq. (4.8a), are called forward-facing, and those corresponding to the − characteristics, Eq. (4.8b), are called backward-facing. Discontinuities in the slope of the dependent variables, for example those at the head of a rarefaction wave, also propagate along + and − characteristics. (Discontinuities in the variables themselves form shocks and cross the characteristics.) Discontinuities in ρ and λ (but not in p and u) may be maintained along a particle-path characteristic, as for example at an interface between two different materials or at some other contact discontinuity.

To this point the principal assumptions, other than the neglect of transport properties, have been (1) that a state of local partial thermodynamic equilibrium exists everywhere, with only the chemical reactions not in equilibrium, and (2) that E and r are functions of p, ρ, and λ.

We now extend this assumption to include the local applicability of all of equilibrium thermodynamics (the usual assumption of irreversible thermodynamics). For this to be useful we have to assume that the temperature function $T(p,v,\lambda)$ is also given as part of the equation-of-state information. With our analytical apparatus augmented in this way, we can obtain an equation for the rate of entropy production by the chemical reaction, and relate quantities such as c and σ to more familiar ones.

To get the entropy-production equation, we begin with the thermodynamic differential relation for the internal energy:

$$\dot{E} = T\dot{S} - p\dot{v} + \Sigma_i \mu_i \dot{m}_i \qquad (4.9)$$

$$\mu_i = (\partial F/\partial m_i)_{T,p},$$

where F is the specific Gibbs free energy, S is the specific entropy, and μ_i is the specific chemical potential of species i. Using the relation $\dot{m}_i = \nu_i \dot{\lambda}$, the last term can be transformed to

$$\Sigma_i \mu_i \dot{m}_i = (\Delta F)\dot{\lambda} \qquad (4.10a)$$

$$\Delta F \equiv (\partial F/\partial \lambda)_{T,p} = \Sigma_i \nu_i \mu_i \qquad (4.10b)$$

$$\Delta \equiv (\partial/\partial \lambda)_{T,p}; \qquad (4.11)$$

combining the above expression for $\dot{E}$ with the energy equation of motion $\dot{E} + p\dot{v} = 0$ gives the rate of entropy production

$$\dot{S} = -(\Delta F/T)r = -(\Delta F/T)\dot{\lambda}. \qquad (4.12)$$

The quantity ΔF measures the deviation from chemical equilibrium, and is negative for excess of reactants and positive for excess of products. The condition for chemical equilibrium is that F be a minimum as a function of λ at constant T and p, that is, that

$$\Delta F = 0.$$

The coefficient ΔF and the rate r must have opposite signs, so that the entropy production is always positive, except at a point of chemical equilibrium, where both ΔF and r vanish. About a point of chemical equilibrium, the rate can be expanded in a Taylor's series in ΔF. The linear term is positive and nonvanishing {for more than one reaction it is the positive definite Onsager matrix [Wood and Salsburg (1960), Eq. (2.21)]}.

Sec. 4A THE MODEL

We next express c and σ in terms of more familiar quantities, and discuss their significance. The desired relations are obtained by the usual thermodynamic transformations, the details of which are given in Appendix 4D. We find

$$c^2 = (\partial p/\partial \rho)_{S,\lambda} \tag{4.13a}$$

$$\sigma \equiv (\partial p/\partial \lambda)_{E,v}/\rho c^2 \tag{4.13b}$$

$$= -(\beta/C_p)(\partial E/\partial \lambda)_{p,v} \tag{4.13c}$$

$$= \rho(\partial v/\partial \lambda)_{H,p} \tag{4.13d}$$

$$= \Delta v/v - \beta \Delta H/C_p, \tag{4.13e}$$

where the constant pressure heat capacity C_p and thermal expansion coefficient β are defined by

$$C_p = (\partial H/\partial T)_{p,\lambda} \; ; \; \beta = (\partial v/\partial T)_{p,\lambda}/v.$$

The velocity c is called the frozen sound speed because the derivative is taken at fixed composition. In some references it is denoted by c_o or a_o. Also of interest is the equilibrium sound speed

$$c_e^2 = (\partial p/\partial \rho)_{S,\Delta F=0}, \tag{4.14}$$

for which the derivative is taken at equilibrium composition, that is, with the composition shifting during the displacement so as to maintain chemical equilibrium. The relations between c and c_e, derived in Appendix 4D for n reactions, are, for one reaction

$$c_e^2 = c^2[1 - c^2\sigma^2/(\partial \Delta F/\partial \lambda)_{S,v}], \tag{4.15}$$

$$c_e \leq c.$$

It is the frozen sound speed c which appears in the characteristic equations. Thus an acceleration wave (slope discontinuity in p and u such as a rarefaction wave head) moves into a stationary medium with speed c. This result, while formally correct, does not tell the whole story. What happens is that after a time the bulk of the disturbance moves with equilibrium sound speed c_e, but is preceded by a precursor wave moving at speed c, whose amplitude decays exponentially with time. The phenomenon is similar to that seen in the propagation of electromagnetic waves in a dispersive medium (Stratton, 1941). Some examples are given in Sec. 4C. Typical differences in the values of frozen and equilibrium sound speeds are given in Sec. 5A5.

As stated earlier the product σr is a measure of the rate of transformation of chemical bond energy to molecular and bulk translational energy. In an unreacted explosive, the material consists entirely of reactants in metastable equilibrium with the reaction waiting to be triggered by some agency. Hence r is initially positive and σ must be positive if the material is to be an explosive. The last form of σ given above, Eq. (4.13e), is important because it shows that both the volume change and energy change of the reaction contribute to the explosive performance. Even a material with zero heat of reaction but with $\Delta v/v > 0$ and thus a positive σ would support a detonation and have the usual properties of an explosive. All useful explosives do in fact have a positive heat-release term (negative ΔH), but the volume-change term is often large enough to have important practical consequences in work with condensed explosives. An example of this may be found in the search for high-energy compounds which might make better explosives. At one time, boron compounds were thought to be quite promising because, with reasonable assumptions about the composition of the reaction products, quite large values of ΔH could be calculated. However, the volume-change term, while positive, is significantly smaller than in conventional explosives because the products consist of fewer molecules with larger molecular weight. Detailed calculations (Mader, 1961) showed that the improvement in performance for the intended applications is marginal at best.

For a mixture of two polytropic gases with the same (constant) specific heat capacity, transformed into each other by a single chemical reaction, the pressure, enthalpy, and σ are

$$p = nRT/v$$

$$H = C_p T - \lambda q,$$

$$\sigma = \Delta n/n + q/C_p T, \tag{4.16}$$

where n is the number of moles per unit mass and Δn is the mole change in the reaction.

Sec. 4A
THE MODEL

As noted earlier, we have written our results for a single reaction, but generalization to the n-reaction case is straightforward in most cases. Quantities like λ, σ, r, ΔF, etc. become n-component vectors. The products σr and $(\Delta F)r$ become dot products $\boldsymbol{\sigma}\cdot\mathbf{r}(= \Sigma_j\sigma_j r_j)$ and $\Delta\mathbf{F}\cdot\mathbf{r}$. The definition of $\boldsymbol{\sigma}$ would be written $(\partial p/\partial \boldsymbol{\lambda})_{E,v}$, meaning a vector with components $\sigma_i = (\partial p/\lambda_i)_{E,v,\lambda'}$, the prime on λ indicating constant λ_j, $j \neq i$. The general form of the relation between the frozen and equilibrium sound speeds, Eq. (4.15), is given in Appendix 4D.

4A5. Steady Solutions

For a uniform initial state the equations of motion have one-dimensional steady (time-independent) solutions of the form discussed in Chapter 2. These consist of a plane shock moving with constant velocity followed by a flow region in which the chemical reaction proceeds to completion. In a coordinate frame moving with the shock the flow is steady between the shock and the plane at which the reaction terminates.

The frame in which the flow is steady will be called the *shock frame*, or sometimes the *steady frame*, in contrast to the *laboratory frame* in which the undisturbed material is at rest. These are compared in Fig. 4.2. In the laboratory frame the shock moves toward positive x with velocity D, the velocity relative to the material ahead. In the shock

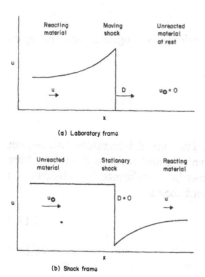

Fig. 4.2. Laboratory and shock coordinate frames. The numerical values of u_o and D are the same. Positive x is to the right and velocities are positive throughout.

frame, the shock is at rest at x = 0 and the undisturbed material moves into the shock with positive velocity u_o equal in sign and numerical magnitude to D. The direction in which the shock faces is reversed in the two frames so as to have positive velocities throughout. The velocity of a shocked particle with respect to the shock, a commonly occurring quantity, is u in the shock frame and D − u in the laboratory frame; it has of course the same numerical magnitude in either. We do not distinguish the frames by the symbols used for velocity, as is sometimes done (that is, here the symbol u stands for the particle velocity in the given frame in either case), and the frame being used must therefore be specified in each instance. An exception to this convention is the symbol D, which will sometimes be used interchangeably with u_o to denote the velocity of the undisturbed material in the steady frame.

Since by definition the steady solution does not depend on the time, we obtain the equations for it by dropping the time derivative terms from the flow equations. From Eqs. (4.5) we thus obtain

$$u\rho_x + \rho u_x = 0$$

$$\rho u u_x + p_x = 0$$

$$u E_x + p u v_x = 0$$

$$u \lambda_x = r,$$

with u the particle velocity in the shock frame. Together with the equation of state and the rate

$$E = E(p,\rho,\lambda)$$

$$r = r(p,\rho,\lambda),$$

these determine the steady solution. The partial derivatives with respect to x now become just ordinary derivatives and are so written from here on. The first three flow equations may be transformed by substitutions among themselves into the equivalent forms

$$d(\rho u)/dx = 0 \tag{4.17a}$$

$$d(\rho u^2 + p)/dx = 0 \tag{4.17b}$$

$$d(E + pv + 1/2\, u^2)/dx = 0. \tag{4.17c}$$

Sec. 4A   THE MODEL

These can be integrated immediately to give the Rankine-Hugoniot relations, which are usually transformed into

$$\rho u = \rho_0 u_0$$

$$\mathcal{R} = (\rho_0 u_0)^2 - (p - p_0)/(v_0 - v) = 0$$

$$\mathcal{H} = E(p,v,\lambda) - E(p_0,v_0,0) - 1/2\,(p + p_0)(v_0 - v) = 0.$$

As the reaction proceeds in the shocked material, λ changes from zero at the shock to the final equilibrium value, so that each point of the steady flow has an associated value of λ. With the appropriate value of λ inserted, these equations connect any point in the flow with the initial state. As we have seen in Chapter 2, for a given large enough value of u_0, they determine at most two flow states for each value of λ. Disregarding for the present the solution on the weak branch of the Hugoniot curve, whose accessibility will occupy much of our attention later, we see that they give a unique solution for the state variables p, u, and ρ as functions of λ. It remains to determine the dependence of λ on distance. This is obtained by integrating the rate equation beginning at $\lambda = 0$, $x = 0$ at the shock. Since all of the flow variables are now functions of λ alone, it becomes just as ordinary differential equation for $\lambda(x)$:

$$d\lambda/dx = r(\lambda)/u(\lambda).$$

It is often convenient to display the steady solution in the form of the state history seen by a moving observer. The only such observer we will have occasion to consider is one riding on a particle and recording its state as a function of time. All particles passing through the steady solution of course have the same history. Along a steady-solution particle path we may change from the independent variable x to the independent variable t through the transformation

$$dt = dx/u.$$

We make the convention that whenever the steady-state derivative is written with t as independent variable, a particle history is implied. The rate equation, for example, becomes

$$d\lambda/dt = r.$$

We use the derivative notation d/dt everywhere in writing the steady-state equations. Its physical meaning is of course identical to that of the dot used for the Lagrangian derivative in the time-dependent equations; both denote the change of a particle property with time.

Making use of the differential relation between p, ρ, and λ for an adiabatic change, Eq. (4.7a), we can obtain single equations for the time derivatives of each of the state variables along a steady-state particle path. These are

$$dp/dt = - \rho u^2 \sigma r/\eta \qquad (4.18a)$$

$$du/dt = u\sigma r/\eta \qquad (4.18b)$$

$$d\rho/dt = - \rho \sigma r/\eta \qquad (4.18c)$$

$$dE/dt = - pv\sigma r/\eta \qquad (4.18d)$$

$$dS/dt = - (\Delta F)r/T \qquad (4.18e)$$

$$d\lambda/dt = r \qquad (4.18f)$$

$$\eta = 1 - M^2;\ M = u/c. \qquad (4.18g)$$

These equations show that in the steady flow the state point in the p-v plane moves with increasing time down the Rayleigh line if $\sigma r/\eta > 0$, and up the Rayleigh line if $\sigma r/\eta < 0$. Immediately behind the shock wave the flow is subsonic and $\eta > 0$, so that if chemical energy is released to the flow, i.e., if $\sigma r > 0$, the pressure decreases, and if energy is absorbed from the flow, the pressure increases. If at any point η becomes zero, then σr must also become zero if the derivatives of the flow variables are to remain finite. The importance of the points at which these quantities vanish will be seen in the examples to follow.

Finally, we remark that it is sometimes convenient to write the equations for the steady solution in terms of velocities in the laboratory frame instead of using velocities in the shock frame. The change-over is easily accomplished by replacing u by $(D - u)$ and u_0 by D. As examples consider the integral mass conservation relation and the equation just given for du/dt:

Steady Frame	Laboratory Frame
$\rho_0 u_0 = \rho u$	$\rho_0 D = \rho(D - u)$
$du/dt = u\sigma r/\eta$	$-du/dt = (D - u)\sigma r/[1 - (D - u)^2/c^2]$

4A6. The Shock-Change Equation

A very useful relation, which we call the *shock-change* equation, relates the time rate of change of the strength of a shock to the velocity or pressure gradient, the reaction rate, and the sonic parameter η just behind it. The original idea goes back to Jouguet (1917), who obtained a rather complicated version without chemical reaction, and used it to analyze the decay of (instantaneous-reaction) spherical detonation waves. There is a large literature on its application to a great variety of materials; see, for example, Chapter 5 by Herrman and Nunziato in Chou and Hopkins (1972), or Chen and Gurtin (1971). We give a derivation in Appendix 4E. Specialized to plane flow, the result is (in the laboratory frame)

$$(dp/dt)_S = \rho c^2 (\sigma r - \eta u_x)/[1 + \rho_0 D (du/dp)_{\mathcal{H}}], \qquad (4.19)$$

where $(dp/dt)_S$ is the rate of change of shock pressure with time, $(du/dp)_{\mathcal{H}}$ is the rate of change of particle velocity with pressure along the Hugoniot curve for the unreacted material, u_x is the gradient of particle velocity, and $\eta = 1 - (D-u)^2/c^2$ is the sonic parameter. The denominator will be positive for most equations of state. In the numerator the term σr is related to the net rate at which chemical bond energy is released to the flow by the chemical reaction, and the term ηu_x is related to the rate at which energy is transported to the rear by the flow.

The condition that the flow be steady is that these two terms exactly balance, so that the numerator vanishes, and the shock pressure is constant in time. For positive σr and subsonic flow ($\eta > 0$), the gradient u_x (and therefore also p_x in the laboratory frame) must be positive to achieve the balance required for the steady state. Thus, in moving back away from the shock in a detonation, the particle velocity and the pressure fall as energy is released.

The equation has an additional application to steady reactive flow. Since the Rankine-Hugoniot equations apply at any point in the flow, it also applies at any point, provided that $(dp/dt)_S = 0$ is interpreted to mean that the pressure is constant with time along any line parallel to the path of the shock front in x-t space, and that $(du/dp)_{\mathcal{H}}$ is evaluated on the appropriate partial-reaction Hugoniot curve. The terms in the parentheses must therefore be of equal magnitude at every point in the steady flow. Setting the parentheses on the right to zero, in fact, and recalling $dx = udt$, gives the steady-flow equation (4.18b). Roughly speaking, we may say that the gradient acts like a rarefaction, transporting energy to the rear, and that this loss must be compensated by the energy supplied by the chemical reaction.

Adams and Cowperthwaite (1965) and Nunziato, Kennedy, and Hardesty (1976) have used the shock-change equation to study the initiation of explosives by shock waves. Pack and Warner (1965) used an assumed approximation to the shock-change equation, called Whitham's approximation (see Whitham, 1974, Chapter 8), in a similar study. The approximation doesn't work well for any but the earliest stages of the problem, because it can have no steady solutions. Pack and Warner recognize the difficulties, and get interesting results in spite of them. It is instructive to compare the methods and results. Additional work has been done by Cowperthwaite and Adams (1967) and Cowperthwaite (1969a).

4B. MATERIAL PROPERTIES

Simple models of the material have been much used in studying the solutions of the reactive flow equations. We make extensive use of them in our exposition, particularly in Chapter 5. Ideally, such a model should contain the principal features of interest, but be sufficiently simple to allow qualitatively correct results to be obtained analytically, thus exposing in optimum fashion the most interesting features of the solution. If approximate numerical results of useful accuracy can also be obtained in this way, so much the better. The motivation for constructing such a model is nicely described by Lighthill (1957) in his account of the construction of what has come to be called the "Lighthill gas."

The two perhaps most widely used simple models are described below. We have been able to use some variation of these wherever we have had occasion to work out a simple example for purposes of illustration. Both are based on the simple polytropic gas form for the equation of state of a pure component

$$E = pv/(\gamma - 1), \quad \gamma \text{ constant,} \tag{4.20}$$

and use some form of the forward Arrhenius reaction rate

$$r = k(1 - \lambda)e^{-E^\dagger/RT}. \tag{4.21}$$

The first form is the simplest and can be applied to both gaseous and condensed-phase explosives (with quite different values of the constants), but does not contain all of the desired properties. The second is an appropriate generalization of the Lighthill gas which contains in some form all the desired features but which applies directly only to ideal gases. The description of this second form also serves to illustrate in more complete fashion the application of the thermodynamic formulae of Sec. 4A4.

When an idealized fluid is needed for study, a single reaction is often assumed with its rate given by the mass-action law with Arrhenius rate constants. This is a severe oversimplification of real systems, which ordinarily have many elementary reactions. These can of course be replaced by a smaller number (but usually more than one) of stoichiometric reactions, but only at the cost of having the more complicated effective rates which are linear combinations of the elementary rates. As examples, see the ozone system described in Appendix 4A and the hydrogen-oxygen system described in Chapter 5.

4B1. The Polytropic Gas

This simplest model treats the system as an ideal mixture of polytropic gases all having the same (constant) heat capacity. The components A, B, C, ..., are transformed into each other by reactions with no mole change

$$A \leftrightarrows B, A \leftrightarrows C, B \leftrightarrows C, \ldots .$$

Thus both the heat capacity and the number of moles of the mixture are constant. There is, of course, no allowance for different sizes of the different molecules.

The equation of state is

$$E = pv/(\gamma - 1) - Q \tag{4.22a}$$

$$Q = \Sigma_i \lambda_i q_i \tag{4.22b}$$

with λ_i and q_i the progress variable and (constant) heat of (stoichiometric) reaction i. The isotherms are

$$pv = RT, \tag{4.23a}$$

and the constant-composition isentropes are

$$pv^\gamma = \text{constant}. \tag{4.23b}$$

Other quantities of interest are

$$H = E + pv = \gamma pv/(\gamma - 1) - Q \tag{4.23c}$$

$$c^2 = \gamma pv \tag{4.23d}$$

$$\sigma_1 = (\gamma - 1)q_1/c^2. \tag{4.23e}$$

For a binary mixture, with mole fractions $x_A = (1 - \lambda)$, $x_B = \lambda$, the equilibrium composition is given by

$$x_B/x_A = \exp[q/RT + (S_B - S_A)/R], \tag{4.24a}$$

with $S_B - S_A$ the (constant) entropy difference between the pure components. The frozen and equilibrium sound speeds for a binary mixture are related by

$$c_e^2/c^2 = 1 - \gamma\sigma^2/[\gamma^2\sigma^2/(\gamma - 1) + x_A^{-1} + x_B^{-1}]. \tag{4.24b}$$

Eqs. (4.24) are obtained by specializing the results of Sec. 4B2.

With appropriate values of the constants, this equation of state can be applied to both gases and condensed-phase explosives. For gases γ is the heat capacity ratio, always less than or equal to 5/3. For condensed-phase materials, we must abandon $pv = RT$, so γ can have larger values, determined by comparison with experiment, and typically about three. The equation of state is usually applied only to the detonation products, not the unreacted liquid or solid, and temperature and entropy are not available.[2] For many purposes the temperature is not required. Most hydrodynamic calculations, for example, need only an equation of state of the form $E(p,v)$. The limitation does, however, preclude consideration of reversible reactions and the accompanying phenomena of chemical equilibrium, as in Eqs. (4.24).

Some restrictions must be observed in applying this form to condensed-phase explosives. Consider the single-reaction case. For the products, represented by $\lambda = 1$, it has been widely used with reasonable success. For the reactants, represented by $\lambda = 0$, it is probably fairly accurate for pressures of the order of the detonation pressure, but is poor at low pressures — less than a few gigapascals. For example, the calculated initial sound speed is several orders of magnitude too small. All the partial-reaction Hugoniot curves have the same asymptote $v/v_o = (\gamma - 1)/(\gamma + 1)$, and the $\lambda = 0$ Hugoniot curve lies very close to the asymptote for all but very low pressures. (See Sec. 2C). For CJ detonation in a condensed-phase explosive the calculated von Neumann spike

---

[2]The problem of determining the temperature from the internal energy given in the form $E(p,v)$ is taken up in Appendix 4B. The function $T(p,v)$ is the solution of a partial differential equation whose coefficients are functions of p,v, and the partial derivatives of E. With physically reasonable choices for the initial conditions of integration, the polytropic gas form for E gives grossly inaccurate results for the temperature of the detonation products of condensed explosives.

pressure is twice the CJ pressure (for any value of γ), a not unreasonable result. Choosing q to match the measured CJ pressure and velocity usually gives a reasonable value in agreement with thermochemical estimates. For a steady-state calculation based on the von Neumann model, where the state point moves down the Rayleigh line and remains in a region of relatively high pressure, this equation of state is probably reasonably accurate.

The neglect of p_o often gives a considerable simplification of analytic results. It is clearly justified for condensed-phase explosives under the conditions of interest here. For gases the resulting error is also small; the calculated CJ pressures are typically in error by only a few percent.

Simple rate laws, like

$$r = k(1 - \lambda),$$

or

$$r = k(1 - \lambda)^{1/2},$$

which do not have any temperature dependence, are often used with this equation of state.

4B2. Binary Mixture of Different Polytropic Gases

This form allows for a mole change in the reaction and differences in the heat capacity, but is restricted to gaseous explosives. It consists of an ideal mixture of two polytropic gases with a single reversible reaction, represented as

$$A \rightleftharpoons (\delta + 1)B, \tag{4.25}$$

in which species A and B may have different (constant) heat capacities. In this equation the symbols A and B denote a mole of the species, and δ the mole change. The equation, then, is in the usual chemist's notation for a reaction, and δ is an integer or rational fraction.

All the expressions to follow are written using specific units, i.e., all quantities are stated for a system of unit mass. The only exceptions to this rule are the mole change δ and the molar gas constant $\tilde{R}$. The subscripts A and B indicate that the specific quantity subscripted is that for unit mass of the indicated (pure) species. The subscript o indicates that the quantity is to be evaluated at the initial pressure and temperature. The equations are written assuming $\lambda_o = 0$.

The initial number of moles per unit mass of A is $1/M_A$, where M_A is the molecular weight of species A, and the initial state ($\lambda = 0$) consists of pure A. At any stage of the reaction the total number of moles per unit mass is

$$n = (1 + \lambda\delta)/M_A \qquad (4.26)$$

The mole fractions x_A and x_B are

$$x_A = n_A/n = (1 - \lambda)/(1 + \lambda\delta) \qquad (4.27)$$

$$x_B = n_B/n = \lambda(1 + \delta)/(1 + \lambda\delta), \qquad (4.28)$$

where n_A and n_B are the number of moles of A and B. The equation of state is

$$pv = n\widetilde{R}T = (1 + \lambda\delta)RT \qquad (4.29)$$

with $R = \widetilde{R}/M_A$, so that R is the specific gas constant for species A. We use subscript zero to denote an arbitrary reference state (T_o, p_o) for each pure component. Letting

$$q = -\Delta H_o = -(H_B - H_A)_o \qquad (4.30)$$

$$\Delta C_p = C_{pB} - C_{pA}, \qquad (4.31)$$

the frozen heat capacities, internal energy, and enthalpy are

$$C_p = C_{pA} + \lambda\Delta C_p \qquad (4.32)$$

$$C_v = C_p - n\widetilde{R} = C_p - (1 + \lambda\delta)R \qquad (4.33)$$

$$E = E_{oA} + C_v(T - T_o) - \lambda q \qquad (4.34)$$

$$H = H_{oA} + C_p(T - T_o) - \lambda q = H_{oA} + C_{pA}(T - T_o)$$
$$+ \lambda[-q + \Delta C_p(T - T_o)]. \qquad (4.35)$$

The entropy and free energy are

$$S = S_{oA} + \lambda(S_B - S_A)_o + C_p \ln(T/T_o) - (1 + \lambda\delta)R \ln(p/p_o)$$
$$- (1 + \lambda\delta)R (x_A \ln x_A + x_B \ln x_b) \qquad (4.36)$$

$$F = H - TS. \tag{4.37}$$

For the quantities c^2, σ, and ΔF occurring in the flow equations, we need expressions for the quantity γ, and the partial derivatives of volume, enthalpy, and entropy. These are

$$\gamma = C_p/C_v \tag{4.38}$$

$$\Delta v = (\partial v/\partial \lambda)_{T,p} = [\delta/(1 + \lambda\delta)]v \tag{4.39}$$

$$\Delta H = (\partial H/\partial \lambda)_{T,p} = -q + \Delta C_p(T - T_0) \tag{4.40}$$

$$\Delta S = (\partial S/\partial \lambda)_{T,p} = (S_B - S_A)_0 + \Delta C_p \ln T/T_0$$
$$- \delta R \ln p/p_0 - R \ln (x_B^{1+\delta}/x_A). \tag{4.41}$$

In terms of these the desired quantities are

$$c^2 = \gamma p v \tag{4.42}$$

$$\sigma = \Delta v/v - \Delta H/(C_p T) \tag{4.43}$$

$$\Delta F = \Delta H - T\Delta S. \tag{4.44}$$

The condition for chemical equilibrium is

$$\Delta F = 0$$

or

$$x_B^{1+\delta}/x_A = (T/T_0)^{\Delta C_p/R}(p/p_0)^{-\delta} \exp\{[q - \Delta C_p(T - T_0)]/RT$$
$$+ (S_B - S_A)_0/R\}. \tag{4.45}$$

The mass-action rate is given by Eq. (4.3). The rate multipliers k_f and k_b are not independent since r must vanish at equilibrium. Taking advantage of this fact, and assuming that Eq. (4.25) is the elementary reaction, the mass-action rate can be transformed (see, for example, Fickett, Jacobson, and Wood, 1970, p. 49) to

$$r = k(1 - \lambda)e^{-E^{\dagger}/RT}(1 - e^{\Delta F/RT}). \tag{4.46}$$

We call this the "Arrhenius rate." The "irreversible-reaction" model that we use in many places has the back-reaction arbitrarily suppressed. It is the limit of Eq. (4.46) as $\Delta F \to -\infty$. The actual limit is, of course, physically unattainable, because it would require something like $(S_B - S_A)_o \to \infty$ in Eq. (4.45). However, the true equilibrium point, particularly in condensed-phase explosives, can lie so far to the right that the difference is of no practical consequence.

The equilibrium and frozen sound speeds are related by

$$\frac{c_e^2}{c^2} = 1 - \frac{(1 + \lambda\delta)\gamma\sigma^2}{(C_V/R)[\gamma\sigma - \delta(1 + \lambda\delta)^{-1}]^2 + (1 + \lambda\delta)^{-1}[x_A^{-1} + (1 + \delta)^2 x_B^{-1}]}, \quad (4.47)$$

where the right-hand side is to be evaluated at the values of the variables corresponding to the equilibrium composition at which the relation is to be applied. This expression is obtained from Eq. (4.15) by straightforward but lengthy manipulation of the partial derivatives. The Lighthill gas can be obtained from the above equations by setting $\delta = 1$ and $\Delta C_p = R$.

4C. REPRESENTATIVE FLOWS

We discuss here some representative one-dimensional flows, other than detonations, which illustrate some of the effects of adding chemical reaction to the effects considered. The first three sections are brief summaries of topics with which we are not directly concerned, included for the sake of completeness. Sec. 4C1 is just a reminder that the reader needs some background in a few of the elements of supersonic compressible flow. In Sec. 4C2 we point to the existence of hydrodynamic instability in a spatially uniform reacting system. Microscopic fluctuations give rise to local hot spots that continue to grow if the reaction is sufficiently sensitive to temperature. For this problem, heat conduction is an important effect. Sec. 4C3 is concerned with the acoustics of a reactive medium, an extensive subject with which we are not much concerned.

In Sec. 4C4 we consider the propagation of a shock into a reactive material at equilibrium. This differs from a detonation in that the initial state is a true equilibrium state, and not a state far from equilibrium preserved by a vanishingly small reaction rate. Because the initial state is one of true equilibrium, both the frozen and equilibrium Hugoniot curves pass through it, in contrast to the detonation case, so shocks of all strengths are possible and there is nothing like a CJ condition. If the shock is generated and supported by an instantaneously started constant-velocity piston, the initial shock is to a point on the frozen

Hugoniot curve. This initial configuration decays to the final steady solution consisting of a shock to a lower pressure point on the frozen Hugoniot curve, followed by a steady reaction zone. The lead shock jump, of course, displaces the state instantaneously, so the original composition is no longer the equilibrium composition, and reaction begins immediately to bring the system back to equilibrium. An important result is that, regardless of the sign of σ, the thermicity product σr is always negative, so the reaction process is effectively endothermic, and the pressure rises through the steady reaction zone.

Another interesting result comes from considering small piston velocities. If the piston velocity is so small that the steady-solution propagation velocity is less than the initial-state frozen sound speed, the shock decays to zero strength in finite time, and becomes the head of an acoustic precursor, running at the frozen sound speed, and decaying in strength exponentially with time. The bulk of the disturbance travels at a lower velocity that, in the limit of zero piston velocity, approaches the equilibrium sound speed. The steady solution extends to infinity in both directions, and is called a "diffuse" or "fully dispersed" shock. The roles of the two sound speeds are now clear; in the acoustic limit the equilibrium sound speed becomes the group velocity and the frozen sound speed becomes the phase velocity.

The rarefaction wave into a reactive equilibrium state, discussed in Sec. 4C5, has some of the same properties. At early time we have a self-similar frozen wave with the head propagating at frozen sound speed, and at late time an equilibrium wave with the head propagating at the slower equilibrium sound speed, preceded by an exponentially decaying precursor propagating at frozen sound speed.

4C1. Flow Without Reaction

The equation of motion (4.6) are, when $r = 0$, the usual equations for plane, time-dependent flow in a nonreacting medium, with the neglect of all transport effects. Since there are many excellent standard works on hydrodynamics which treat these flows, they will not be discussed here. It will be assumed, however, that the reader has some familiarity with them.

4C2. Reaction Without Flow

The case of a spatially uniform perturbation of material initially in a spatially uniform state is described by the equations (4.6) with initial values of u, u_x, and p_x all zero. This type of perturbation might be accomplished by heating by radiation which is very weakly absorbed by the medium to produce a uniform nonequilibrium state. The resulting sequence of states will be spatially uniform only if the system remains hydrodynamically stable at all times.

With large enough activation energy the system is unstable; local hot spots form and initiate flames, so that the flow becomes three-dimensional. Jones (1975) has treated this stability problem in the short wave-length limit for the Euler equations, with the result that the system is unstable for nonzero activation energy. Borisov (1974) gives an extensive qualitative discussion of the problem, with a review of the experimental work.

4C3. Sound Waves in a Reactive Mixture

Probably the simplest case of the interaction of the motion of a substance with the chemical reaction is in a mixture of reacting species in chemical equilibrium through which a sound wave is propagating. The infinitesimal variations in pressure and temperature caused by the sound wave cause infinitesimal shifts in the chemical equilibrium with delays characterized by the chemical reaction rate, and these infinitesimal shifts give rise to finite, and sometimes large, changes in the sound velocity, which is a ratio of infinitesimals. At a sufficiently low sound frequency the shift in equilibrium follows the variations caused by the sound wave, and the wave propagates at the equilibrium sound speed, Eq. (4.14). At sufficiently high sound frequency the equilibrium does not shift appreciably in the half period of the sound wave, and the wave propagates at the frozen sound speed, Eq. (4.13a). The energy of the sound wave is absorbed by a material which reacts chemically, as is indicated by the fact that the entropy production term is always positive, Eq. (4.12). The equations which describe the propagation of a plane sound wave in a reacting medium can be obtained by linearizing Eqs. (4.6) about the equilibrium composition, but for some applications, such as investigating material properties by measuring sound dispersion and velocity, transport properties must also be considered; see Herzfeld and Litovitz (1959). The complicated problem of the amplification of sound waves propagating in non-equilibrium reacting mixtures has also received considerable attention; see, for example, Garris et al. (1975).

4C4. Shock Wave in a Reactive Mixture

A shock wave propagating in a reactive mixture initially in thermodynamic equilibrium provides a reasonably simple illustration of the interaction of flow and reaction variables. The Hugoniot curves in the p-v plane are shown in Fig. 4.3. The equilibrium Hugoniot curve is the representation of Eq. (2.5) with $\Delta F(p,v,\lambda) = 0$ at every point, so that λ

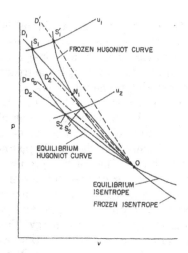

Fig. 4.3. Diagram of equilibrium and frozen Hugoniot curves and isentropes through an initial state of chemical equilibrium O. The Rayleigh line with propagation velocity $D = c_0$ shown as a chain line (— · — · —) is the dividing line between sharp and diffuse shocks.

changes along the curve. The frozen Hugoniot curve is the representation of Eq. (2.5) with λ held fixed at its value for equilibrium of the initial state. The frozen and equilibrium isentropes, which are tangent to their respective Hugoniot curves at the initial state point, are also shown in Fig. 4.3 for volumes greater than that of the initial state. For all substances having a thermodynamically stable equilibrium state, regardless of the sign of σ, the frozen isentrope and Hugoniot curve will be steeper than the corresponding equilibrium curves at the initial state O. For the simple substance of Sec. 4B2, consideration of the equilibrium relation $\Delta F = 0$ shows that these curves cross only at the initial state point. This is typical for a well-behaved equation of state. Thus the processes are endothermic, having $\sigma\tau < 0$ throughout.

If Fig. 4.3 had been drawn for a nonreacting material, only the frozen Hugoniot curve would appear. A shock wave, in such a medium, has a final state defined by the intersection of the Hugoniot curve with the Rayleigh line, Eq. (2.3), through the initial state. The slope of the Rayleigh line is proportional to the square of the shock velocity, and the shock velocity approaches the sound velocity as the shock pressure is reduced to zero. Thus the minimum shock velocity corresponds to the Rayleigh line tangent to the Hugoniot curve at the initial point, and less steep Rayleigh lines are physically meaningless.[3] For a shock in a reacting medium we have both the frozen and equilibrium Hugoniot curves. A Rayleigh line which intersects the equilibrium Hugoniot curve at both the initial point and some higher pressure point, representing a shock transition in the material, will, for small enough propagation

[3] We ignore intersections in the regions $v > v_0$.

velocity, intersect the steeper frozen Hugoniot curve only at the initial state. Because of this, there are two distinctly different types of shock solutions. They are divided by the Rayleigh line with propagation velocity $D = c_o$, shown as a chain line in Fig. 4.3, where c_o is the frozen velocity of sound, Eq. (4.13a), in the material at the initial state.

For $D > c_o$, say D_1, the steady solution consists of a shock discontinuity from O to N_1, on the Rayleigh line D_1, followed by a zone of chemical reaction carrying the state point from N_1 to S_1. The material velocity is raised to u_1, Eq. (2.4), which is therefore the steady piston velocity required to maintain the steady solution. The pressure profile is diagramed in Fig. 4.4a, together with some earlier profiles in the time-dependent flow produced by a piston instantaneously accelerated to this constant velocity. The shifting of composition due to reaction always results in a net shift from translational to chemical bond energy. The same piston velocity u_1 in the frozen material (i.e., with the reaction inhibited) would produce a shock, represented by the Rayleigh line D_1', to the state S_1' which would extend back to the piston. In the reacting material, a piston instantaneously accelerated to velocity u_1 at $t = 0$ produces, for times much less than the reaction time, just this shock to state S_1' because the material has not changed from the initial composition. As time goes on and the reaction proceeds, the flow pattern approaches the steady solution with final state S_1. Numerical time-dependent calculations, i.e., the integration of Eqs. (4.6), can be found in Fickett, Jacobson, and Wood (1970), p. 30.

For a propagation velocity $D < c_o$, say D_2, such as is represented by the Rayleigh line D_2 in Fig. 4.3, a complication arises. Signals moving at the frozen sound velocity in the initial state c_o will run ahead of the steady flow profile which propagates at the velocity D_2. No shock discontinuity is involved; all the points along the Rayleigh line between O and S_2 can be attained in sequence. The pressure profile of the steady solution is approximately profile 4 of Fig. 4.4b. It is similar to that for a viscous shock.

The steady piston velocity required to maintain this configuration is u_2. The corresponding frozen shock is the state S_2' on the Rayleigh line D_2'. This is the shock which a piston discontinuously accelerated to velocity u_2 would produce in the reacting material for times in which no appreciable reaction takes place. The shock strength decreases exponentially with time, and its velocity approaches the frozen sound speed c_o. The bulk of the disturbance moves at velocity D_2; the configuration eventually approached is the steady solution in a frame moving with this velocity. It is called a fully dispersed shock, or sometimes a diffuse shock, and was first described by Bethe and Teller (1942). A numerical

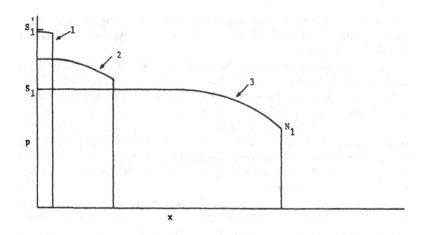

(a) $D > c_o$

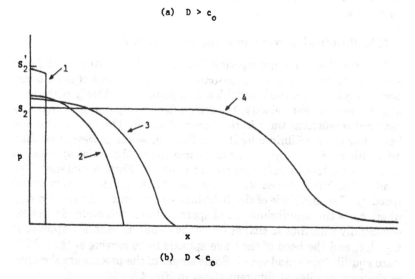

(b) $D < c_o$

Fig. 4.4. Shock waves in a gas in chemical equilibrium. The laboratory frame is used, and the shock is traveling to the right. (a) Sharp shock, $D > c_o$. (b) Diffuse shock, $D < c_o$. The numbered curves are diagrams of successive profiles in the flow generated by pistons instantaneously accelerated to the constant velocities u_1 and u_2, respectively, of Fig. 4.3. The last profile in each case is the steady configuration.

calculation of a similar time-dependent problem can again be found in Fickett, Jacobson, and Wood (1970), p. 31. Typical profiles are sketched in Fig. 4.4b.

It is important to notice how the steady-state shock wave in a reacting medium differs from the simpler steady shock in a nonreacting medium. In the nonreactive case the shock wave is localized at a point in the moving reference frame and is followed by a constant state. In the reactive case, the steady diffuse shock is always infinite in extent, and the sharp-fronted wave has a large extent which may be infinite or not depending on the form for the reaction rate. The steady solution is not a configuration which the wave has at any finite time, but only one which is approached in infinite time.

Diffuse shocks are ordinarily not of great practical importance since their strength is limited by the ratio of the frozen and equilibrium sound speeds. For typical gas systems the maximum pressure ratio for a diffuse shock is about 1.3.

4C5. Rarefaction Wave in a Reactive Mixture.

A rarefaction wave propagating into a reactive mixture initially at equilibrium provides another reasonably simple example of the interaction of flow and reaction variables. If a piston is suddenly withdrawn with finite constant velocity, the initial flow configuration is that of a centered rarefaction wave in the frozen material. The frozen isentrope lies below the equilibrium isentrope, Fig. 4.3, so the process is exothermic, with $\sigma r > 0$ throughout. At late times the configuration approaches that of a centered rarefaction wave in the equilibrium material. The head of the initial frozen wave moves at the initial-state frozen sound speed c_o. The amplitude of the disturbance between $x = c_o t$ and $x = c_{eo} t$, where c_{eo} is the equilibrium sound speed in the initial state, decays exponentially with time so that at $t = \infty$ the slope discontinuity appears at $x = c_{eo} t$, and the head of the wave appears to be moving at the initial-state equilibrium sound speed. Some stages of the process are sketched as pressure profiles at different times in Fig. 4.5.

Numerical calculations for flows of this type have been performed by Wood and Parker (1958), and by Arkhipov (1962). In these examples the approach to the equilibrium flow configuration is found to be very slow compared to the reaction rate.

APPENDIX 4A. CHEMICAL REACTION EQUATIONS

For the theoretical discussion of detonation we need a set of independent progress variables which define the chemical composition, and for

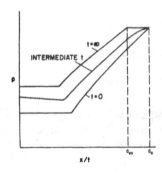

Fig. 4.5. Diagrams of profiles of a rarefaction wave generated in a gas in chemical equilibrium by suddenly withdrawing a piston with finite velocity. The points labeled c_∞ and c_0 are the equilibrium and frozen sound speeds in the initial state.

each of these the effective rate of the corresponding formal reaction. In this appendix we describe the method, with some examples, for the reduction of the elementary chemical reaction equations to an equivalent independent set of formal reactions.

In Sec. 4A1 we showed that a set of t elementary chemical reaction equations involving s species can be written

$$\sum_{i=1}^{s} \nu_{ji}\{X_i\} = 0, \quad j = 1, \ldots, t \tag{4A-1}$$

and also that

$$\sum_{i=1}^{s} \nu_{ji} = 0. \tag{4A-2}$$

For each of the t reactions of Eq. (4A-1) there is a progress variable λ_j, and in general they are not independent variables. For example, in the simple case of

$$A \rightleftharpoons B \tag{4A-3}$$

the elementary reactions are usually

$$A + A \rightleftharpoons A + B, \tag{4A-4}$$

which means that an A molecule collides with another A molecule and that one of them is transformed to a B molecule, and

$$A + B \rightleftharpoons B + B, \tag{4A-5}$$

in which A collides with B to make the transformation.

In the usual mass action approximation of chemical kinetics, the rates for Eqs. (4A-3), (4A-4), and (4A-5) can be written, as in Eq. (4.3),

$$r_3 = M_A(\vec{k}_3 m_A/M_A - \overleftarrow{k}_3 m_B/M_B)$$

$$r_4 = 2M_A\rho[\vec{k}_4(m_A/M_A)^2 - \overleftarrow{k}_4(m_A/M_A)(m_B/M_B)]$$

$$r_5 = (M_A + M_B)\rho[\vec{k}_5(m_A/M_A)(m_B/M_B) - \overleftarrow{k}_5(m_B/M_B)^2].$$

In the notation of Eq. (4A-1), all three of Eqs. (4A-3), (4A-4), and (4A-5) are identical, and thus obviously not independent. For our purposes the chemical transformation by all three reactions can be represented by the single formal reaction

$$A \leftrightarrows B,$$

whose rate of transformation is just the sum

$$r = r_3 + r_4 + r_5,$$

and whose progress variable is

$$\lambda = \lambda_3 + \lambda_4 + \lambda_5.$$

Notice how complicated the rate is for this very simple (unphysically simple) set of chemical reactions. The six $\vec{k}_i$ and $\overleftarrow{k}_i$ are all functions of temperature and pressure, λ enters through the m_i in various ways, and the density dependence in r_3 is different from that in r_4 and r_5.

Another kind of reaction which can also be represented by Eq. (4A-3) is the set

$$2A \leftrightarrows A_2$$

$$A_2 + A \leftrightarrows 2A + B,$$

if one of the reactions is much faster than the other. For this set

$$m_A = 1 - 2\lambda_1 + \lambda_2$$

$$m_{A_2} = \lambda_1 - \lambda_2$$

$$m_B = \lambda_2.$$

If the first reaction is very fast, it will maintain equilibrium, and is described by

App. 4A   CHEMICAL REACTION EQUATIONS

$$K = c_{A_2}/c_A^2 = \rho n_{A_2}/(\rho n_A)^2$$

$$= \rho^{-1}(m_{A_2}/M_{A_2})/(m_A/M_A)^2$$

$$= (M_A/2\rho)(\lambda_1 - \lambda_2)/(1 - 2\lambda_1 + \lambda_2)^2,$$

where K is the equilibrium constant, usually a function of temperature and pressure. We assume that K is small, because if it is not, the initial composition is not even approximately pure A. In this case

$$\lambda_1 \simeq \lambda_2 = \lambda$$

and $m_A = 1 - \lambda$, $m_{A_2} = (2\rho K/M_A)(1 - \lambda)^2$ from the equilibrium equation, and $m_B = \lambda$. The rate for the reaction set is the rate for the second reaction, because the first is, by assumption, so fast that it stays in equilibrium. With the mass action approximation the rate is

$$r = 3M_A[\rho \bar{k}(m_{A_2}/M_{A_2})(m_A/M_A) - \rho^2 \bar{k}(m_A/M_A)^2(m_B/M_B)]$$

$$= 3(\rho/M_A)^2[2K\bar{k}(1 - \lambda)^3 - \bar{k}\lambda(1 - \lambda)^2].$$

This rate must be added to the other rates to describe the progress of the formal reaction

$$A \leftrightarrows B.$$

The examples given above show several interesting features. The representation of the reactions as $A \leftrightarrows B$ is not as far from reality as might be thought. The rates, however, are very complicated. The number of elementary reactions, and therefore the number of parts to the rate expression, is large, and is usually reduced to the smallest number which will give the accuracy desired. There are many more possibilities, for example the termolecular reactions, than we have considered here.

As an illustration of further properties of the set of elementary reactions, consider the following set of reactions for collisions among three species of oxygen, elementary reactions written in the usual molar units on the left, and in the form for transformation of unit mass on the right:

(1) $O_3 + O_3 \leftrightarrows O + O_2 + O_3 \quad -1\{O_3\} + 1/3\{O\} + 2/3\{O_2\} = 0$

(2) $O_3 + O \leftrightarrows 2O + O_2 \quad\quad\quad -1\{O_3\} + 1/3\{O\} + 2/3\{O_2\} = 0$

(3) $O_3 + O_2 \rightleftharpoons O + 2O_2$ $\quad -1\{O_3\} + 1/3\{O\} + 2/3\{O_2\} = 0$

(4) $O_3 + O \rightleftharpoons 2O_2$ $\quad -3/4\{O_3\} - 1/4\{O\} + 1\{O_2\} = 0$

(5) $O_2 + O \rightleftharpoons 3O$ $\quad -1\{O_2\} + 1\{O\} = 0$

(6) $O_2 + O_3 \rightleftharpoons 2O + O_3$ $\quad -1\{O_2\} + 1\{O\} = 0.$

Writing $i = 1,2,3$ for O, O_2, and O_3 respectively, and $\mathbf{v}_j$ for the collection $(\nu_{j1}, \nu_{j2}, \nu_{j3})$, the reaction equations become

$$\mathbf{v}_1 = \mathbf{v}_2 = \mathbf{v}_3 = (1/3, 2/3, -1)$$

$$\mathbf{v}_5 = \mathbf{v}_6 = (1, -1, 0)$$

$$\mathbf{v}_4 = (-1/4, 1, -3/4) = 3/4\, \mathbf{v}_1 - 1/2\, \mathbf{v}_5.$$

Clearly there are only two independent reactions, and we might choose

$$\mathbf{v}_a = \mathbf{v}_1 = (1/3, 2/3, -1)$$

$$\mathbf{v}_b = \mathbf{v}_5 = (1, -1, 0)$$

to represent the set. For the rates we must satisfy the fundamental relationship

$$\dot{m}_i = \Sigma_j \nu_{ji} r_j.$$

To accomplish this we just take the expression of each ν_{ji} in terms of ν_{ai}, and ensure that all terms are included, by using the coefficients of the ν_{ai}. The appropriate rates (related to ordinary chemists rates by Eq. (4.4)) are

$$r_a = r_1 + r_2 + r_3 + 3/4\, r_4$$

$$r_b = -1/2\, r_4 + r_5 + r_6.$$

There is no unique choice of the two independent reaction equations, and they may be chosen to satisfy special requirements. One might choose instead

$$\mathbf{v}_A = \mathbf{v}_1 + 2/3\,\mathbf{v}_5 = (1,0,-1)$$

$$\mathbf{v}_B = \mathbf{v}_1 - 1/3\,\mathbf{v}_5 = (0,1,-1),$$

so that the first reaction represents the production of O from O_3 and the second the production of O_2 from O_3. With this choice we have

$$\mathbf{v}_1 = \mathbf{v}_2 = \mathbf{v}_3 = 1/3\,\mathbf{v}_A + 2/3\,\mathbf{v}_B$$

$$\mathbf{v}_4 = -1/4\,\mathbf{v}_A + \mathbf{v}_B$$

$$\mathbf{v}_5 = \mathbf{v}_6 = \mathbf{v}_A - \mathbf{v}_B,$$

so the rates are

$$r_A = 1/3\,(r_1 + r_2 + r_3) - 1/4\,r_4 + r_5 + r_6$$

$$r_B = 2/3\,(r_1 + r_2 + r_3) + r_4 - r_5 - r_6.$$

The coefficients of the r_i in these expressions could equally well be chosen by inspection of the equations in their original form to see how O and O_2 are produced in or removed from the system by the reactions. The progress variables are defined by

$$\dot{\lambda}_A = r_A, \quad \dot{\lambda}_B = r_B,$$

and the mass fractions (assuming the initial state is pure O_3) by

$$m_3 = 1 - \lambda_A - \lambda_B$$

$$m_1 = \lambda_A$$

$$m_2 = \lambda_B.$$

The diagram in λ-space is like that of Fig. 4.1(a).
The other choice, with

$$\mathbf{v}_a = (1/3, 2/3, -1),\ \mathbf{v}_b = (1,-1,0),$$

yields

$m_3 = 1 - \lambda_a$, $m_1 = 1/3\, \lambda_a + \lambda_b$, and $m_2 = 2/3\, \lambda_a - \lambda_b$.

The accessible region of λ-space now includes negative values of λ_b. It is shown in Fig. 4A-1.

The examples are simple enough that the relationships can be seen by inspection, but it is easy to imagine a set of reaction equations which are not so simple. Fortunately there are methods for finding an independent set of equations for any homogeneous system of the form of Eq. (4A-1), and these are described in any textbook of linear algebra, for example Franklin (1968), Shields (1964). If, as in Eq. (4A-1), there are t reactions among s species, the terms ν_{ji} form a t × s matrix, with t rows and s columns. The rank n of the matrix is defined as the dimension of the largest nonzero determinant which can be formed from the rows and columns of the matrix. Determinants are square, and therefore the rank cannot be larger than the smaller of t and s; Eq. (4A-2) states that each row of the matrix has zero as its sum, and reduces the rank at least one. It can be shown that the number of independent reactions needed to represent the set is the rank of the matrix. A possible choice for the independent reactions is the set made up from Eq. (4A-1) of those in the rows of the matrix used to make the nonzero determinant.

Suppose an independent set of equations

$$\sum_{i=1}^{s} \hat{\nu}_{ki}\{X_i\} = 0, \quad k = 1, \ldots, n$$

is chosen. Then it will be possible to express the original ν_{ji} in terms of the $\hat{\nu}_{ki}$, since they are dependent, as

$$\mathbf{v}_j = \sum_k c_{jk} \hat{\mathbf{v}}_k, \quad j = 1, \ldots, t.$$

This vector equation represents t × s equations for the c_{jk}. For each j (that is, for each elementary reaction equation) there are a set of s equations, one for each component of $\mathbf{v}_j$. The rate for independent reaction equation k is then

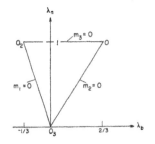

Fig. 4A-1. The accessible region of λ-space for one choice of the minimum set of reaction equations for the transformation of O_3 to O and O_2. Part of the accessible region has negative values of λ_b.

$$\hat{r}_k = \Sigma_{j=1}^j c_{jk} r_j, \quad k = 1, \ldots, n.$$

These n rates are for the n independent reactions, and corresponding to each we have

$$\dot{\lambda}_k = \hat{r}_k$$

$$\dot{m}_i = \Sigma_{k=1}^p \hat{\nu}_{ki} \hat{r}_k$$

$$\dot{m}_i = m_i^0 + \Sigma_{k=1}^p \hat{\nu}_{ki} \lambda_k.$$

The reduction to the minimum number of equations is essential for theoretical discussion, because the λ-space, an n-dimensional hyperspace with the independent λ's along the n axes, is revealing of the important properties of solutions to the detonation problem. Obtaining a set of independent λ's and minimizing the dimension of the space are important simplifications.

APPENDIX 4B. TEMPERATURE FROM INTERNAL ENERGY

In many hydrodynamic problems the natural form for the equation of state is $E = E(p,v)$, and usually other thermodynamic functions are not needed. The temperature can be obtained from the internal energy rather simply if, in addition, some initial data are known.

From the expression

$$dE = TdS - pdv \tag{4B-1}$$

we can write

$$(\partial E/\partial v)_T = T(\partial S/\partial v)_T - p \tag{4B-2}$$

and from the Maxwell equation (taken from $dA = -pdv - SdT$)

$$(\partial S/\partial v)_T = (\partial p/\partial T)_v = T_p^{-1}, \tag{4B-3}$$

where the subscript notation for partial differential implies $T = T(p,v)$. From $dE = E_p dp + E_v dv$ we find

$$(\partial E/\partial v)_T = E_p (\partial p/\partial v)_T + E_v \tag{4B-4}$$

and from the usual rule for partial derivatives

$$(\partial p/\partial v)_T = -T_v/T_p. \tag{4B-5}$$

Eqs. (4B-4) and (4B-5) can be combined to get

$$(\partial E/\partial v)_T = -E_p T_v/T_p + E_v. \tag{4B-6}$$

Eqs. (4B-2) and (4B-3) can be combined to give

$$(\partial E/\partial v)_T = T/T_p - p. \tag{4B-7}$$

Equating the right-hand sides of (4B-6) and (4B-7) we find

$$(E_v + p)T_p - E_p T_v = T, \tag{4B-8}$$

which is the partial differential equation for $T(p,v)$ with coefficients known if $E(p,v)$ is given.

Equation (4B-8) is hyperbolic, and by putting it in characteristic form we can identify the initial data that are needed to determine temperature. Along the characteristic

$$\left(\frac{dp}{dv}\right)_C = -\frac{E_v + p}{E_p} \tag{4B-9}$$

Eq. (4B-8) becomes

$$-T_p (dp/dv)_C - T_v = T/E_p \tag{4B-10}$$

or, equivalently,

$$(dT/dv)_C = -T/E_p. \tag{4B-11}$$

We recall [see Eq. (4.7b)] that

$$-(E_v + p)/E_p = (\partial p/\partial v)_s \tag{4B-12}$$

and realize that the characteristic curves defined by (4B-9) are isentropes. Thus if $E(p,v)$ is given in some region of p,v space, and T is given along some nonisentrope (noncharacteristic) arc, then $T(p,v)$ is determined over the band between the isentropes through the end points of the arc, as diagrammed in Fig. 4B-1.

These equations apply for any fixed value of λ. For each λ another initial data arc is required.

A good general discussion of this problem has been given by Cowperthwaite (1969b).

This approach can be extended, and the ratio of specific heats can be found. Equation (4B-8) can be divided through by T, and rewritten as

$$(E_v + p)(\partial \ln T/\partial p)_v - E_p(\partial \ln T/\partial v)_p = 1. \qquad (4B\text{-}13)$$

Then the partial derivative with respect to ln p is taken, and the order of differentiation interchanged as needed, remembering that

$$[\partial(\partial \ln T/\partial p)_v/\partial \ln p]_v = [\partial(\partial \ln T/\partial \ln p)_v/\partial p]_v$$

$$- (\partial \ln T/\partial p)_v, \qquad (4B\text{-}14)$$

and using Eq. (4B-13) to eliminate $(\partial \ln T/\partial v)_p$. Then we define

$$Z = (\partial \ln T/\partial \ln p)_v, \qquad (4B\text{-}15)$$

and rearrange to find

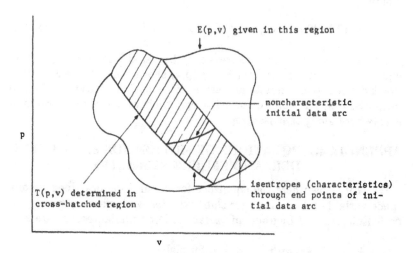

Fig. 4B-1. Temperature integration for E(p,v).

FLOW IN A REACTIVE MEDIUM   Chap. 4

$$-[(E_v + p)/E_p]Z_p + Z_v = (p/v)\{Z[v(E_{vp} + 1)/pE_p$$
$$- vE_{pp}(E_v + p)/pE_p^2 - v(E_v + p)/p^2 E_p] + vE_{pp}/E_p^2\}. \qquad (4B\text{-}16)$$

The left-hand side is just $(dZ/dv)_C$, analogous to Eq. (4B-11). The expression in the square brackets on the right hand side is easily shown, using the identity $\gamma = v(E_v + p)/pE_p$, to be just γ_p. The last term on the right-hand side is found, using $\beta = E_p/v$, to be β_p/β^2. Thus Eq. (4B-16) can be written

$$(dZ/dv)_C = (p/v)(Z\gamma_p + \beta_p/\beta^2). \qquad (4B\text{-}17)$$

Some manipulation of thermodynamic derivatives gives the identity

$$Z^{-1} = \beta\gamma(1 - C_v/C_p). \qquad (4B\text{-}18)$$

An identical procedure, except that the derivative of Eq. (4B-13) is performed with respect to ln v, yields

$$(dY/dp)_C = - (v/p)[Y\gamma_v/\gamma^2 - \alpha_v/(1 + \alpha)^2], \qquad (4B\text{-}19)$$

where

$$Y = (\partial \ln T/\partial \ln v)_p = [\beta(C_p/C_v - 1)]^{-1} \qquad (4B\text{-}20)$$

The application of these expressions, discussed briefly in Sec. 2B, to equations of state of the form $E = E(p,v)$ shows where these equations break down in their thermal properties. If T and C_p/C_v are known (or can be guessed) at any point on an isentrope, their values can be computed everywhere along that isentrope.

APPENDIX 4C. EQUATIONS OF MOTION FOR SLAB, CYLINDER, AND SPHERE SYMMETRY

We write down here the equations for lineal flows described by one space variable x and time t for slab, cylinder, and sphere symmetry in both Eulerian and Lagrangian variables. For Eulerian coordinate x:

$$\dot{\rho} + \rho u_x = - \rho\alpha u/x \qquad \begin{array}{l}\alpha = 0 \text{ for slab} \\ 1 \text{ for cylinder} \\ 2 \text{ for sphere}\end{array}$$

App. 4C EQUATIONS OF MOTION

$$\dot{u} + vp_x = 0$$

$$\dot{E} + p\dot{v} = 0$$

$$\lambda = r.$$

For $\alpha > 0$, x is the distance from the center.

Often in computational work it is convenient to replace x by the Lagrangian coordinate h, defined by

$$dh = \rho x^\alpha (dx - u\,dt),$$

so that h is mass per unit area for slab symmetry, mass per unit angle (radians) per unit length for cylindrical symmetry, and mass per unit solid angle (steradians) for spherical symmetry. Using the usual techniques of partial differentiation to change f(x,t) to f(h,t), it is easy to show that the particle-path (Lagrangian) derivative denoted by a dot over the symbol is $(\partial/\partial t)_h$, so that for any $f = f(h,t)$ it follows that $\dot{f} = f_t$.

After making the transformation we obtain

$$\dot{x} = u$$

$$\dot{\rho} + \rho^2 x^\alpha u_h = -\rho \alpha u/x \quad \text{or} \quad \dot{v} = (x^\alpha u)_h$$

$$\dot{u} + x^\alpha p_h = 0$$

$$\dot{E} + p\dot{v} = 0 \quad \text{or} \quad \dot{\mathcal{E}} + (pux^\alpha)_h = 0, \quad \mathcal{E} = E + 1/2\,u^2$$

$$\lambda = r.$$

The characteristic equations are, in the Eulerian coordinate x,

$$(dx/dt)_\pm = u \pm c; \quad \dot{x} = u$$

$$(dp/dt)_\pm \pm \rho c (du/dt)_\pm = \rho c^2 (\sigma \cdot r - \alpha u/x)$$

$$\dot{E} + p\dot{v} = 0$$

$$\lambda = r.$$

For the Lagrangian coordinate h, replace the first line by

FLOW IN A REACTIVE MEDIUM   Chap. 4

$(dh/dt)_\pm = \pm \rho c x^\alpha$; $\dot{x} = u$.

Throughout, the particle-path energy-conservation relation $\dot{E} + p\dot{v} = 0$ may be replaced by $\dot{p} = c^2 \dot{\rho} + \rho c^2 \boldsymbol{\sigma} \cdot \mathbf{r}$ or by $\dot{S} = -(\boldsymbol{\Delta F}/T) \cdot \mathbf{r}$ with $E(p,\rho,\boldsymbol{\lambda})$ replaced by $S(p,\rho,\boldsymbol{\lambda})$.

APPENDIX 4D. FROZEN AND EQUILIBRIUM SOUND SPEEDS AND σ

In this appendix we derive a relationship between the frozen and equilibrium sound speeds, and some different forms of σ. The manipulation of partial derivatives so common in thermodynamics is complicated here by the fact that when there are n independent reactions there are n + 2 independent variables, say $(S, v, \boldsymbol{\lambda})$ for example, in the expressions. We use the standard vector notation, where

$$\boldsymbol{\lambda} = (\lambda_1, \lambda_2, \ldots, \lambda_n),$$

and extend the notation to derivatives so that, for example, $(\partial T/\partial \boldsymbol{\lambda})_{S,v}$ is a vector whose ith component is $(\partial T/\partial \lambda_i)_{S,v,\lambda_j, j \neq i}$. In the subscript notation for partial differentiation, for which the independent variables must be specified, this vector derivative is written $T_{\boldsymbol{\lambda}}$. The second derivative $T_{\boldsymbol{\lambda\lambda}}$ is an n × n matrix with element ij (the jth component of the ith row) equal to $(\partial^2 T/\partial \lambda_i \partial \lambda_j)_{S,v}$. The dot product $T_{\boldsymbol{\lambda\lambda}} \cdot d\boldsymbol{\lambda}$ is the vector whose ith component is $\Sigma_j (\partial^2 T/\partial \lambda_i \partial \lambda_j)_{S,v} d\lambda_j$. Equation (4.10b) defines $\Delta F = (\partial F/\partial \lambda)_{T,p}$ for one reaction, and this definition must be extended to n reactions where it becomes a vector $\boldsymbol{\Delta F}$ whose ith component ΔF_i is $(\partial F/\partial \lambda_i)_{T,p}$. Notice that in the subscript notation the independent variables are $(S,v,\boldsymbol{\lambda})$, while in defining $\boldsymbol{\Delta F}$ we used $(T,p,\boldsymbol{\lambda})$; thus $\boldsymbol{\Delta F} \neq F_{\boldsymbol{\lambda}}$. The derivative $(\partial \boldsymbol{\Delta F}/\partial \boldsymbol{\lambda})_{S,v} = \boldsymbol{\Delta F}_{\boldsymbol{\lambda}}$ is an n × n matrix with element ij equal to $(\partial \Delta F_i/\partial \lambda_j)_{S,v}$, and the dot product $(\partial \boldsymbol{\Delta F}/\partial \boldsymbol{\lambda})_{S,v} \cdot d\boldsymbol{\lambda}$ is a vector whose ith component is $\Sigma_j (\partial \Delta F_i/\partial \lambda_j)_{S,v} d\lambda_j$.

The frozen sound speed is most simply defined for the independent variables $(S, v, \boldsymbol{\lambda})$

$$v^2 (\partial p/\partial v)_{S,\boldsymbol{\lambda}} = -c^2, \tag{4D-1}$$

and the equilibrium sound speed for the set $(S, v, \boldsymbol{\Delta F})$:

$$v^2 (\partial p/\partial v)_{S, \boldsymbol{\Delta F} = 0} = -c_e^2. \tag{4D-2}$$

The relation between these two sets is

App. 4D   FROZEN AND EQUILIBRIUM SOUND SPEEDS

$$d(\Delta F) = (\partial \Delta F/\partial S)_{v,\lambda} dS + (\partial \Delta F/\partial v)_{S,\lambda} dv$$
$$+ (\partial \Delta F/\partial \lambda)_{S,v} \cdot d\lambda. \quad (4D\text{-}3)$$

Equation (4.9)

$$dE = TdS - pdv + \Delta F \cdot d\lambda \quad (4D\text{-}4)$$

can be used to express the partial derivatives in terms of familiar quantities. Here

$$T = (\partial E/\partial S)_{v,\lambda} = E_S$$
$$-p = (\partial E/\partial v)_{S,\lambda} = E_v$$
$$\Delta F = (\partial E/\partial \lambda)_{S,v} = E_\lambda, \quad (4D\text{-}5)$$

and from the requirement that the order of partial differentiation not affect the result, that is, $E_{S,\lambda} = E_{\lambda,S}$, which gives the Maxwell relations of thermodynamics, we obtain

$$E_{S,\lambda} = (\partial T/\partial \lambda)_{S,v} = T_\lambda \quad (4D\text{-}6)$$

$$E_{\lambda,S} = (\partial \Delta F/\partial S)_{v,\lambda} = \Delta F_S \quad (4D\text{-}7)$$

and therefore

$$T_\lambda = \Delta F_S. \quad (4D\text{-}8)$$

Similarly, from

$$E_{\lambda,v} = E_{v,\lambda},$$

we obtain

$$\Delta F_v = -p_\lambda. \quad (4D\text{-}9)$$

Also, differentiating ΔF with respect to λ in (4D-5), we have

$$\Delta F_\lambda = E_{\lambda\lambda} \equiv \varphi, \quad (4D\text{-}10)$$

where we have defined

$$\phi_{ij} = \left(\frac{\partial^2 E}{\partial \lambda_i \partial \lambda_j}\right)_{S,v} \tag{4D-11}$$

Substituting from Eqs. (4D-8), (4D-9), and (4D-10) into Eq. (4D-3), we have

$$d(\Delta F) = T_\lambda dS - p_\lambda dv + \boldsymbol{\phi} \cdot d\boldsymbol{\lambda}. \tag{4D-12}$$

This is solved for $d\boldsymbol{\lambda}$ by multiplying from the left by the inverse[4] $\boldsymbol{\phi}^{-1}$ of $\boldsymbol{\phi}$, and

$$d\boldsymbol{\lambda} = -\boldsymbol{\phi}^{-1} \cdot T_\lambda dS + \boldsymbol{\phi}^{-1} \cdot p_\lambda dv + \boldsymbol{\phi}^{-1} \cdot d(\Delta F). \tag{4D-13}$$

From this expression we obtain

$$(\partial \boldsymbol{\lambda}/\partial v)_{S,\Delta F} = \boldsymbol{\phi}^{-1} \cdot p_\lambda. \tag{4D-14}$$

Now from the rules for partial differentiation

$$\left(\frac{\partial p}{\partial v}\right)_{S,\Delta F} = \left(\frac{\partial p}{\partial v}\right)_{S,\lambda} + \left(\frac{\partial p}{\partial \boldsymbol{\lambda}}\right)_{S,v} \left(\frac{\partial \boldsymbol{\lambda}}{\partial v}\right)_{S,\Delta F} \tag{4D-15}$$

and the terms in this expression can be replaced to give

$$-\rho^2 \hat{c}_e^2 = -\rho^2 c^2 + p_\lambda \cdot \boldsymbol{\phi}^{-1} \cdot p_\lambda, \tag{4D-16}$$

where $\hat{c}_e$ is defined by

$$-\rho^2 \hat{c}_e^2 = (\partial p/\partial v)_{S,\Delta F}$$

in which ΔF is constant but not necessarily zero. Equation (4D-16) can be rearranged as

$$\hat{c}_e^2 = c^2 - v^2(p_\lambda \cdot \boldsymbol{\phi}^{-1} \cdot p_\lambda). \tag{4D-17}$$

Now we specialize to $\Delta F = 0$. Again from the rules for partial differentiation,

$$p_\lambda = (\partial p/\partial \boldsymbol{\lambda})_{S,v} = (\partial p/\partial \boldsymbol{\lambda})_{E,v} + (\partial p/\partial E)_{v,\lambda}(\partial E/\partial \boldsymbol{\lambda})_{S,v}, \tag{4D-18}$$

[4] In the discussion following Eq. (4D-26) below we show that $\boldsymbol{\phi}^{-1}$ exists.

App. 4D FROZEN AND EQUILIBRIUM SOUND SPEEDS

but at $\Delta F = 0$

$$(\partial E/\partial \lambda)_{S,v} = \Delta F = 0 \qquad (4D\text{-}19)$$

and

$$p_\lambda = (\partial p/\partial \lambda)_{E,v}. \qquad (4D\text{-}20)$$

From the definition of σ

$$(\partial p/\partial \lambda)_{E,v} = \rho c^2 \sigma, \qquad (4D\text{-}21)$$

so that

$$p_\lambda = \rho c^2 \sigma \qquad (4D\text{-}22)$$

and finally, after substituting back into Eq. (4D-17), we find

$$c_0^2 = c^2(1 - c^2 \sigma \cdot \varphi^{-1} \cdot \sigma). \qquad (4D\text{-}23)$$

When this is specialized to one reaction, Eq. (4.15), the term

$$\sigma \cdot \varphi^{-1} \cdot \sigma \qquad (4D\text{-}24)$$

becomes

$$\sigma^2/(\partial \Delta F/\partial \lambda)_{S,v}. \qquad (4D\text{-}25)$$

It remains to be shown that (4D-24) or (4D-25) is always positive. We do this by expanding $E(S,v,\lambda)$ in λ about $\Delta F = 0$, to find

$$E(S,v,\lambda) = E(S,v,\lambda_e) + E_\lambda \cdot d\lambda + d\lambda \cdot \varphi \cdot d\lambda + \ldots. \qquad (4D\text{-}26)$$

Because $E_\lambda = \Delta F$, the second term is zero at equilibrium. Now at equilibrium a closed system must have a minimum in the energy at constant S,v. Therefore the third term in Eq. (4D-26) can never be negative, and this requirement is expressed by saying that the matrix φ must be positive definite. A theorem of matrix algebra states that the inverse of a positive definite matrix is also positive definite, and that therefore

$$\sigma \cdot \varphi^{-1} \cdot \sigma \geq 0. \qquad (4D\text{-}27)$$

Thus it follows from Eq. (4D-23) that we arrive at the result stated in Eq. (4.15)

$$c_e \leq c. \tag{4D-28}$$

For the different forms of σ, Eq. (4.13), we begin by obtaining some preliminary results. We now change our subscript notation to denote partial derivatives for the set of independent variables

(p, v, λ).

The preliminary results needed are

$$E_p = TS_p \tag{4D-29}$$

$$H_v = TS_v = \rho C_p/\beta \tag{4D-30}$$

$$\rho c^2 = C_p/\beta E_p \tag{4D-31}$$

$\beta \equiv (\partial v/\partial T)_{p,\lambda}/v.$

The first and second of these are obtained directly from

$$dE = TdS - pdv + \Delta F \cdot d\lambda$$

$$dH = TdS + vdp + \Delta F \cdot d\lambda$$

and

$$H_v = (\partial H/\partial T)_{p,\lambda}(\partial T/\partial v)_{p,\lambda} = \rho C_p/\beta.$$

The third is obtained as follows, using the first two:

$$\rho^2 c^2 = -(\partial p/\partial v)_{s,\lambda} = S_v/S_p = \rho C_p/\beta E_p.$$

With these in hand, we transform the first form of σ, Eq. (4.13b), to the second, Eq. (4.13c)

$$(\partial p/\partial \lambda)_{E,v}/\rho c^2 = -(E_\lambda/E_p)/(C_p/\beta E_p) = -(\beta/C_p)E_\lambda.$$

and then the second to the third, Eq. (4.13d),

App. 4E SHOCK CHANGE EQUATION

$$-(\beta/C_p)E_\lambda = -\rho H_v^{-1}E_\lambda = -\rho H_v^{-1}H_\lambda = \rho(\partial v/\partial \lambda)_{H,p},$$

and finally the third to the fourth, (Eq. 4.13e), using $v = v(\lambda T, p)$

$$\rho(\partial v/\partial \lambda)_{H,p} = \rho[(\partial v/\partial \lambda)_{T,p} + (\partial v/\partial T)_{p,\lambda}(\partial T/\partial \lambda)_{H,p}]$$

$$= \Delta v/v - \beta(\partial H/\partial \lambda)_{T,p}/(\partial H/\partial T)_{p,\lambda}$$

$$= \Delta v/v - \beta\Delta H/C_p.$$

APPENDIX 4E. SHOCK CHANGE EQUATION

We work in the laboratory frame. The equations of motion in one space variable are (Appendix 4C)

$$p_t + up_x + \rho c^2 u_x = \rho c^2 \psi \tag{4E-1}$$

$$u_t + uu_x + vp_x = 0 \tag{4E-2}$$

$$\psi = \sigma \cdot r - \alpha u/x, \tag{4E-3}$$

where we have used Eq. (4.7a) to replace $\dot\rho$ by $\dot p$. Derivatives along the shock path are given by

$$(dp/dt)_S = p_t + Dp_x \tag{4E-4}$$

$$(du/dt)_S = u_t + Du_x. \tag{4E-5}$$

Add (4E-1) to (4E-4), and (4E-2) to (4E-5) to obtain

$$(dp/dt)_S + (u - D)p_x + \rho c^2 u_x = \rho c^2 \psi \tag{4E-6}$$

$$(du/dt)_S + (u - D)u_x + vp_x = 0. \tag{4E-7}$$

Eliminate p_x by multiplying (4E-7) by $-\rho(u - D)$ and adding to (4E-6), to obtain

$$(dp/dt)_S - \rho(u - D)(du/dt)_S - \rho(u - D)^2 u_x + \rho c^2 u_x = \rho c^2 \psi. \tag{4E-8}$$

The derivatives of p and u along the shock path are related by the Hugoniot curve (in p-u space) for the material as

$$(dp/dt)_S = (du/dt)_S (dp/du)_\mathcal{H}, \qquad (4E\text{-}9)$$

where the subscript $\mathcal{H}$ denotes a derivative along the Hugoniot curve. Using (4E-9), the conservation of mass relation

$$\rho(D - u) = \rho_0 D, \qquad (4E\text{-}10)$$

and some rearrangement, one obtains the shock-change equation

$$(dp/dt)_S = \rho c^2 (\psi - \eta u_x)/[1 + \rho_0 D (du/dp)_\mathcal{H}] \qquad (4E\text{-}11)$$

where η is the sonic parameter

$$\eta = 1 - (D-u)^2/c^2 \qquad (4E\text{-}12)$$

This is the form given as Eq. (4.19). If u_x is eliminated instead of p_x the result is

$$(dp/dt)_S = \rho c^2 [\psi - \eta (\rho_0 D)^{-1} p_x]/[1 + (1 - \eta)^{-1} \rho_0 D (du/dp)_\mathcal{H}]. \qquad (4E\text{-}12)$$

5

STEADY DETONATION

In Chapter 3 we reviewed the failure of the simple theory to predict the experimental results in gaseous systems. As this failure was becoming apparent, a substantial theoretical effort was mounted to see if there were other types of steady solutions which would give better results. Although not the complete answer to the problem, this hope turned out to be well founded. Removing the restrictions of the simple ZND model can produce a new type of steady solution, which we call the *weak* or *eigenvalue* solution, in contrast to the *normal* or *CJ* solution of Chapter 2. The eigenvalue solution has the property suggested by the experimental results: it terminates at a point on the weak or supersonic branch of the detonation Hugoniot curve, below the CJ point.

What changes in the model give rise to the eigenvalue solution? Roughly, introducing any essentially endothermic or dissipative effect opens up the possibility of an eigenvalue solution. This could be a mole decrement in the reaction, a secondary endothermic reaction, slight radial divergence of the flow, viscosity, etc. The earliest example is von Neumann's (1942) pathological detonation due to a mole decrement. Whether we have the CJ or the eigenvalue solution in a particular system depends on the details of all the constitutive properties of the mixture of unreacted and reacted material—reaction rate, equation of state, viscosity, etc. In the eigenvalue solution, the location of the final state on the detonation Hugoniot curve and thus the steady propagation velocity, depend on these constitutive details. This is in sharp contrast to the CJ case, where the final state and steady propagation velocity depend only on the equation of state of the reaction products.

Although the eigenvalue solution shifts the final state in the right direction to improve agreement with experiment, it is not the whole story. Both the hydrodynamic stability theory (Chapter 6) and experimental observation (Chapter 7) show that the steady one-dimensional solution is usually unstable, being replaced by a three-dimensional time-dependent structure. Nevertheless, the steady one-dimensional solutions are of considerable interest. They presumably represent the actual flow in some (transverse average) approximation. They are of interest in their own right, as part of the theory of reactive flow. They are a necessary preliminary to the study of hydrodynamic stability in Chapter 6, in that a thorough knowledge of the mathematical object to be perturbed is a necessary preliminary to the study of the perturbation itself. Finally, their study should have general application to many fluid flow problems with a kinetic process, for example the whole field of shock-induced phase changes, including polymorphic transitions in crystals.

The results of the investigations of these steady solutions have unfortunately not received the attention they deserve. The rather formal mathematical arguments and the unfamiliar geometrical spaces in which the equations are most conveniently studied have left most experimentalists with a limited appreciation of the work and its implications. We simplify the presentation as much as possible by starting with the simplest models and introducing complications one at a time, working out simple examples as we go. We concentrate on concepts and representative results, leaving the details to the original papers. Even so, there is much material to be covered.

The basic assumptions remain those of the ZND model. We seek one-dimensional (or quasi-one-dimensional) plane, laminar solutions, steady in a frame moving with a constant velocity D. Except in Sec. 5F, where we include transport effects, the solution begins with a lead shock (moving at the same constant velocity) which triggers the reaction.

Under the steady assumption, the problem reduces to a set of ordinary differential equations. Except in Sec. 5F, the shock wave is assumed to be of infinitesimal thickness, so the state of the material changes instantaneously from the initial state to the shocked state. The state immediately behind the shock (called the von Neumann point) is thus the initial point for the integration of the equations. This solution is, for mathematical rather than physical reasons, conveniently considered in a space where some variable, usually pressure or particle velocity, is represented as a function of the reaction progress variable(s). The set of ordinary differential equations has, as usual, a one-parameter family of solutions, or *integral curves*. The solution to the problem posed is the particular integral curve selected from the family by imposing the initial condition that it start from the von Neumann point N. We call this integral curve the *solution*, the *reaction path*, or I_N. It terminates at the end of the region of steady flow, which is followed by a constant state or by a time-dependent flow. We call the state at the end of the steady region the *final* or *terminal* state. In considering the character of the solutions the approach described in Chapter 2 is used. The nature of the solutions is first studied with the propagation velocity D as a parameter (the D-discussion). With this knowledge in hand, the problem of satisfying a given rear boundary condition and the corresponding determination of an appropriate following flow is taken up (the piston problem).

As stated above, the solution may be of either the *CJ* or *normal* type, or the *weak* or *eigenvalue* type, depending on the properties of the fluid. The CJ type is similar to the familiar solution for the ZND model discussed in Chapter 2. The unsupported wave propagates at CJ velocity with sonic flow at the end of the reaction zone, and the overdriven wave propagates at a velocity above CJ velocity with subsonic flow at the end of the reaction zone and extending back to the following piston. Although the pressure profile and the length of the reaction zone depend on the rate of the chemical reaction, the propagation velocity and the final state depend only on the equation of state and not on the reaction rate. This is a consequence of the mathematical result that the strong (or CJ) point in which such solutions terminate is a nodal-type critical point of the differential equations.

The weak type of solution is quite different. The solution for the unsupported detonation can terminate satisfactorily only for a unique value of D, the eigenvalue, for which it can pass through a sonic saddle point (called the pathological point P) and proceed on down to the weak nodal critical point W. This eigenvalue velocity and the final state of this solution, in contrast to the normal case, depend on the rate function. The final state is supersonic, with pressure and particle velocity less than the CJ values. The unsupported detonation runs at this special velocity, with the flow at the end of the reaction zone supersonic. The following rarefaction therefore falls farther and farther behind the terminal state, so the reaction zone is followed by an ever-widening zone of constant state. For piston velocities between the particle velocities at the weak and strong points for the special value of D, a double wave structure exists and the steady reaction zone is followed by a slower-moving shock wave. For piston velocities above this range the solutions are similar to the usual overdriven ones.

The cases treated are summarized in Table 5.1, to which it is suggested that the reader refer as he proceeds through the chapter. Throughout, the equation of state is assumed to be well-behaved. Transport effects are neglected, except in Sec. 5F; and the flow is assumed to be one-dimensional and plane (constant-area channel) except in Sec. 5G.

Table 5.1 CASES CONSIDERED

Section	Number of Reactions	Reversible?	Is Any σ_1 Negative?	Other	Property Allowing Weak Solutions
5A	1	Yes	No	---	No weak solution
5B	2	No	(1) No	---	No weak solution
			(2) Yes ($\sigma_2 < 0$)	---	$\sigma_2 < 0$
5C	1	No	Yes	---	Mole change gives von Neumann pathological case.
5D	2	Yes	No	---	Second reaction driven past equilibrium by fast first reaction.
5E	Arbitrary	Yes	---	---	Combination of 5B2, 5C, and 5D.
5F	1	No	No	Viscosity	Viscous effects, if reaction is very fast.
5G	1	No	No	Radial Divergence	Radial divergence

5A. ONE REACTION, $\sigma > 0$

In this section we consider an explosive which undergoes a single reaction for which σ is always positive. Some of the illustrations are qualitative diagrams with certain features exaggerated to better display the properties of interest. In making them we have in mind the ideal gas of Sec. 4B2 with $\delta = 0$, $C_{pA} = C_{pB}$, and $q > 0$.

We discuss both the irreversible case (with some repetition of Chapter 2) and the reversible case. The principal features of interest are the λ-space diagrams, which serve as an introduction to the later sections, and, for the reversible reaction, the effect of the equilibrium sound speed.

Secs. 5A1 and 5A2 describe the equation-of-state and Hugoniot-curve properties for both fixed and equilibrium composition. Secs. 5A3 and 5A4 describe detonations for irreversible and reversible reactions. The correspondence with the first two sections is that the final state for the irreversible reaction is the fixed-composition state of complete reaction, whereas that for the reversible reaction is the equilibrium state of incomplete reaction. Sec. 5A5 gives some numerical examples showing the magnitude of the effect of making the reaction reversible, and Sec. 5A6 displays a complete numerical solution for a realistic gas system, dilute hydrogen/oxygen.

5A1. Properties at Fixed Composition

We list some properties of the material at fixed composition. Derivations follow from the assumption of steady flow and the assumptions about the equation of state, and may be found in Evans and Ablow (1961) or the standard references given there.

As in Sec. 4A2, we require for the state function $E = E(p,v,\lambda)$,

$$E_v > -p \quad \text{and} \quad E_p > 0.$$

Recall that these assumptions are equivalent, taking $S = S(p,v,\lambda)$, to

$$S_v > 0 \quad \text{and} \quad S_p > 0.$$

These assumptions are sufficient to ensure that the sound speed is real, and further that the isentropes never cross in the p-v plane. Here we also assume that

$$(\partial^2 p/\partial v^2)_{s,\lambda} > 0,$$

ensuring that the isentropes and Hugoniot curves are everywhere concave upward. A typical set of isentropes and Hugoniot curves are shown in Fig. 5.1a.

The properties which we emphasize are

1. Each isentrope has one and only one point at which a Rayleigh line through p_0, v_0 is tangent to it in the p-v plane. The locus of all such points, referred to as the locus of tangents, is a continuous curve passing through p_0, v_0.

2. Along a Rayleigh line, Eq. (2.3), the entropy has a single maximum which occurs at the point where the Rayleigh line intersects the locus of tangents.

3. The way in which the isentropes intersect the Rayleigh line determines the sonic character of the steady flow. If the isentrope is steeper than the Rayleigh line in the p-v plane, then in the steady reference frame we have subsonic flow with $c > u$ (in the laboratory frame $c > D - u$). If the Rayleigh line is steeper than the isentrope, then the flow is supersonic with $c < u$.

4. It follows from property 3 that the locus of tangents separates the p-v plane into subsonic and supersonic regions, and that on the locus the flow is exactly sonic, with $c = u$ and $\eta = 0$. The locus of tangents is therefore usually called the sonic locus.

5. The entropy along an Hugoniot curve, Eq. (2.6), with λ fixed, has a single minimum at the point where the Hugoniot curve crosses the locus of tangents (sonic locus). The Hugoniot curve, Rayleigh line, and isentrope passing through such a point all have a common tangent, as shown in Sec. 2C. In the p-v plane this common tangent coincides with the Rayleigh line. Above and below the common point, the isentrope lies between the Rayleigh line and the Hugoniot curve.

6. Jouguet's Rule: When a Rayleigh line crosses an Hugoniot curve in two points, the flow at the upper point is subsonic and that at the lower point is supersonic.

7. When a Rayleigh line crosses an Hugoniot curve in two points, then a new shock Hugoniot curve centered at the lower intersection also passes through the upper intersection. This new shock Hugoniot curve does not in general coincide with the original Hugoniot curve centered at p_0, v_0 except at the two points in question, as shown in Fig. 5.1b. This

Sec. 5A   ONE REACTION, $\sigma > 0$

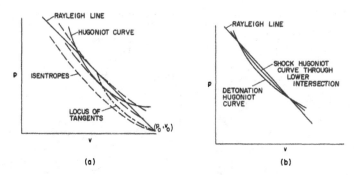

Fig. 5.1. Curves at fixed composition: (a) Hugoniot curve and isentropes. (b) Shock connection between upper and lower intersections.

property is important to the double-wave structure which appears in later sections of this chapter.

5A2. Properties at Equilibrium Composition

We next consider the *equilibrium equation of state* $p^e(v,E)$, defined by the requirement that the (varying) composition λ at any given v,E be the equilibrium value λ_e, that is

$$p^e(v,E) = p(v,E,\lambda_e)$$

$$\Delta F(v,E,\lambda_e) = 0,$$

the second equation determining λ_e.

The isentropes are less steep than those at fixed composition (the equilibrium sound speed being less than the frozen). The equilibrium Hugoniot curve, Fig. 5.2, is also less steep, and is of course not a member of the family of (fixed-compositon) partial-reaction Hugoniot curves.

The equilibrium Hugoniot curve has two points at which a form of the Chapman-Jouguet condition is satisfied. One is the point at which the locus of tangents (to the frozen isentropes) crosses the equilibrium Hugoniot curve. This point is called the "frozen CJ point" or CJ_0. The Rayleigh line through it is tangent to the frozen isentrope there and the flow is thus frozen-sonic. The equilibrium Hugoniot curve is *not* tangent to the frozen isentrope here. The other is the point at which an equilibrium isentrope is tangent to the equilibrium Hugoniot curve. This point is called the "equilibrium CJ point" or CJ_e. It is more like the CJ point of the simple theory of Chapter 2. The equilibrium Hugoniot curve, an equilibrium isentrope, and the Rayleigh line through the point are all tangent there, and the entropy along the equilibrium Hugoniot

139

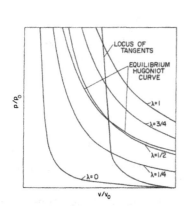

Fig. 5.2. Partial-reaction and equilibrium Hugoniot curves for the case of a single reversible reaction, calculated for the fluid parameters of Sec. 5A5.

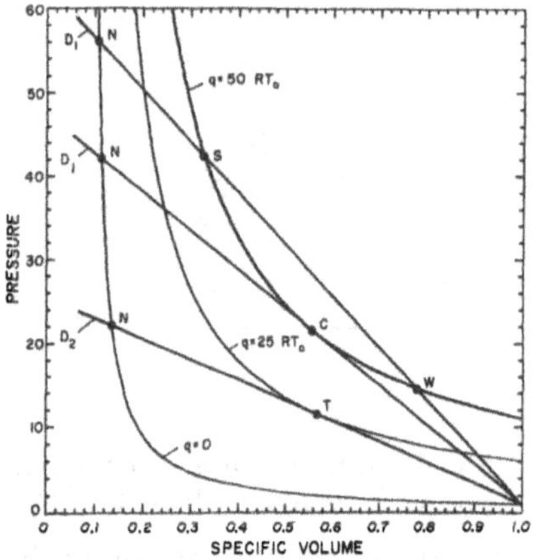

Fig. 5.3. Detonation with one irreversible reaction in the polytropic gas $pv = (\gamma - 1)(E + \lambda q)$, $\gamma = 1.2$, $q = 50\, RT_0$.

curve is a minimum there. Jouguet's rule in terms of the equilibrium sound speed holds on the equilibrium Hugoniot. Above CJ_e the flow is equilibrium subsonic, with $c_e > u$, below CJ_e it is equilibrium supersonic, with $c_e < u$, and at CJ_e it is equilibrium sonic, with $c_e = u$.

For any well-behaved equation of state, the frozen CJ point always lies below the equilibrium CJ point in the p-v plane. The frozen CJ point is by definition frozen sonic so that $c = u$ there; but since the frozen sound speed is everywhere greater than the equilibrium sound speed, that is $c > c_e$ (Eq. 4.15), we must have $c_e < u$ at CJ_0. Thus CJ_0 is equilibrium supersonic and must therefore lie on the supersonic branch of the equilibrium Hugoniot curve, that is below CJ_e.

5A3. Detonation with One Irreversible Reaction

We are now ready to consider the steady solution for the simplest case: a single irreversible reaction. In the p-v plane, Fig. 5.3, all states of the steady solution lie on the Rayleigh line for the given D. As a particle passes through the shock at the front, the state is first changed instantaneously, without reaction, from the initial point O to the von Neumann point N. It then moves down the Rayleigh line as the reaction proceeds. Although the steady solution does not contain all points of the Rayleigh line lying between points N and O, it is convenient to consider the properties of the entire set of points comprising this segment, excluding only those (if any) lying outside the stoichiometric boundary.

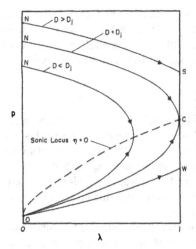

Fig. 5.4. Variation of p with λ along three different Rayleigh lines for a single irreversible reaction with positive σ.

Solutions $p = p(\lambda)$ are obtained by eliminating v from the equations for the Rayleigh line and the Hugoniot curve, using the explicit equation of state $E = E(p,v,\lambda)$, with D as a parameter. Figure 5.4 shows the solutions in p-λ (variation of p with λ along the Rayleigh line) for different values of D. Each solution has a strong (subsonic) and a weak (supersonic) branch. The arrows show the direction in which the state changes with time as a particle traverses the steady solution. This is, of course, always toward increasing λ, since the back reaction is excluded by assumption. Each solution has a vertical tangent where it crosses the sonic locus. At this point the Rayleigh line is tangent to a partial-reaction Hugoniot curve, so that λ has its maximum value.

The weak branch of each curve is not of physical interest because it cannot be reached from point N for any value of D under the assumptions about the mechanism of initiation of reaction. A weak continuous solution could be achieved by having initiation (by some outside agency) take place on a plane moving with velocity D without appreciable pressure rise.

5A3.1. D-Discussion

The steady solutions $p = p(\lambda)$ with D as a parameter are obtained from the conservation equations without consideration of the rate equation. Obviously it must also be satisfied, and its effect enters as a restriction on the values of the parameter D.

For $D < D_j$, Fig. 5.4 suggests that there is a steady solution starting from point N and terminating on the sonic locus (i.e., consisting of the upper branch of the inner parabola). Such a solution is immediately suspect, for the steady flow equations (4.18) give infinite derivatives at the sonic point ($\eta = 0$ by definition, $r > 0$ since $\lambda < 1$, and $\sigma > 0$ by assumption). A more rigorous argument excludes such a solution by showing that it fails to satisfy the characteristic equation (4.8d)

$$(dp/dt)_+ + \rho c (du/dt)_+ = \rho c^2 \sigma r.$$

The argument proceeds as follows. The path of the sonic locus in x-t space has slope $dx/dt = u + c = D$. The equation holds on this path since it has the characteristic direction $dx/dt = u + c$. The state is constant on this path since in a solution steady in a frame moving with velocity D any line $dx/dt = D$ is by definition a constant-state locus. Hence the left-hand side vanishes while the right-hand side does not (since $\sigma r > 0$ as shown above), and we have a contradiction.

Thus the only acceptable steady solutions are those which terminate with $\sigma r = 0$ and thus $\lambda = 1$, i.e., on the $q = 50\,RT_0$ Hugoniot curve of Fig. 5.3 and the right-hand edge $\lambda = 1$ of Fig. 5.4. This excludes all values of D less than D_j, leaving $D \geq D_j$ as the range of acceptable solutions.

5A3.2. The Piston Problem

By the "piston problem", Sec. 2A3, we mean the construction of a complete solution to the flow equations, containing a steady reaction zone at the front and having a prescribed (piston) velocity u_p at the rear.

The problem, then, is to construct, for the given piston velocity, some flow spanning the space between the terminal state of the steady solution and the piston. At the same time we must also determine the value of D having that steady solution for whose terminal state this is possible.

Before getting down to cases, we pause to clarify one point. For the common forms of the reaction rate, an infinite time is required for the attainment of chemical equilibrium, so that the length of the reaction zone is formally infinite, and whatever flow we attach behind it starts at infinite distance behind the front. Of course, any given small deviation from equilibrium is attained at a finite distance behind the front and after a few characteristic reaction times the state of any particle is in practice indistinguishable from the equilibrium one. In any actual experiment there are other effects to be considered. For example, the lateral flow expansion discussed in a later section, present in any finite-diameter charge, in effect terminates the reaction zone at a finite distance from the front.

For $u_p \geq u_j$ (in the laboratory frame) the solution consists of that steady zone whose terminal state has $u = u_p$ (thus $D \geq D_j$) followed by a uniform state (the same as this terminal state) extending back to the piston. For $u_p > u_j$ we have overdriven detonations, and for $u_p = u_j$ (thus $D = D_j$) a supported CJ detonation. For $u_p < u_j$ we have an unsupported CJ detonation, consisting of the CJ steady solution followed by a (time-dependent) rarefaction wave which reduces (increases in the steady frame) the particle velocity to that of the piston. The rarefaction wave is followed by a uniform state extending back to the piston. The greater the difference between u_p and u_j, the stronger the rarefaction.

Can the head of the time-dependent rarefaction wave legitimately be joined to the end of the steady zone? At the joining surface between these two regions the x-derivatives must be discontinuous because they are changing with time on one side and constant on the other. Such discontinuities can occur only along a characteristic curve of the flow. In the steady zone, constant states occur, by definition, along lines of slope D in the x-t plane, while characteristics have slope $u + c$. Therefore, along such a joining surface the condition $D = u + c$, equivalent to $\eta = 0$, must be satisfied. Since the CJ point is sonic, this condition is satisfied and the joining may take place there.

We are thus able to exhibit complete solutions for the entire range of piston velocities.

5A4. Detonation with One Reversible Reaction

The reversible case (both forward and backward reactions) is more complicated. In our discussion of it we follow unpublished work of Z. W. Salsburg (1959). Figure 5.5 shows the integral curves and the equilibrium Hugoniot curve $\mathcal{H}_e$. Points S and W are the intersections of the Rayleigh line with the equilibrium Hugoniot curve. The frozen CJ point C_0 is the intersection of the sonic locus with the equilibrium Hugoniot curve. The equilibrium CJ point C_e is formed by the coalescence of S and W at $D = D_e$. Note that in the range $D_e < D < D_0$, point W lies on the subsonic[1] branch. The weak branch is disposed of in the same way as before. Again the arrows show the change in the state point with time. Their direction follows from the sign of the rate—positive to the left of the equilibrium Hugoniot curve and negative to the right. Again the curves have a vertical tangent where they cross the sonic locus.

The S-v diagram, Fig. 5.6, offers another view of the same system. Since the entropy must increase with time in our adiabatic fluid element, the state point in this diagram always moves upward to higher S.

---
[1] Where the unmodified words "subsonic" or "supersonic" are used, the adjective "frozen" is to be understood.

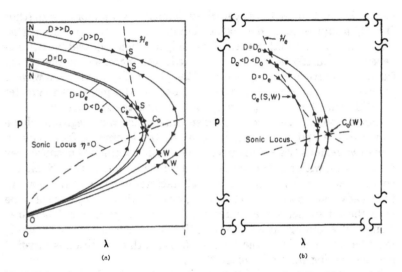

Fig. 5.5. Solutions in p-λ for a single reversible reaction with positive σ. (a) Complete diagram, (b) expanded view near CJ points. $\mathcal{H}_e$ is the equilibrium Hugoniot curve, and C_0 and C_e are the frozen and equilibrium CJ points.

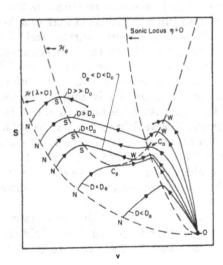

Fig. 5.6. Variation of S with v along Rayleigh lines for the steady detonation of Fig. 5.5.

The variable v serves as a measure of distance along the Rayleigh line. The slope of the curves may be found from Eqs. (4.18c, e):

$$(dS/dv) = -(\eta \Delta F)/(\sigma v T). \qquad (5.1)$$

They have extrema at equilibrium points $\Delta F = 0$ and at sonic points $\eta = 0$. At point N, ΔF is negative and dS/dv is positive. The first extremum at point S on the equilibrium Hugoniot is a maximum (or inflection point) and the next extremum, which is either at the sonic locus or at point W on the equilibrium Hugoniot, is a minimum. When S and W coalesce to form C_e, an inflection point of zero slope is formed, and similarly for C_0, formed by the coincidence of W and the sonic locus. Notice that the extremum at the sonic point changes from a minimum above the equilibrium Hugoniot to a maximum below as ΔF changes sign, and that the extremum at point W changes from a maximum above the sonic locus to a minimum below as η changes sign.

Before proceeding to the remainder of the discussion, we pause to list those properties of the solutions which turn out to be general results for the more complicated systems treated in the following sections:

1. The vertical slope of the integral curves at the sonic point, Fig. 5.5. Most other thermodynamic variables (ρ, T, E, r) plotted against λ also have a vertical tangent at this point. The entropy reaches a local maximum or minimum in a cusp at the sonic locus.

2. The uniqueness of the solution for given D. That is, if we begin the integration of the differential equations anywhere on the integral curve, its subsequent course is determined unless it begins at the sonic point, where the simple type of double-valuedness seen in Fig. 5.5 obtains (subsonic and supersonic branches both entering or leaving the sonic point).

3. The appearance of point W on the subsonic branch for $D_e < D < D_0$ in the reversible case.

5A4.1. D-Discussion

For $D < D_e$ we have, as before, no steady solution. For $D > D_0$ we have overdriven detonations terminating at the strong point S on the equilibrium Hugoniot. The portion (frozen strong, equilibrium weak) of the equilibrium Hugoniot from just below C_e down to and including C_0, having $D_e < D < D_0$ is inaccessible since the state point cannot proceed beyond the equilibrium point S. This property is special to the one-reaction case; for two or more reactions the more complicated geometry of the λ-space allows this segment to be reached under some conditions. For $D = D_e$ we have the unsupported CJ detonation terminating at C_e. Its property of not being a frozen-sonic point causes some difficulty which will be discussed in the next section. Note that if the state point were somehow displaced to a point slightly beyond C_e on this integral curve it would proceed on to $\eta = 0$ (in Fig. 5.6 the entropy increases to the right of the zero-slope inflection point C_e). Presumably this possibility has no physical significance since C_e itself is reached only in infinite time.

5A4.2. Piston Problem

For $u_p > u_e$ (with u_e the particle velocity at the equilibrium CJ point C_e) the complete solution again consists of the steady reaction zone followed by a uniform state extending back to the piston. For $u_p > u_e$ (thus $D > D_e$) we have overdriven detonations, and for $u_p = u_e$ (thus $D = D_e$) we have a supported "CJ" detonation. The situation to this point is like the irreversible case, with D_e the minimum propagation velocity. The unsupported case offers some difficulty, however, because all the admissible steady solutions down to and including that for $D = D_e$ have (frozen) subsonic terminal states, so that a rarefaction cannot be simply joined on at this point. The frozen CJ point C_0, which is sonic and would allow this joining, is inaccessible to any solution beginning at point N.

Strictly speaking, no exact solution to the complete flow problem for this case has been found. But Wood and Salsburg (1960) offer an approximate one which constitutes a plausibility argument in favor of D_e as the unsupported detonation velocity. To see what this is, recall the discussion in Sec. 4C of a rarefaction wave moving into a material at chemical equilibrium (in the present case that issuing from the end of the steady zone with state C_e). At late time, deviations from equilibrium within the rarefaction wave are small, so that to first approximation the characteristic velocity is $u + c_e$ and the head of the wave may be joined to the terminus of the steady zone at C_e. In the next approximation the rarefaction wave has an exponentially decaying precursor penetrating the reaction zone, which is thus slightly perturbed from the steady solution, while the rarefaction wave is not quite that described in Sec. 4C, since it is moving into material not quite at equilibrium. The reaction and rarefaction regions are separated by a small zone in which the flow derivatives, although not discontinuous, change rapidly. The characteristic diagram, Fig. 5.7, for unsupported detonations in the irreversible

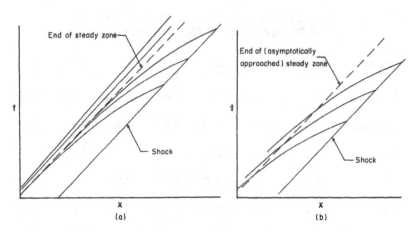

Fig. 5.7. Characteristic diagram at Chapman-Jouguet velocity for (a) irreversible reaction, (b) reversible reaction.

and reversible cases makes them appear quite different, for in the reversible case, Fig. 5.7b, the characteristics from the rarefaction penetrate the reaction zone. The amplitude of the signals which propagate along them, however, decays exponentially with time.

5A5. Magnitude of the Effect of Reversibility

Differences between frozen and equilibrium sound speeds in gases typically range from 5 to 10%. In the neighborhood of the CJ point, the difference is about 7% for the idealized reversible-reaction system described below, and 10% for the dilute hydrogen/oxygen system described in Sec. 5A6. Other numerical examples are given by Strehlow (1968a), p. 157, Vincenti and Kruger (1965), p. 260, and Soloukin (1966), p. 137. In condensed explosives, typical differences are 1 to 2% in calculations we have done, one of which is given in Table 5.2.

For detailed comparison in a gas, we have calculated two idealized systems with exactly the same heat release at the CJ and CJ_e points, attained in one case by an irreversible reaction and in the other by a reversible reaction. The first system, with irreversible reaction and $q = 50\ RT_0$, is that of Fig. 2.11. The second is the same except that $q = 100\ RT_0$ but the reaction is reversible with the equilibrium constant chosen to give $\lambda = 1/2$ at the equilibrium CJ_e point, so that the heat release at this point, $\lambda q = 50\ RT_0$, is unchanged. These are compared in Table 5.2. The CJ pressures differ by about 5% for the gaseous system, compared to 2% for the condensed system.

147

STEADY DETONATION   Chap. 5

Table 5.2 CJ STATES TO ILLUSTRATE EFFECTS OF REVERSIBLE CHEMICAL REACTION

System	State	p	D	c	c_o	γ [a]	γ_o [a]
Condensed	CJ_e	32.45	8.7959	6.829	6.746	3.373	3.292
Condensed	CJ_o	31.86	8.7970	6.785	6.701	3.371	3.289
Gas 1	CJ	21.53	6.2162	3.464	---	1.2	---
Gas 2	CJ_e	23.29	6.2347	3.487	3.255	1.2	1.046
Gas 2	CJ_o	21.77	6.2508	3.482	3.250	1.2	1.045

[a] $\gamma = (\partial \ln p / \partial \ln v)_{s,\lambda}$; $\gamma_o = (\partial \ln p / \partial \ln v)_{s,\Delta F\, =\, 0}$.

Condensed System: Solid explosive RDX at $\rho_o = 1.8$ Mg/m³ calculated with the Lennard-Jones-Devonshire equation of state, using the parameters of the third ("central point") entry of Table 4.2, p. 88, and the $\rho_o = 1.8$ entry of Table D1, p. 139, of Fickett (1962).
Gas 1: Irreversible reaction. $A \rightarrow B$, $\gamma = 1.2$, $q = 50\,RT_o$, $r = k(1-\lambda)e^{-E^\dagger/RT}$, the fluid of Fig. 2.11.
Gas 2: Reversible reaction. $A \rightleftarrows B$, $\gamma = 1.2$, $q = 100\,RT_o$, $r = k(1-\lambda)e^{-E^\dagger/RT}(1 - e^{\Delta F/RT})$. The entropy difference between A and B is chosen to make $\lambda = \lambda_o = 1/2$ at the CJ_o point. The equilibrium composition is $(1-\lambda)/\lambda = K$; $K = K_o \exp[(q/R)(T^{-1} - T_o^{-1})]$. The required value of $\ln K_o$ is 91.777. The rate multiplier $k = 1166.6$ is chosen to place $t = 1$ at $\lambda = 1/2$.
Units: For the condensed system, p is in GPa, and velocities are in km/s. For the gases, p is in units of p_o and velocities in units of c_o.

Additional details for the two gas systems are given in Figs 5.8 and 5.9. The system with reversible reaction has about 0.3% higher detonation velocity. Figure 5.8 shows the Hugoniot curves, and the steady profiles for p and T. The frozen Hugoniot curves are, of course, steeper than the equilibrium Hugoniot curve. In the steady profile, the temperature maximum is almost nonexistent in the reversible case because the slope of the equilibrium isentrope is almost the same as that of the isotherm. Figs. 5.9a and 5.9b show the variation of the frozen and equilibrium sound speeds and Mach numbers along the $\lambda = 1/2$ Hugoniot curve for the system with reversible reaction. The equilibrium sonic point (CJ_e point) is marked on the curves for c_e and M_e and the frozen-sonic point (which is the same as the frozen CJ point for the irreversible system) is

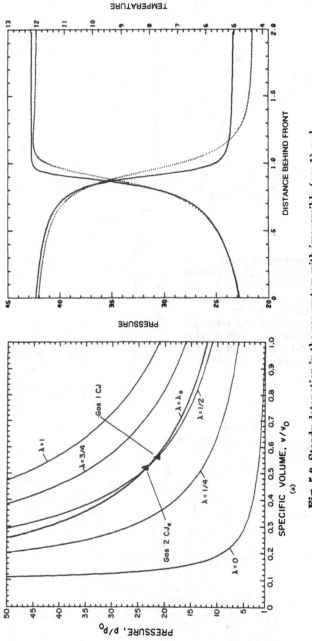

Fig. 5.8. Steady detonation in the gas system with irreversible (gas 1) and reversible (gas 2) reaction described in Table 5.2. (a) Hugoniot curves; triangles mark the CJ points. (b) Steady solution profiles: irreversible (······), reversible (———).

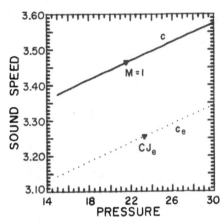

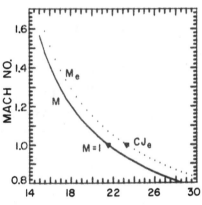

Fig. 5.9. Additional properties of the irreversible and reversible systems. (a), (b) Variation of c_e and c, and of M_e ($=[D-u]/c_e$) and M ($=[D-u]/c$) along the $\lambda = 1/2$ Hugoniot curve *of the reversible system.* (c) Loci of CJ points in the p-v plane for the two systems, obtained by varying q (see text).

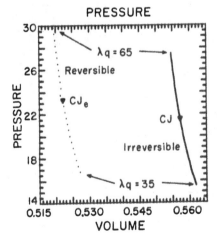

marked on the curves for c and M. Fig. 5.9c shows the loci of CJ points for the two systems generated by varying q. In the variation for the reversible system, K_0 (see Table 5.2) is chosen for each q to make $\lambda = \lambda_0 = 1/2$ at the CJ_0 point.

5A6. A Realistic Example: Hydrogen/Oxygen

The hydrogen/oxygen system has been extensively studied and the kinetics are well understood. A short discussion and some calculations are presented here for contrast with the simplified systems we have used for illustration. Strehlow (1968a) gives an elementary treatment; and Browne, White, and Smookler (1969) list most of the possible elementary reactions.

The branching-chain reactions of this system give the reaction zone a structure not found in our simpler examples. A small subset of the elementary reactions illustrates the main features:

(1a) $\quad H + O_2 \rightleftarrows OH + O$

(1b) $\quad O + H_2 \rightleftarrows OH + H$

(1c) $\quad OH + H_2 \rightleftarrows H_2O + H$

(2) $\quad H + OH + M \rightleftarrows H_2O + M$.

The reaction process takes place in two stages. The first is the almost thermally neutral *induction stage* in which the pressure, temperature, and number of moles remain macroscopically constant at the von Neumann values while the atom and radical (H, O, and OH) concentrations increase approximately exponentially with time from very small initial values via the branching-chain dissociation reactions (1a), (1b), and (1c). The second stage involves the usually slower exothermic recombination of the atoms and radicals through a number of three-body processes like reaction (2). These, of course, make a negligible contribution to most of the induction stage because their rates are proportional to products like [H][OH].

The induction stage starts from an initial concentration of atoms and radicals which is quite small but known to be significantly larger than their equilibrium concentrations in the undisturbed material. The mechanism which produces these is not understood in detail; it may be that most of the dissociation takes place within the shock. The effective activation energy for the induction stage is that of reaction (1a), about 18 kcal/mole, for it is the slowest and thus the rate-determining step.

The exponential nature of the growth may be easily seen by considering the extreme case in which reactions (1b) and (1c) are much faster than (1a) so that the OH and O produced by (1a) are immediately converted to H_2O and H, that is, the O produced by (1a) is converted by (1b) to OH and H, and the OH thus produced by (1a) and (1b) is then converted by (1c) to H_2O and H. The net effect is

$$H + O_2 + 3H_2 \leftrightarrows 2H_2O + 3H,$$

so that the net H-atom concentration doubles at each step and the early growth is proportional to e^{2kt}, where k is the forward rate constant for reaction (1a).

As the products accumulate to the point where terms in the rate equations such as [H][OH] are no longer negligible compared to those such as [H][H_2], the recombination reactions like reaction (2) come into play and the recombination stage begins. In this stage the pressure decreases and the temperature (and here the mole number) increase in the conventional way for an exothermic reaction in steady flow. The recombination rates are relatively insensitive to temperature but more sensitive to density because they require three-body collisions. Thus the ratio of the recombination to induction-zone lengths decreases with increasing density.

A rough stoichiometric summary of the course of the reaction is

(1) $2H_2 + O_2 \leftrightarrows H_2O + H + OH$, $\Delta H = 0$ kcal/mole; $\Delta n = 0$

(2) $H + OH \leftrightarrows H_2O$, $\Delta H = -116$ kcal/mole; $\Delta n = -1$.

As stated earlier, the heat release takes place in the recombination zone. As will be seen in the computer results presented below, the principal oversimplification is the neglect of O and the setting of OH equal to H. Actually O and OH are about equal throughout and are considerably smaller than H.

The negative mole change for the recombination stage raises the question of the possibility of a pathological detonation, that is, one in which σ becomes negative at some point. Although the λ-space has not been explored in detail, this possibility appears to be unlikely, for the negative mole change term in σ, Eq. (4.16), amounts to only 3% of the total value of σ for the stoichiometric recombination reaction (2).

In passing we comment that the much studied hydrocarbon systems have a similar but less well-known reaction mechanism. The observed over-all activation energy for the induction zone is about the same as for hydrogen/oxygen, probably because reaction (1a) is still the rate-determining step. The induction-zone reactions are mildly exothermic, so that the mechanism, while predominantly a branching chain, takes on some of the character of a thermal explosion, with a more rapid temperature rise at the transition to the recombination zone. The recombination zone is still the strongly exothermic part of the reaction process, with smaller mole decrement and faster rate than in hydrogen/oxygen, so that at the same density the reaction zone has more nearly a square-wave shape.

Using computer programs[2] available at the Los Alamos Scientific Laboratory, we have calculated the steady CJ reaction zone for the dilute hydrogen/oxygen system described in Table 5.3. The results, Fig. 5.10, are plotted on three different distance scales to properly display the induction and recombination segments. The tabulated CJ point is the equilibrium CJ point.

5B. TWO IRREVERSIBLE REACTIONS

We work out here the simplest case of more than one reaction. Despite its simplicity, it gives interesting results and introduces many of the concepts needed for more complicated systems.

The equation of state is the polytropic gas with no mole change

$$pv = (\gamma - 1)(E + Q) \tag{5.2}$$

$$Q = \lambda_1 q_1 + \lambda_2 q_2. \tag{5.3}$$

The two reactions are taken to be consecutive

$$A \to B$$

$$B \to C,$$

with progress variables λ_1 and λ_2 related to the mass fractions by

$$\lambda_1 = 1 - x_A$$

$$\lambda_2 = x_C,$$

[2]The computer programs are HUG (Bird et al. 1964) which calculates the shock and CJ states, and KIN which integrates the rate equations for the steady solution. KIN was constructed by LASL by C. Hamilton, with a plotting capability added by J. D. Jacobson. The calculations it performs are essentially those described by Duff (1958); the program is similar to that of Garr and Marrone (1963) except that the thermodynamic functions are entered as polynomial fits of tabulated values.

Table 5.3 CJ DETONATION IN $2H_2 + O_2 + 9Ar$
($p_o = 10.1$ kPa, $T_o = 300$ K $D = 1572$ m/s, $c_o = 344.9$ m/s, $D/c_o \equiv M_o = 4.558$)

State	p/p_o	ρ/ρ_o	T/T_o	u/c_o	c/c_o	γ
Shock	25.47	3.991	6.383	1.142	2.482	1.517
CJ	15.19	1.769	9.157	1.981	2.821	1.460

Particle velocity u is in the laboratory and c_o, c, and γ are frozen.
At CJ, $\gamma_e = 1.215$, $c_e/c_o = 2.577$.

Reaction	k (cm^3 $mol^{-1)c-1}$ sec^{-1}	$E^\dagger$ kcal/mole	ΔH kcal, $T = T_o$
(1) $H + O_2 = OH + O$	9.54×10^{13}	14.7	16.16
(2) $H_2 + O = OH + H$	5.60×10^{13}	10.2	-1.74
(3) $H_2 + OH = H_2O + H$	6.15×10^{13}	5.6	-14.40
(4) $OH + OH = H_2O + H$	3.80×10^{13}	1.0	-12.66
(5) $H_2 + O_2 = OH + OH$	2.5×10^{12}	39.0	14.42
(6) $H + H + Ar = H_2 + Ar$	7.5×10^{14}	0	-101.23
(7) $H + H + H_2O = H_2 + H_2O$	1.0×10^{16}	0	-101.23
(8) $H + O + Ar = HO_2 + Ar$	2.2×10^{14}	0	-45.62
(9) $H + O_2 + H_2O = HO_2 + H_2O$	5.4×10^{15}	0	-45.62
(10) $H + OH + Ar = H_2O + Ar$	3.3×10^{20}	0	-115.63
(11) $H + OH + H_2O = 2H_2O$	6.6×10^{20}	0	-115.63
(12) $OH + HO_2 = H_2O + O_2$	3.0×10^{14}	0	-70.01

The rate of depletion of reactant i by the forward reaction in question is

$$\dot{\rho n}_i = a_i k_f \rho^a \Pi_j n_j^{a_j}; \quad a = \Sigma_j a_j$$

$$k_f = k e^{-E^\dagger/\alpha T}$$

where n_i is the number of moles of species i per unit mass, the a_i are the reactant stoichiometric coefficients in the reactions as written above, and the sum and product extend over the reactants (see Sec. 4A). The backward reaction rates are calculated from the forward rates and the equilibrium constants.

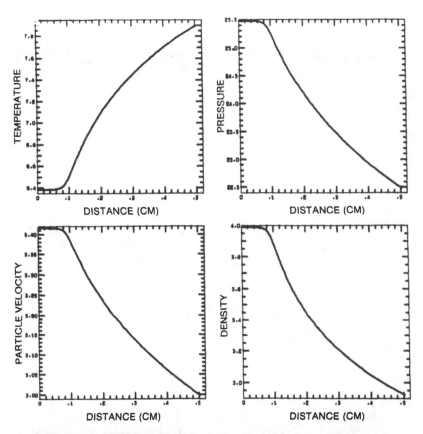

Fig. 5.10. Calculated steady reaction zone for the CJ detonation of Table 5.3. A selection of plots for different ranges of the distance x is shown, in order to display the induction and recombination zones, which have very different lengths. The dimensionless axis labels are temperature T/T_0, pressure p/p_0, particle velocity u/c_0 (laboratory frame), density ρ/ρ_0, specific enthalpy $(H - H_0)/RT_0$, characteristic speed $\rho c/\rho_0 c_0$, and specific mole number n/n_0. Note that, for maximum resolution, different vertical scales are used for the composition of each species.

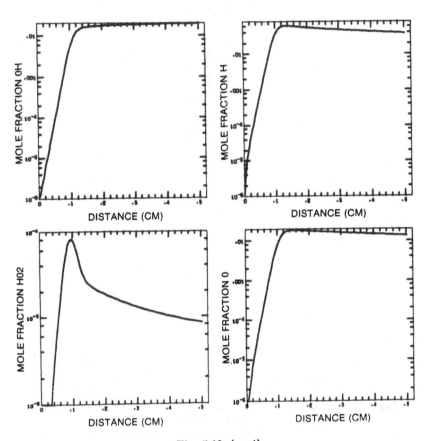

Fig. 5.10. (cont)

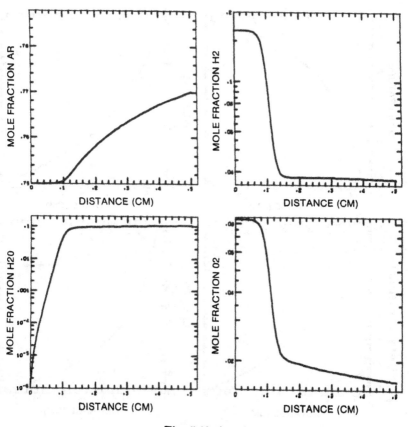

Fig. 5.10. (cont)

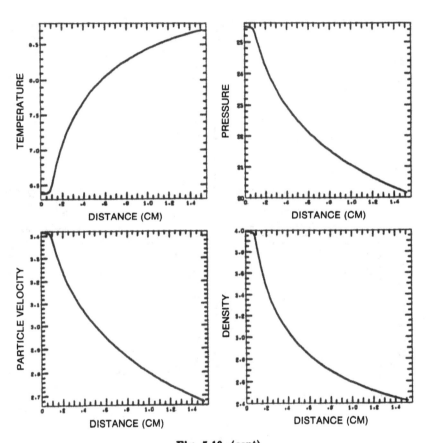

Fig. 5.10. (cont)

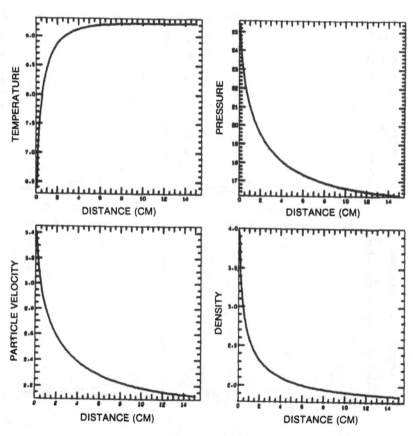

Fig. 5.10. (cont)

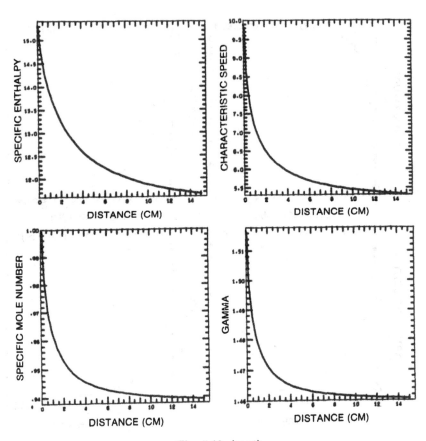

Fig. 5.10. (cont)

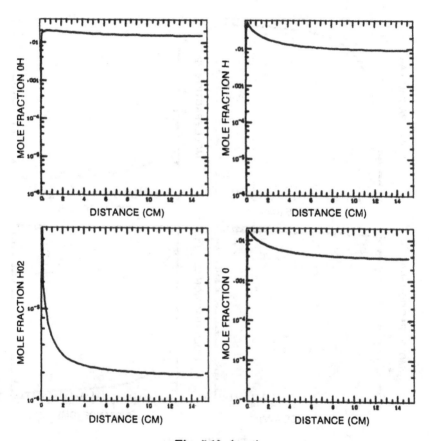

Fig. 5.10. (cont)

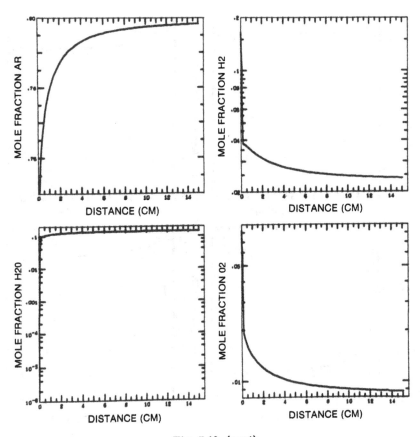

Fig. 5.10. (cont)

Sec. 5B   TWO IRREVERSIBLE REACTIONS

and first-order rates depending on composition only

$$d\lambda_1/dt = r_1 = 1 - \lambda_1 = x_A \tag{5.4a}$$

$$d\lambda_2/dt = r_2 = \lambda_1 - \lambda_2 = x_B. \tag{5.4b}$$

We consider two cases

(1) both reactions exothermic: $q_1 > 0$, $q_2 > 0$

(2) second reaction endothermic: $q_1 > 0$, $q_2 < 0$.

The first case turns out to be identical to that of one reaction (with the same total heat release), Sec. 5A3. The second is more interesting. The thermicity

$$\boldsymbol{\sigma}\cdot\mathbf{r} = \sigma_1 r_1 + \sigma_2 r_2 \tag{5.5}$$

passes through zero as the reaction proceeds, giving our first example of an eigenvalue detonation. Physically, one may think of a solid explosive with an endothermically decomposing binder.

5B1. Both Reactions Exothermic

We take

$$q_1 = q_2 = 25\ RT_0.$$

The family of partial-reaction Hugoniot curves, Fig. 5.11, is still a one-parameter family, as in Fig. 5.3, with the parameter λq replaced by $Q = \lambda_1 q_1 + \lambda_2 q_2$, Eq. (5.3), so that along any one Hugoniot curve the total heat release is constant, and the composition may vary over pairs of λ_1 and λ_2 satisfying Q = constant.

Figures 5.12 through 5.14 show the full p-λ_1-λ_2 base plane or λ-space. The pressure surface, Fig. 5.12, resembles the p-λ space for one reaction, Fig. 5.4, being a one-parameter (D) family of surfaces instead of curves. Each surface is a cylinder whose generators (light lines) are horizontal lines with base plane slope $d\lambda_2/d\lambda_1 = -q_1/q_2$; that is, they are lines of constant Q. Each surface has a vertical tangent plane along one generator; as we shall see below, this generator is the *sonic locus*. Each reaction path (heavy line) has a subsonic branch beginning at point N

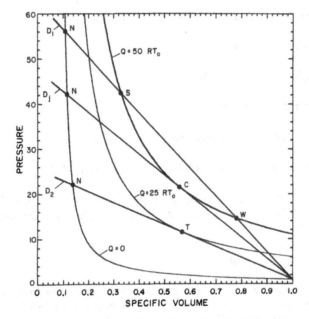

Fig. 5.11. Diagram of the partial-reaction Hugoniot curves in the p-v plane for two irreversible reactions, both exothermic.

and a supersonic branch beginning at point O, meeting at a point on the sonic locus.

In the λ-plane, Figs. 5.13 and 5.14, lines of constant x_A, x_C, and x_B are horizontal, vertical, and parallel to AC, respectively. Loci of constant Q are lines of slope $-q_1/q_2 = -1$, with that for the maximum Q of 50 RT_0 passing through the upper right corner at point C, and that for Q = 0 through the lower left corner at point A. To keep the mass fractions positive, λ_1 and λ_2 must be confined to the stoichiometric triangle ABC:

$$0 \leq \lambda_1 \leq 1 \tag{5.6a}$$

$$0 \leq \lambda_2 \leq \lambda_1. \tag{5.6b}$$

The arrows show the *phase portrait*, or family of all possible integral curves of the rate equations (5.4), which we may write as

$$d\lambda_2/d\lambda_1 = r_2(\lambda_1,\lambda_2)/r_1(\lambda_1,\lambda_2) = (\lambda_1 - \lambda_2)/(1 - \lambda_1). \tag{5.7a}$$

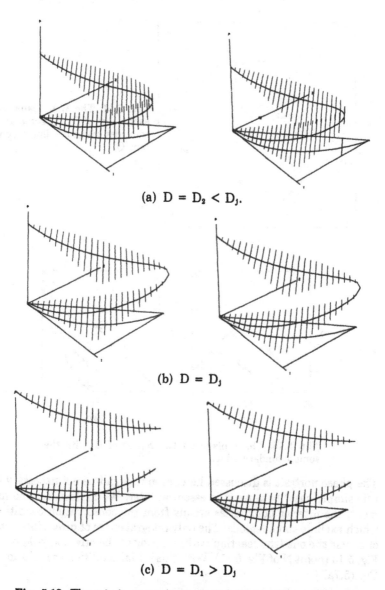

Fig. 5.12. The p-λ_1-λ_2 space (stereo pairs) for the three values of D shown in Fig. 5.11. The generators of the surfaces are lines of constant Q. The sonic locus is shown in the base plane where it lies within the stoichiometric boundary; this occurs here only in case (a).

STEADY DETONATION Chap. 5

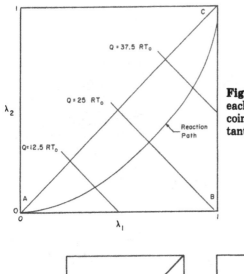

Fig. 5.13. The λ_1-λ_2 plane. For each value of D, the sonic locus coincides with some line of constant Q.

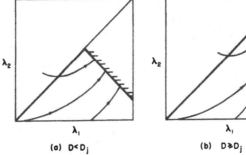

Fig. 5.14. The λ-plane for two ranges of D, with the sonic boundary added.

The phase portrait is discussed here, even though it is not necessary for this simple case, because it is essential later. We neglect, for the moment, the interior boundaries arising from the conservation conditions which exclude some regions. The only integral curve of direct interest is of course the one (the reaction path) starting at the origin $\lambda_1 = \lambda_2 = 0$ of Fig. 5.13 (point N of Fig. 5.11). With this initial condition the solution of Eq. (5.7a) is

$$\lambda_2 = \lambda_1 + (1 - \lambda_1) \ln (1 - \lambda_1), \tag{5.7b}$$

which is the reaction path plotted in Figs. 5.12-5.14. Other integral curves are indicated because they help to determine its properties. The critical points, where both rates vanish so that $d\lambda_2/d\lambda_1 = 0/0$, are of special interest. Their topology can be determined by a standard analysis (see, for example, Davis [1960], Chapter 11) which we will not describe here. Away from the critical points the qualitative phase portrait is easily constructed from the differential equation (5.7a). For example, on the stoichiometric boundary AC the derivative $d\lambda_2/dt$ is zero, and just inside it both $d\lambda_2/dt$ and $d\lambda_1/dt$ are positive, so this boundary is a locus of *turning points* at which integral curves have zero slope and are concave upward. The remainder of the phase portrait is easily seen to be as shown. On the right boundary $\lambda_1 = 1$ the integral curves have infinite slope. Point C turns out to be a nodal critical point. Critical points of nodal and saddle type are discussed in more detail later; here we simply note that all of the integral curves end in this point, and enter it with infinite slope.

The conservation conditions which confine the steady solution to the Rayleigh line impose an additional boundary in the λ-plane whose position depends on D. On a given Rayleigh line, Q has a maximum value at the point where the Rayleigh line is tangent to a partial-reaction Hugoniot curve, for example point T of Fig. 5.11. Call this maximum value $Q_{max}(D)$. The corresponding restriction in the λ-plane is

$$Q = \lambda_1 q_1 + \lambda_2 q_2 \leq Q_{max}(D). \tag{5.8}$$

This adds the additional boundary $Q = Q_{max}(D)$ shown as a hatched line in Fig. 5.14a. This boundary is called the *sonic line* or *sonic boundary* because it is the image of a tangent, and therefore sonic, point in p-v. As D increases, it moves up and to the right in the λ-plane, passing through the upper right corner C for $D = D_j$. In Fig. 5.12, it is the projection onto the base plane of that generator of the cylindrical surface along which the surface has a vertical tangent.

We remark that our assumption that the reaction rates depend only on λ makes this example degenerate. Here both the super- and subsonic branches of the reaction path in the pressure surface have the same projection onto the base plane. Were we to let the reaction rates depend on other state variables as well, this would not be the case. A more general case is treated in Sec. 5D.

5B1.1 D-Discussion

For $D < D_j$ the segments NT and OT of Fig. 5.11 map into the upper and lower branches of the reaction path of Fig. 5.12a. As shown there and in Fig. 5.14, the reaction path terminates on the sonic boundary, with nonzero rate, so that, as shown earlier, it cannot be a steady solution.

For $D = D_j$, the sonic boundary is at $\lambda_1 = \lambda_2 = 1$, the upper right corner of Fig. 5.14b, and the reaction is complete at the complete-reaction sonic point, that is, the CJ point. The upper branch, Fig. 5.12b, is the steady CJ solution.

For $D > D_j$ the sonic point has moved outside the stoichiometric triangle, the complete reaction point $\lambda_1 = \lambda_2 = 1$ is subsonic, and the two branches, Fig. 5.12c, no longer meet. The upper branch is the steady overdriven solution.

The steady solutions are thus exactly those of the one-irreversible-reaction detonation of Sec. 5A3 having the same total heat release. There are steady solutions terminating in sonic and subsonic points for $D = D_j$ and $D > D_j$ respectively, and no steady solution for $D < D_j$.

5B1.2. Piston Problem

The piston problem is identical to that of the corresponding one-reaction detonation, with the overdriven solutions susceptible to degradation by rarefactions from the rear, and with the CJ solution the unsupported detonation. The important property, that the final states of the steady solutions are independent of the reaction rates, still holds.

5B2. Second Reaction Endothermic (Eigenvalue Detonation)

With the second reaction endothermic, that is, $q_2 < 0$, we have our first example of an eigenvalue detonation. Our assumption of irreversibility is now less realistic for the second reaction, for the heat absorption shifts the equilibrium composition to a point of less complete reaction. But this is unimportant for the purpose of illustrating the main features. The practical interest in this case arises because many solid explosives consist of a mixture of an explosive and an inert binder, and a possible course for the reaction is the exothermic reaction of the explosive followed by the endothermic decomposition of the binder.

We take the system of the preceding example with the values of q_1 and q_2 changed to

$$q_1 = 100 \, RT_0 \tag{5.9}$$

$$q_2 = -75 \, RT_0. \tag{5.10}$$

The p-v diagram, p-λ_1-λ_2 space, and λ-plane (λ_1-λ_2 base plane) are shown in Figs. 5.15-5.18. The projection on the p-λ_1 plane, not used in the preceding section, is shown in Fig. 5.19.

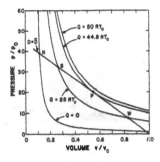

Fig. 5.15. Diagram of the partial-reaction Hugoniot curves in the p-v plane for two irreversible reactions with the second reaction endothermic. The $Q = 44.8\,RT_0$ Hugoniot curve is the highest attainable. Rayleigh line $\tilde{D}$ is tangent to this curve.

The negative sign of q_2 changes the slope of the straight lines which are the constant-Q contours, Eq. (5.3), from negative to positive, Fig. 5.17, with profound consequences. Each of these contours is still the sonic boundary, Eq. (5.8), for some value of D. Let $\tilde{Q}$ be the value of $Q_{max}(D)$ for which the sonic boundary is tangent to the reaction path, Fig. 5.17. For $Q > \tilde{Q}$ the sonic boundary lies to the right of the contour $Q = \tilde{Q}$, so the entire reaction path is accessible and along it Q increases from zero at A to the maximum value $\tilde{Q} = 44.8\,RT_0$ at the point where the reaction path is tangent to $Q = \tilde{Q}$, and then decreases to $25\,RT_0$ at C. The locus of neutral thermicity $\sigma \cdot r = 0$, Eq. (5.5), is shown as a dashed line in Fig. 5.17. Its intersection with the sonic boundary, called the pathological point P, is the sonic point of neutral thermicity. It plays an important role in the problem. For $Q = \tilde{Q}$ it coincides with the tangent point mentioned above.

The reaction-path topology in the pressure surface has also changed. Fig. 5.16 shows the p-λ_1-λ_2 space, and the projections, for three representative values of D, of the reaction paths onto the p-λ_1 plane. The special value of D, which we call $\tilde{D}$, for which the sonic boundary is tangent to the reaction path, Fig. 5.17, is the analog of D_j in the normal detonation. For $D > \tilde{D}$ the reaction path has two vertically separated segments, the upper entirely subsonic and connecting points N and S, and the lower supersonic and connecting points O and W. For $D < \tilde{D}$ there are two horizontally separated segments, each having both a subsonic and a supersonic branch. Only the left segment is accessible from points N and O; it is like the p-λ curves for one reaction, Fig. 5.4. For $D = \tilde{D}$ we have the saddle configuration centered at point P. Note that only at $D = \tilde{D}$, with the reaction path passing through P, is the weak point W accessible from N.

5B2.1. D-Discussion

For $D < \tilde{D}$ only the upper branch of the left segment (Fig. 5.19) can be reached from point N, and it is excluded as before by its termination at

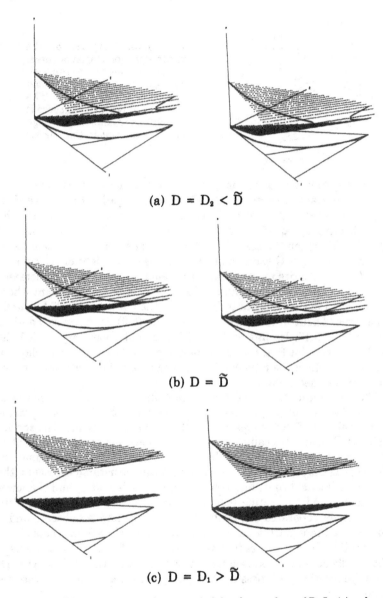

Fig. 5.16. The p-λ_1-λ_2 space (stereo pairs) for three values of D. In (c) only that portion of the surface with Q less than the maximum is shown.

Sec. 5B TWO IRREVERSIBLE REACTIONS

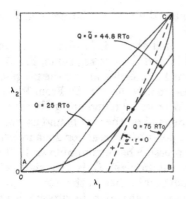

Fig. 5.17. The λ-plane. The thermicity $\boldsymbol{\sigma}\cdot\mathbf{r} = \sigma_1 r_1 + \sigma_2 r_2$ is positive to the left and negative to the right of $\boldsymbol{\sigma}\cdot\mathbf{r} = 0$.

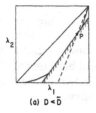 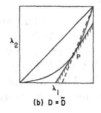 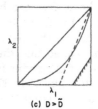

(a) $D < \tilde{D}$ (b) $D = \tilde{D}$ (c) $D > \tilde{D}$

Fig. 5.18. The λ-plane for three values of D.

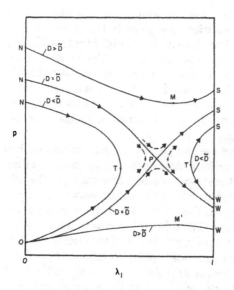

Fig. 5.19. The p-λ_1 diagram for a two-reaction detonation ith second reaction endothermic. The dashed curves show integral curves in the neighborhood of the pathological point (see text).

the unacceptable sonic point T where the rate is nonzero and $\sigma \cdot r$ is positive, as seen from Fig. 5.17.

For $D = \tilde{D}$ the reaction path, as noted above, is tangent to the sonic boundary at the pathological point P, and the two segments, Fig. 5.19, have now joined at the center in point P. Referring also to the p-v plane, Fig. 5.15, the state point first moves from N down to P. From P it can proceed on down to the weak point W or turn around and go back up to the strong point S. The choice will depend on the rear boundary condition. Note that P is an interior sonic point of the reaction zone in either case, and that in the strong case the pressure has a minimum and a first-derivative discontinuity at P. To summarize: $\tilde{D}$ is the eigenvalue velocity, the unique velocity for which point P lies on the reaction path, and P is a saddle point, the only point through which the supersonic part of the surface can be reached.

Finally, for $D > \tilde{D}$, with the reaction path entirely within the sonic boundary, the maximum value of Q on the reaction path remains unchanged, so that the state point in the p-v plane turns around at this value of Q, moving from point N down to point M and back up to S (the upper segment in Fig. 5.19). The reaction zone, though entirely subsonic, continues to have an interior pressure minimum.

5B2.2. Piston Problem

The piston problem presents some new and interesting possibilities. Diagrams of a possible set of solutions are given in Fig. 5.20, where $\tilde{u}_S$ and $\tilde{u}_W$ are the final strong and weak particle velocities for $D = \tilde{D}$. Consider first any piston velocity $u_p > \tilde{u}_S$, Fig. 5.20a. The solution is the usual overdriven one followed by a uniform state between the end of the reaction zone and the piston. As discussed above there is a minimum in the pressure profile within the reaction zone. The entire flow is subsonic with respect to the shock front, so any disturbance in the rear can propagate to affect any region.

For $u_p = u_S$, Fig. 5.20b, there are two possible configurations, each of which is followed by the same uniform state back to the piston. The first, the upper one in the figure, is the strong solution terminating at the subsonic point S. In an experiment in which the piston velocity is slowly decreased, it is expected to appear first because it is the continuous extension of the overdriven solution. The other solution with the same terminal state is a double-wave solution consisting of the steady reaction zone solution terminating at the supersonic point W, followed by a shock with final state S. This shock moves with the same velocity as the front since its Rayleigh line SW coincides with that for $\tilde{D}$ (see property 7, Sec. 5A1).

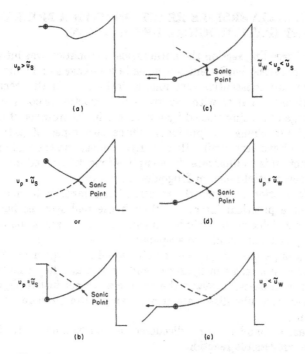

Fig. 5.20. Pressure-distance profiles for possible steady solutions of the two-reaction detonation with second reaction endothermic. The circle marks the end of the steady zone.

For piston velocities between $\tilde{u}_S$ and $\tilde{u}_W$, the terminal state W can be matched to the piston by means of a shock which recedes from the reaction zone, Fig. 5.20c. The shock connects points W on $D = \tilde{D}$, Fig. 5.15, with the piston by increasing the particle velocity.

As the piston velocity is decreased still further, the strength of the following shock decreases and it recedes more rapidly from the reaction zone. Finally, at piston velocity $\tilde{u}_W$, Fig. 5.20d, the reaction zone is connected to the piston by a uniform flow at state W. For lower piston velocities, Fig. 5.20e, a rarefaction wave is required to reduce the particle velocity to that of the piston. Since point W is supersonic with respect to the front, the head of this rarefaction wave will recede from the front. This configuration then consists of, from front to back: the steady reaction zone terminating at the supersonic point W, an ever-widening region of uniform state, a forward-facing rarefaction wave, and finally, an ever-widening region of uniform state between the tail of the rarefaction wave and the piston.

STEADY DETONATION   Chap. 5

5C. ONE IRREVERSIBLE REACTION WITH A MOLE DECREMENT (PATHOLOGICAL DETONATION)

In the preceding section the pathological detonation was introduced. The energy release coefficient σ was made to change sign by making the second of two consecutive reactions endothermic. In this section the pathological detonation with one irreversible reaction is considered, the sign change of σ being caused by a decrease in the number of moles as the reactants change to products. These two types of pathological detonation have very similar flow configurations, and the main purpose of this section is to illustrate the complications introduced into the p-v plane when partial-reaction Hugoniot curves cross.

The detonation in which the reaction has a large mole change is of considerable practical interest, although the real systems have complications which are not included in this example. Detonation in a dust of finely powdered aluminum suspended in gaseous oxygen is expected to show a large mole decrement combined with a large heat release. A stoichiometric mixture of hydrogen and oxygen also has a mole decrement. In both cases, however, the products are not just Al_2O_3 or H_2O, since these molecules dissociate at high temperature to reduce the mole decrement.

The case at hand was first discussed by von Neumann (1942). Consider the irreversible reaction

$$A \rightarrow (1 + \delta)B, \tag{5.11}$$

where δ is negative so there is a mole decrement. The energy and volume terms in the energy release coefficient σ, Eq. (4.16), have opposite sign, so σ may change sign at a point in the steady flow for some choice of the values of those terms. The case in which the sign of σ does change is, of course, the one of interest in this section. It can be shown that the partial-reaction Hugoniot curves then cross in the p-v plane, and that they have an envelope on which $\sigma = 0$. This can be seen by recalling that for $\sigma > 0$ the successive Hugoniot curves lie above those for lesser degree of reaction, and for $\sigma < 0$ below those for less reaction, so that the envelope is the locus $\sigma = 0$. The partial-reaction Hugoniot curves are shown in Fig. 5.21, and the solutions in p-λ in Fig. 5.22.

Let $\tilde{D}$ denote the value of D for which the Rayleigh line is tangent to the $\sigma = 0$ envelope. For $D > \tilde{D}$ the subsonic and supersonic branches are separated as shown in Fig. 5.22, since part of the Rayleigh line lies above the envelope and therefore above all of the partial-reaction Hugoniot curves. Starting at the shock point N, the state point moves down the

Sec. 5C ONE IRREVERSIBLE REACTION

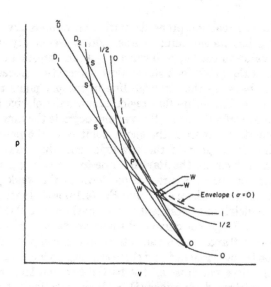

Fig. 5.21. The pathological one-reaction detonation in the p-v plane. In the example sketched the tangent point on the envelope is at $\lambda = 1/2$.

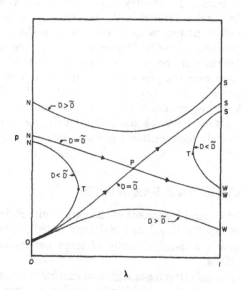

Fig. 5.22. The pathological one-reaction detonation in the p-λ plane.

175

Rayleigh line as reaction proceeds until it encounters the envelope. At this point p has its minimum value, and consequently u also has a minimum value so that $u_x = 0$ when $\sigma = 0$, in agreement with the requirements of Eq. (4.19) for a steady flow. As reaction proceeds the state point moves back up the Rayleigh line in the p-v plane to the $\lambda = 1$ Hugoniot curve. The entire flow region is subsonic relative to the shock and therefore rarefaction from the rear can degrade the flow. For $D = \tilde{D}$, the state point moves from the shock point N to the point where the Rayleigh line is tangent to the envelope, and then may either turn around and go back up the Rayleigh line to the strong point S on the $\lambda = 1$ Hugoniot curve, or proceed on down to the weak point W. In neither case is $u_x = 0$ where $\sigma = 0$, but Eq. (4.19) is still satisfied because where the Rayleigh line is tangent to the envelope it is also tangent to a partial-reaction Hugoniot curve and therefore we have $\eta = 0$ there. The point where $\sigma = 0$ and $\eta = 0$ is called the pathological point P. For $D < \tilde{D}$ the envelope is never reached and there is no steady solution because the Rayleigh line does not cross all the partial-reaction Hugoniot curves.

For all overdriven detonations ($D > \tilde{D}$) the reaction zone terminates at the strong point S where the flow is subsonic. For $D = \tilde{D}$ the steady flow terminates at either the strong point, where the flow is subsonic, or at the weak point, where the flow is supersonic. The character of the flow at these end points is exactly that found for the two-reaction pathological detonation treated in the preceding section, and therefore the discussion of the piston problem is the same as that given there, and the pressure profiles following the reaction zone are the same. The pressure profiles through the reaction zone are also more or less similar, but in the present case the pressure minimum in p-λ moves closer and closer to the shock front as the degree of overdrive (detonation velocity) is increased. Also the pressure at the strong end point S is always greater than the pressure at the shock front, for the simple case treated here. The unsupported detonation has a pressure profile which is similar in the two cases.

5D. TWO REVERSIBLE REACTIONS

Following the general treatment of the case of an arbitrary number of reversible reactions by Wood and Salsburg (1960), Erpenbeck (1961, 1964a) undertook the detailed study of some particular cases of two reversible reactions.

The system is a polytropic gas with two exothermic reactions, having no mole change and thus nominally similar to that of Sec. 5B1. However, the reactions are reversible and the rates (first-order

Sec. 5D   TWO REVERSIBLE REACTIONS

Arrhenius) are temperature-dependent. Both of these features greatly complicate the problem. Also, the reversibility can produce an eigenvalue detonation if the first reaction is sufficiently faster than the second. The first reaction may displace the composition beyond the equilibrium point of the second, which then becomes endothermic by running backwards.

The added complexity appears as a more complicated topology of the λ-plane. With reversible reaction there are interior critical points of different types. With temperature-dependent rates the reaction paths and the locus of neutral thermicity $\sigma \cdot r = \sigma_1 r_1 + \sigma_2 r_2 = 0$ depend on the detonation velocity D, rather than being the same for all values of D when projected onto the λ-plane as in Figs. 5.12, 5.13, 5.16, and 5.17. Also, the projections of the supersonic and subsonic branches are no longer superposed as they are in these figures.

The essential parameters are the same as those of Sec. 5B2: detonation velocity D, relative heat release q_2/q_1, and relative rate r_2/r_1. The case discussed is $q_1 \geq 2q_2 > 0$, for all values of r_2/r_1.

Because the reactions are reversible, the following rarefaction or shock wave in the piston problem runs into a reactive medium and will have the forms discussed in Secs. 4C4 and 4C5. This property and the more complicated topology require the discussion of a great many cases; but the essential features of the piston problem are those already discussed in Sec. 5B2.

5D1. The λ-Plane

The fluid is a polytropic gas with two parallel reversible reactions $A \leftrightarrows B$ and $A \leftrightarrows C$, with the initial state pure A. Transport properties are neglected, and the solutions sought are steady, plane, and laminar. The Arrhenius form is used for both reaction rates, similar to Eq. (4.46) but generalized to two reactions, so that there are two non-trivial parameters: the difference in activation energies and the ratio of the frequency factors

$$k = k_2/k_1.$$

The problem is conveniently treated by considering variation of this ratio, with the difference in activation energies held fixed.

Most of the discussion will involve the λ-plane [$\lambda = (\lambda_1, \lambda_2)$] and the reaction path in it. As before, to each partial-reaction Hugoniot curve in the p-v plane, which is similar to Fig. 5.11, there corresponds a particular value of Q, and this value of Q defines a sonic line in the λ-plane.

The differential equations to be studied are

$$d\lambda_1/dt = r_1 \quad (5.12)$$

$$d\lambda_2/dt = r_2 \quad (5.13)$$

or

$$d\lambda_1/d\lambda_2 = r_1/r_2 \quad (5.14)$$

We discuss only the case considered in greatest detail by Erpenbeck (1964a): $q_1 \geq 2q_2 > 0$. He also considered the consecutive reaction case with the second reaction endothermic ($q_2 < 0$); its piston problem is similar.

Figure 5.23 is a diagram of the λ-plane for a detonation velocity between the equilibrium and frozen CJ values. (The stoichiometric boundaries are different from those of Sec. 5B because the reactions are parallel instead of consecutive.) Recall that any dependent variable plotted on a third vertical axis defines a double-valued surface with a vertical tangent plane at the sonic locus $\eta = 0$ (the entropy has a cusp instead). In Fig. 5.23 the portions of the curves lying in the upper (subsonic) branch of the surface are shown with solid lines, and those lying in the lower (supersonic) branch are shown with dashed lines.

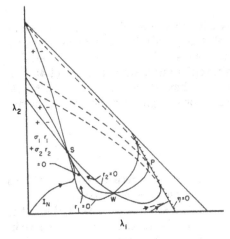

Fig. 5.23. The λ-plane for the steady detonation with two reversible reactions for $D_e < D < D_0$. The + and − signs show that $r_1 > 0$ to the left of its partial equilibrium curve $r_1 = 0$, etc.

Sec. 5D   TWO REVERSIBLE REACTIONS

The surface in question is defined solely by the equation of state and conservation conditions for the steady flow at velocity D, but the integral curves depend on the reaction rates. Other curves depending on the reaction rates are of interest. The partial-equilibrium curves $r_1 = 0$ and $r_2 = 0$ are important because where $r_1 = 0$ the integral curve (projected on the λ-plane) must be parallel to the λ_2-axis, and similarly for $r_2 = 0$. Also each reaction reverses direction where an integral curve crosses its partial-equilibrium curve, that is λ_i increases with time for $r_i > 0$ and decreases with time for $r_i < 0$. The intersections of the two partial equilibrium curves are points of chemical equilibrium and critical points for the differential equation (5.14). The integral curves near $r_1 = 0$ and $r_2 = 0$ can be drawn qualitatively using these properties, and it is easy to see that the intersection S is a node and the intersection W is a saddle for the curves drawn in Fig. 5.23.

The existence of solutions to Eq. (5.14) is assured by suitable conditions of continuity of r_1 and r_2, which are certainly satisfied in any physically reasonable case. To study the uniqueness of a solution from a given initial point it is convenient to consider r_1 and r_2 as thermodynamic functions over the λ-plane. As is the case for most thermodynamic functions, these rate surfaces have a vertical tangent plane above the sonic locus $\eta = 0$ in the base plane. To ensure that the solutions of Eq. (5.14) are unique, it is sufficient that the rate of change of either rate along any line in the surface be finite, that is, that the λ-gradient of the rate be finite, or that the Lipschitz condition be satisfied [see, for example, Davis (1960), p. 85]. For physically reasonable choices of the rate functions this condition is satisfied except on the sonic locus $\eta = 0$. Therefore, provided the two branches are considered separately, solutions of Eq. (5.14) exist and are unique except at the sonic locus.

The locus $\boldsymbol{\sigma} \cdot \mathbf{r} = \sigma_1 r_1 + \sigma_2 r_2 = 0$ divides the plane into exothermic and endothermic parts. The point at which it is tangent to $\eta = 0$ is the pathological point P. As in the earlier examples, integral curves reaching the sonic locus at any point other than P terminate there, as shown near the lower right corner of Fig. 5.23. Only by passing through point P can an integral curve go from the subsonic branch to the supersonic branch, and even this possibility is sometimes excluded in the present reversible case by the critical-point nature of P.

The integral curve I_N representing the solution of interest begins at the initial state point N (the von Neumann point) where $\lambda_1 = \lambda_2 = 0$. In Fig. 5.23 it is shown terminating at point S. For sufficiently small values of the second reaction rate relative to the first it will terminate on the sonic

STEADY DETONATION   Chap. 5

locus instead. In some cases it can pass through the pathological point P into the supersonic branch.

5D2. D-Discussion

Let D_e and D_0 denote the equilibrium and frozen CJ detonation velocities, and D_1 a particular velocity, defined below, peculiar to this problem. The λ-plane phase portrait is sketched in Fig. 5.24 for several values of D. The portion of the plane in which the interesting structure occurs has been exaggerated for clarity. For typical values of the parameters, the critical points are actually quite close to each other. The integral curves are sketched to correspond to a relatively large value of k. Their behavior for small values of k is described in the discussion but not shown in the figure.

5D2.1. $D < D_e$

As sketched in Fig. 5.24a, the partial-equilibrium curves do not intersect, so that there is no point of chemical equilibrium satisfying the conservation conditions. The integral curve I_N may intersect $r_2 = 0$ once, twice, or not at all, but in any case terminates on the sonic locus, so that there is no steady solution.

5D2.2. $D = D_e$

For this D, the partial equilibrium curves $r_1 = 0$ and $r_2 = 0$ have a point of tangency which is the equilibrium CJ point C_0. We digress briefly here to discuss the structure of the resulting higher-order singularity. The critical points of a differential equation such as Eq. (5.14) with parameter D

$$\frac{d\lambda_2}{d\lambda_1} = \frac{r_2(\lambda_1,\lambda_2,D)}{r_1(\lambda_1,\lambda_2,D)} \tag{5.15}$$

are those points at which both r_1 and r_2 vanish. The behavior of the integral curves in the neighborhood of such a point is found by separately expanding the functions r_1 and r_2 in powers of the displacements of λ_1 and λ_2 from the point. The procedure is described in standard texts, such as Davis (1960), Chapter 11. The two types of critical points most often encountered here—a node and a saddle—are shown in Fig. 5.24g in proximity to each other for a value of D slightly greater than D_e. The separatrices I_1, I_n, I_{SW} separate families of integral curves which approach or leave the critical points in particular directions. As D *decreases* toward D_e from above, the nodal and saddle points approach

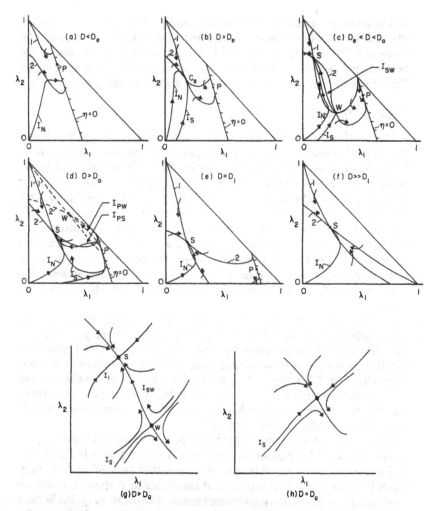

Fig. 5.24. Integral curves in the λ-plane for different values of D. The numbers 1 and 2 denote the curves along with $r_1 = 0$ and $r_2 = 0$, respectively.

each other, and at D_0 a higher-order singularity which has both nodal- and saddle-type sectors is formed as shown in Fig. 5.24h. The portion of the separatrix I_{SW} between S and W has vanished and the separatrices I_1 and I_s have coalesced. I_1 is never important to the discussion, but I_s is.

Depending on the value of k, the separatrix I_s may intersect either the λ_1 or the λ_2 axis. For large k, it intersects the λ_1 axis and I_N lies above it and must therefore cross $r_2 = 0$ above C_e and go down into C_e as shown in Fig. 5.24b. As k is decreased, I_s moves to the left and, at a particular value of k, becomes I_N. For smaller k, I_G intersects the λ_2 axis, and I_N lies below it and is forced to pass below C_e and terminate on the sonic locus. Thus for sufficiently small k there is no steady solution for $D = D_e$.

5D2.3. $D_e < D < D_0$

The partial-equilibrium curves, Fig. 5.24c, cross at two points S and W, both of which lie on the (frozen-)subsonic branch of the surface. Point S is a node and point W a saddle. Again the terminus of I_N depends on the position of the separatrix I_s and thus on the value of k. For large k, I_N terminates at the strong point S, as shown in the Figure. As k decreases, I_s moves to the left, until at some value of k it passes through the origin. For this k, the solution I_N consists of two pieces: the separatrix I_s from the origin to point W, and the curve I_{SW} (another separatrix of the critical point W) connecting points W and S. For k smaller than this value, I_N is forced to the sonic locus and no steady solution exists, with the possible exception of the case in which it reaches point P.

For this range of D the behavior of the integral curves in the neighborhood of the pathological point P may be quite complicated, and has been carefully investigated by Erpenbeck (1961). Without going into detail, it is sufficient to state here that, in the projection onto the p-λ_1 plane (like Fig. 5.19), the pathological point P may have the character of either a node or a focus (including the possibility of limit cycles), except that the direction of reaction changes sign across the sonic locus. Point P cannot be a saddle in this range of D.

Only if it is a node stable from the subsonic surface can P possibly be reached by I_N. Point P neeed not lie above W as it is shown in Fig. 5.24c, and it appears to be possible that in some rather unusual cases I_N may reach P. Such a steady solution must terminate at P since there are no equilibrium points on the supersonic branch of the surface (point W here being on the subsonic branch). The possible complete solution to the piston problem would require some more complicated type of time-dependent continuation beyond P than is usually considered. If I_N, having missed point W, can reach point P, then a possible solution might be such a time-dependent continuation.

5D2.4 $D > D_0$

At $D = D_0$ points W and P coincide and together become the frozen CJ point. For D larger than this value, Fig. 5.24d, point W moves down onto

Sec. 5D   TWO REVERSIBLE REACTIONS

the supersonic branch of the surface and becomes a stable node. In this range the pathological point P can have only the character of a saddle point. For some value of k, the separatrix I_s through the saddle point P becomes I_N. For larger k, I_N terminates at point S; for smaller k it is forced to the sonic locus. Therefore, at least for values of k within some range, there is, for any given k, the important eigenvalue of D for which I_N and I_s coincide. For this D the integral curve I_N proceeds to the (sonic) point P, which it reaches in finite time because P is not a point of chemical equilibrium. From here there are two possibilities: it may proceed back up (along the Rayleigh line) to the subsonic point S along I_{PS}, or it may go on down to the supersonic point W along I_{PW}.

As D increases still further, at some velocity D_1 the minimum in the $r_1 = 0$ curve touches the λ_1 axis. This value of D is the same as the CJ value of D for a material in which the rate of reaction 2 is identically zero, because it is the minimum velocity for which chemical equilibrium can be attained with $\lambda_2 = 0$.

5D2.5. $D > D_1$

For $D > D_1$, Fig. 5.24e, the integral curve I_N can go only to point S, since the partial-equilibrium curve $r_1 = 0$ intersects the λ_1 axis. As D is increased still further, Fig. 5.24f, the sonic locus moves out of the stoichiometric triangle and the partial-equilibrium curve $r_2 = 0$ terminates at $\lambda_1 = 1$, $\lambda_2 = 0$. Point S remains the only possible terminal point for I_N.

5D3. The Piston Problem

There are three ranges of values of k with quite different behavior:

Case I. Values of k large enough that the integral curve I_N always lies to the left of the separatrix and goes to the strong point S or to the equilibrium CJ point C_0.

Case II. Intermediate values of k such that I_N and I_6 coincide for some value of D in the range $D_e < D < D_0$.

Case III. Smaller values of k such that the coincidence of I_N and I_6 takes place for a value of $D > D_0$.

For a given k there will be, in Cases II and III, an eigenvalue of D (which makes I_N and I_s coincide), which we denote by $\tilde{D}(k)$. The function $\tilde{D}(k)$ is monotone decreasing as k increases for the simple fluid considered here. Similarly, we denote by $\tilde{k}$ the value of k which brings I_N and I_s into coincidence at a given D. The subscripts e and o have their usual meaning of equilibrium and frozen CJ points [as $\tilde{k}_0 = \tilde{k}(D_0)$]. For any $\tilde{D}(k)$ there are two particle velocities $\tilde{u}_S$ and $\tilde{u}_W$ denoting the particle

STEADY DETONATION
Chap. 5

velocity at points S and W for that $\tilde{D}$. Figure 5.25 shows these particle velocities as functions of k. The numerical values, from Erpenbeck (1964a) were obtained by numerical integration.

5D3.1 Case I. $k \geq \tilde{k}_e$

In Case I the integral curve I_N is forced to point S for all $D > D_e$ and to C_e for $D = D_e$. The piston problem is therefore the same as that for the single reversible reaction in Sec. 5A4.

5D3.2 Case II. $\tilde{k}_0 \leq k < \tilde{k}_e$

For $D = D_e$ the integral curve I_N is to the right of the separatrix I_s and is therefore forced to the sonic locus, so no steady solution exists. It is obvious then that the steady solutions must be found, if any exist, for $D > D_e$.

For $u_p < \tilde{u}_W$ the suggested solution, Fig. 5.26a, is the eigenvalue detonation moving with velocity $\tilde{D}(k)$ and terminating at point W. Point W is frozen-subsonic but equilibrium-supersonic. A reactive rarefaction wave, the main portion of which recedes from the front as in Sec. 5B2, follows the detonation and in it the particle velocity decreases from $\tilde{u}_W$ to u_p. The presence of the precursor of the rarefaction wave, which attenuates with time, precludes a strictly steady solution; the problem is similar to that of the detonation in material with one reversible reaction discussed in Sec. 5A4.

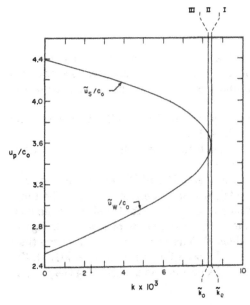

Fig. 5.25. The u_p-k plane used in the discussion of the piston problem. The curves $\tilde{u}_S$ and $\tilde{u}_W$ are the particle velocities on the strong and weak branches of the detonation Hugoniot curve for $D = \tilde{D}(k)$, the eigenvalue detonation velocity. Values are from Erpenbeck (1964a) for $q_1/RT_0 = 60$, $q_2/RT_0 = 20$, $\gamma = 4/3$, $(E_1^\dagger - E_2^\dagger)RT_0 = -45$.

Sec. 5D TWO REVERSIBLE REACTIONS

For $\tilde{u}_W < u_p < \tilde{u}_S$ the suggested configuration, Fig. 5.26b, is the same eigenvalue detonation followed by a slower moving diffuse shock which raises the particle velocity from $\tilde{u}_W$ to u_p. The shock is diffuse because point W is frozen-subsonic, as discussed in Sec. 4C4. The pertinent Hugoniot curves in the p-v plane are shown in Fig. 5.27. The integral curve for the diffuse shock connects point W to the state on the equilibrium shock Hugoniot curve originating at W and having $u = u_p$. On a λ-plane drawn for the diffuse shock velocity this integral curve is similar to I_{WS} of Fig. 5.24c, but goes to the proper point W rather than point S. For $u_p = \tilde{u}_S$, I_e passes through the origin and becomes I_N, so that both the eigenvalue detonation (I_e) and the diffuse shock (I_{SW}) have the same velocity and can be shown in a single diagram like Fig. 5.24c.

For higher piston velocities $u_p > \tilde{u}_S$, I_N lies above the separatrix and goes to point S, and the detonation velocity is greater than $\tilde{D}(k)$. The detonation is overdriven and more or less normal in behavior, except that the pressure profile, Fig. 5.26c, has a minimum for values of u_p near $\tilde{u}_S$, since the integral curve passes close to point W.

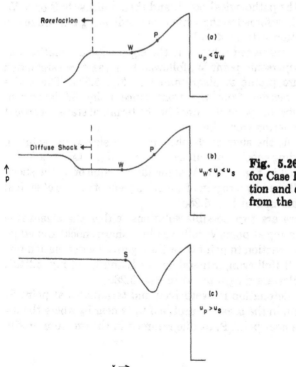

Fig. 5.26. Pressure profiles for Case II. Both the rarefaction and diffuse shock recede from the reaction zone.

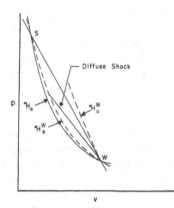

Fig. 5.27. Diagram of the Hugoniot curves for Case II. $\mathcal{H}_o$ is the detonation Hugoniot curve centered at the initial state. $\mathcal{H}_o^W$ and $\mathcal{H}_o^{eW}$ are the equilibrium and frozen Hugoniot curves centered at point W. The slope of the diffuse-shock Rayleigh line is less than the slope of $\mathcal{H}_o^W$ at point W. The effects of the frozen and equilibrium sound speed differences are exaggerated.

5D3.3 Case III. $k < \tilde{k}_0$

When $k < \tilde{k}_0$ the eigenvalue detonation occurs for $D > D_0$. The eigenvalue detonation looks in the λ-plane like Fig. 5.24d, but with I_s becoming I_N, passing from the origin to point P. The integral curve passes through point P, the pathological point, and thence either to S or to W. For all $D < \tilde{D}(k)$, I_N is forced to the sonic locus, indicating that there can be no steady solution with $D < \tilde{D}(k)$.

For $u_p < \tilde{u}_W$ the suggested solution is the eigenvalue detonation terminating at the supersonic point W, followed by a reactive rarefaction wave. The pressure profile is diagrammed in Fig. 5.28a. The entire rarefaction wave recedes from the front since point W is frozen-supersonic. The subsonic point S cannot be the terminal state because it is subject to degradation from the rear.

For $\tilde{u}_W < u_p < \tilde{u}_S$ the suggested solution is the eigenvalue solution followed by a receding shock. For values of u_p near $\tilde{u}_W$, this shock is diffuse, as diagrammed in Fig. 5.28b. At some larger value of u_p the shock becomes sharp, followed by a region of effectively endothermic chemical reaction, as diagrammed in Fig. 5.28c.

At $u_p = \tilde{u}_S$ there are two possible solutions: either the eigenvalue detonation terminating at point W followed by a sharp shock[5] and effectively endothermic reaction to point S, or the eigenvalue detonation terminating at point S (following integral curves I_S and I_{PS} of Fig. 5.24d). The pressure profiles are diagrammed in Fig. 5.28d.

For $u_p > \tilde{u}_S$ the detonation is overdriven and terminates at point S. There is a minimum in the pressure profile if u_p is near $\tilde{u}_S$ where the integral curve passes near point P, as diagrammed in the pressure profile curve, Fig. 5.28e.

[5]It can be shown that the diffuse- to sharp-shock transition takes place at some $u_p < u_s$.

Sec. 5D   TWO REVERSIBLE REACTIONS

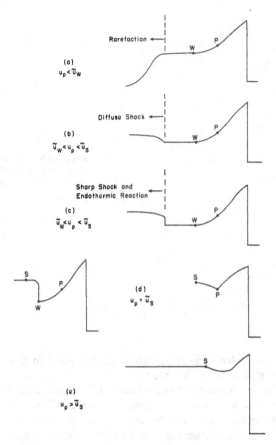

Fig. 5.28. Pressure profiles for Case III.

5D3.4 Summary of Piston Problem Results

In Fig. 5.29a the pressure profiles are diagrammed in their regions of the u_p-k plane. The plane is that of Fig. 5.25, but is greatly distorted to make room for the profiles. Below the (curved) line $u_p = \tilde{u}_s$, and the line extending to the right from C_e, the velocity of the front is independent of the piston, and above those lines the detonations are the overdriven type. To the right of $k = \tilde{k}_o$ the behavior is like that discussed in Sec. 5A4. To the left, the eigenvalue detonations all have the sonic point P reached in finite time (the reaction rate is not zero as the integral curve passes through the sonic point), and the steady reaction zone continues to the weak point W (after infinite time). Behind the weak point there is an ever-increasing constant state, followed by a rarefaction or shock (a

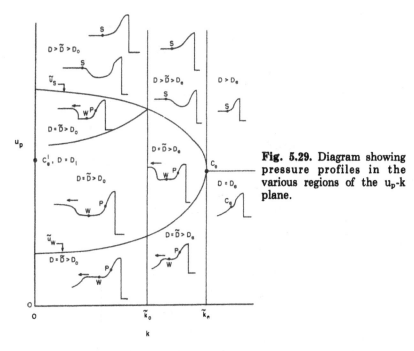

Fig. 5.29. Diagram showing pressure profiles in the various regions of the u_p-k plane.

reactive wave shifting the equilibrium as discussed in Sec. 4C) which moves more slowly than the front.

On the k = 0 axis there is a point marked C_e^1 which is the end point for the detonation in which only reaction 1 is allowed. There is an apparent discontinuity in behavior here, because it does not tie in with the rest of the cases shown. The reason is that as k becomes very small, the wave can be considered in two distinct parts. Near the shock front only reaction 1 goes on to a large extent, and reaction 2 is very far behind. In the limit we ignore reaction 2 altogether, describing the steady zone differently from the descriptions shown in the figure. In Sec. 5D4 the details are elaborated further.

The line extending down and to the left from point A on the upper branch of $u_p = \tilde{u}_S$ marks the diffuse-sharp shock transition.

In the sketches on the figure the profiles are drastically distorted, because the time or distance from the front to point P is very short compared to the time from P to W, and the following waves are not steady and have gradients much less than those in the reaction zone. In Sec. 5D4 some actual calculations for the part of the reaction zone near the front are given, and the pressure difference between $u_p = \tilde{u}_S$ and $u_p = \tilde{u}_W$ is shown.

5D4. Examples

We display here portions of some steady solutions for particular fluids, with the principal object of comparing the forms of normal (CJ type) and eigenvalue (weak type) solutions. The results make use of information kindly supplied to us by Dr. Erpenbeck in addition to that given in Table I of Erpenbeck (1964a).

The fluids used have in common the parameters

$$\gamma = 1.2, \, q_1 = 50 \, RT_0, \, q_2 = 25 \, RT_0, \, E_1^t = E_2^t = 10 \, RT_0,$$

for which

$$D_e = 6.0361 \, c_0, \, D_0 = 6.0365 \, c_0, \, D_1 = 6.1733 \, c_0.$$

Recall that D_1 is the velocity at which the $r_1 = 0$ partial-equilibrium curve is tangent to the λ_1 axis at one point, that is, just the equilibrium CJ velocity for the single-reaction system having only the first of the above two reactions (obtainable from the above by taking the limit $k_2/k_1 = 0$).

We consider three different fluids corresponding to three different values of k ($\equiv k_2/k_1$) with the above set of parameters, and for each a particular value of D:

1. The eigenvalue solution for $k = \widetilde{k}(D) = 0.009183$ with D slightly less than D_1: $D = 6.1597 \, c_0$, corresponding to $(D^2 - D_e^2) = 0.9 \, (D_1^2 - D_e^2)$.

2. The equilibrium CJ solution for the one-reaction system having the first of the above two reactions, that is, the two-reaction system with $k = 0$, $D = D_1$.

3. The normal equilibrium CJ solution for two reactions with $k = 0.1109$, slightly greater than $\widetilde{k}(D_e)$, and with $D = D_e$.

The calculated particle-pressure histories are given in Fig. 5.30. In the eigenvalue solution (1) the pathological point P at $p = 21.1$ is reached in about nine half-reaction times. Beyond this point there is a branch of continuous slope to the weak point at $p = 17.2$ and a branch of discontinuous slope to the strong point at $p = 25.5$. The solution has not been calculated beyond the pathological point except for the initial slopes and final states, but the effective length of this region is estimated by various methods to be several hundred times the distance of the pathological point from the front. The one-reaction equilibrium CJ solution is nearly

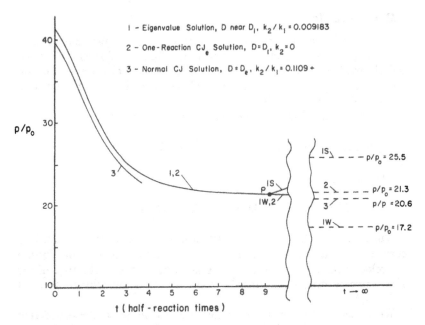

Fig. 5.30. Calculated steady solutions: (1) eigenvalue solution for D slightly less than D_1, (2) equilibrium CJ solution for one reaction ($k_2 = 0$, $D = D_1$), (3) normal equilibrium solution (for two reactions), k slightly larger than $k(D_e)$. The time scale for the particle pressure histories shown is chosen to make $\lambda = 1/2$ at $t = 1$ for case (1), which requires that the unit of time be $3.3/k_1$. With this choice λ is not far from $1/2$ at $t = 1$ for the other two cases. The first part of curve 2 lies slightly above curve 1 but too close to show separately.

indistinguishable, on the scale of Fig. 5.30, from the part of the eigenvalue solution shown, but of course contains no upper branch. The final state of this solution lies about halfway between the two final states of the eigenvalue solution. Finally, the equilibrium CJ solution (for two reactions) of normal type is very similar to the equilibrium CJ solution for one reaction, but lies somewhat below it.

5E. MORE THAN TWO REACTIONS

Detonation in a material having an arbitrary number of reversible chemical reactions has been studied by Wood and Salsburg (1960). Their paper is a general treatment of steady one-dimensional detonation, and forms the basis for all the preceding sections of this chapter.

The examples in the preceding sections illustrate the results of the general treatment, and were of course chosen for that purpose. It is perhaps worth noting that the work of Wood and Salsburg preceded almost all the work on the details of the particular cases.

When there is an arbitrary number n of chemical reactions, λ-space is n-dimensional. The analogue of the pathological point P is the intersection of the surfaces $\eta = 0$ and $\sigma \cdot r = 0$ and is thus an $n - 2$ dimensional locus of critical points, so that the topology is more complicated.

It is our conjecture that all the physically significant detonation solutions are qualitatively similar to those found for two reactions which are discussed in the preceding section. Thus there is nothing new to present in this section, with the exception of a case whose physical significance is obscure and that could conceivably provide an exception.

This case arises from the possibility that a portion of the pathological locus has the character of a collection of nodes stable from the subsonic side in the range $D > D_0$. If it has this character, a two-parameter family of integral curves passes through each nodal point. For purposes of illustration consider the three-reaction case. The pathological locus in λ_1-λ_2-λ_3 space consists of a curve segment lying in the sonic surface. Each point of the nodal portion of this segment is a node for a two-parameter family of integral curves passing from the subsonic (solid lines) to the supersonic (dashed lines) branch of the space, as diagrammed in Fig. 5.31. Thus if the pathological locus has this character, there is a finite range of D (there is no unique eigenvalue D in this range) for which I_N reaches a portion of the pathological locus, and, because of the nodal character, the continuation to the supersonic branch is not unique. The physical meaning is obscure. In the words of the authors, "We know of no physical requirement which could result in unique selection at the point P, and so conclude that the steady flow would probably not continue beyond such a point."

5F. INCLUSION OF TRANSPORT EFFECTS

The effects of transport processes are considered in this section. When these are included, the equations for nonreactive flow have solutions representing shocks of finite thickness. In a detonation the shock region and the reaction zone thus overlap to some extent. For realistic reaction rates the profiles are similar to those of the ZND model, with peak pressures reduced somewhat by the overlapping of shock and reaction zones. For very fast rates of reaction, too fast to have physical significance, the equations have steady detonation solutions characterized by an eigenvalue propagation velocity and terminating on the weak branch of the detonation Hugoniot curve.

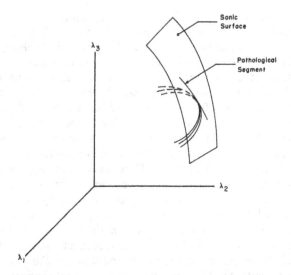

Fig. 5.31. One of a possible continuous sequence of nodes on the pathological locus in the three-reaction case.

The equations used are the Navier-Stokes equations with chemical reaction, which include the effects of viscosity, heat conduction, and diffusion. For simplicity, only a single irreversible chemical reaction with σ positive throughout is considered. The assumption of local thermodynamic equilibrium which is used in all the preceding sections is still made, and may well be a non-negligible source of error in the shock region where the flow gradients are very large. The problem is simplified by assuming certain special but reasonable values for some of the physical parameters. As throughout this chapter, steady solutions are sought. Work on this problem has been done by Friedrichs (1946), Hirschfelder and coworkers (1958), Adamson (1960), Oppenheim and Rosciszewski (1963), Spalding (1963), Koumoutsos and Kovitz (1963), Wood (1961, 1963), and Bowen (1967). The later paper of Wood (1963) is the pattern for discussion of this section.

When the transport effects are added, the steady profile no longer consists of a discontinuous shock transition followed by a zone of chemical reaction. Instead the entire solution is continuous and infinite in spatial extent. It is not possible to identify one portion strictly as shock and the other as reaction zone, but the initial state is often referred to as the cold boundary and the final state as the hot boundary.

Sec. 5F INCLUSION OF TRANSPORT EFFECTS

The essential features of this problem are displayed by the simpler case in which, among the transport processes, only viscosity is considered, and heat conduction and diffusion are omitted. It is not without interest in its own right, for it is the continuous analog of those numerical schemes for the solution of flow problems in which an artificial viscosity is added to "smear out" the shock transition to a size manageable in the discrete steps of the computer program. Unfortunately, however, no solutions for it have been published.[4] We therefore use it to discuss the equations, but turn to the full problem to display the (qualitatively similar) calculated solutions.

In the steady frame (shock at rest) with incoming particle velocity D, the equations for one-dimensional steady flow are

$$\rho u = \rho_0 D \tag{5.16a}$$

$$\nu \, du/dx = p - p_0 - \rho_0 D(D - u) \tag{5.16b}$$

$$\nu v \, du/dx = E - E_0 + pv - p_0 v_0 + 1/2 \, (u^2 - D^2) \tag{5.16c}$$

$$d\lambda/dx = u^{-1} r, \tag{5.16d}$$

where x is distance and ν is the coefficient of viscosity. An equation of state and chemical rate law must also be given: for these we take

$$E = pv/(\gamma - 1) - \lambda q \tag{5.16e}$$

$$r = k(1 - \lambda). \tag{5.16f}$$

It is instructive to note that, using the mass conservation relation, the second and third equations can be put in the form

$$\rho_0^2 D^2 - (p + \pi - p_0)/(v_0 - v) = 0 \tag{5.17a}$$

$$E - E_0 - 1/2 \, (p + \pi + p_0)(v_0 - v) = 0 \tag{5.17b}$$

or

---

[4] Cameron (1963) assumed a constant chemical rate, but this is an oversimplification for the present purpose of illustrating the general case. A rate which vanishes as the reaction approaches completion must be used to obtain the desired character of the result, that is, the presence of a critical point.

STEADY DETONATION Chap. 5

$$\pi/(v_0 - v) = \rho_0^2 D^2 - (p - p_0)/(v_0 - v) = \mathcal{R}(p,v) \tag{5.17c}$$

$$1/2\ \pi(v_0 - v) = E - E_0 - 1/2\ (p - p_0)(v_0 - v)$$

$$= \mathcal{H}(p,v), \tag{5.17d}$$

where π is the "viscous pressure" given by

$$\pi = -\nu\ du/dx. \tag{5.17e}$$

The symbol q is often used for the viscous pressure. These relations, which are just the usual frictionless conservation relations with the pressure p replaced by $p + \pi$, give the deviations of the solution from the Rayleigh and Hugoniot curves.

Equations (5.16) can be reduced to a pair of ordinary differential equations for (say) $v(x)$ and $\lambda(x)$ by eliminating p and u. The fluid parameters ν, γ, q, and k, the initial pressure and density p_0 and ρ_0, and the propagation velocity D remain as parameters. Assuming $p_0 \ll p$, the resulting equations are

$$\nu\ d(v/v_0)/dx = \rho_0 D[1/2\ (\gamma + 1)(v/v_0) + 1/2\ (\gamma - 1)(v_0/v) - \gamma$$

$$+ (\gamma - 1)(q/D^2)(v_0/v)\lambda] \tag{5.18a}$$

$$d\lambda/dx = r[(v/v_0)D]^{-1}. \tag{5.18b}$$

The continuous solution of these equations extends from $-\infty$ to $+\infty$ in x. In the initial and final states the flow gradients vanish and these states are connected by the conservation conditions and are thus identical to those of the frictionless problem with the same propagation velocity. The equations can be put into dimensionless form by defining two characteristic lengths as follows:

$$d(v/v_0)/d\xi = f_1(v/v_0,\lambda,q/D^2,\gamma)$$

$$= 1/2\ (\gamma + 1)(v/v_0) + 1/2\ (\gamma - 1)(v_0/v) - \gamma$$

$$+ (\gamma - 1)(q/D^2)(v_0/v)\lambda \tag{5.19a}$$

$$d\lambda/d\xi = f_2(v/v_0,\lambda,\ell,\gamma) = \ell(v_0/v)(1 - \lambda) \tag{5.19b}$$

$$\ell = \ell_v/\ell_r = k\nu/(\rho_0 D^2) \tag{5.19c}$$

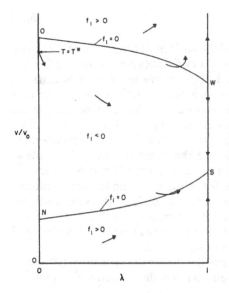

Fig. 5.32. The λ-v/v_0 phase plane for the viscous detonation. The integral curve from point O with the inclusion of an ignition temperature is shown proceeding to T^* before λ begins increasing from zero.

$\ell_v = \nu/(\rho_0 D) =$ characteristic viscous length (5.19d)

$\ell_r = D/k =$ characteristic reaction length (5.19e)

$\xi = x/\ell_v.$ (5.19f)

The characteristic viscous length is used as the distance unit. It would be the natural unit of distance for a nonreactive shock and is of the order of a mean free path. The characteristic reaction length is related through the velocity to the characteristic reaction time k^{-1}. The initial and final states are obtained from the conservation relations (with p_0 omitted)

$$\rho_0^2 D^2 = p/(v_0 - v) \tag{5.20a}$$

$$pv/(\gamma - 1) - \lambda q = 1/2\, p(v_0 - v). \tag{5.20b}$$

In the discussion to follow, the principal variation is to change D for a particular fluid and initial density ($\nu,\gamma,q,k;\rho_0$), but variation of k at fixed D is also considered.

The λ-v/v_0 phase plane shown in Fig. 5.32 is useful for considering the solutions for various values of D. The integral curves of interest are the solutions of the equation obtained from Eqs. (5.19a) and (5.19b) for the reaction in this plane:

$$d(v/v_0)/d\lambda = f_1/f_2. \tag{5.21}$$

The parabola $f_1 = 0$ is the locus of the horizontal turning points of the integral curves. The relation between λ and v obtained by setting $f_1 = 0$ is just that which holds on the usual Rayleigh line in the p-v plane; hence, this locus is referred to as the Rayleigh curve. On the locus $f_2 = 0$, i.e., $\lambda = 1$, $r = 0$, the integral curves are vertical. This locus corresponds to the complete-reaction Hugoniot curve in the p-v plane. The points S and W, where the Rayleigh curve intersects the vertical line $\lambda = 1$, correspond to the strong and weak intersections of the Rayleigh line and complete-reaction Hugoniot curve in the p-v plane. Points O and N, where the Rayleigh curve crosses $\lambda = 0$, are the initial state and the von Neumann point. In this plane the state point in the ZND model jumps discontinuously from point O to point N, then moves up the Rayleigh curve to point S. The analogous solution of the viscous problem is an integral curve leaving point O and terminating at point S.

Points S and W, where $f_1 = 0$ and $f_2 = 0$, are the critical points. Their structure is shown in Fig. 5.33. Point S is a node and point W is a saddle.

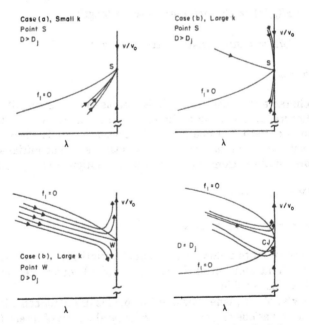

Fig. 5.33. Integral curves in the neighborhood of the critical points S and W for the viscous detonation.

Sec. 5F INCLUSION OF TRANSPORT EFFECTS

The preferred directions of approach to S depend on the values of k and D. At any fixed $D > D_j$, case (a) will obtain at small k and case (b) at large k, as diagrammed in the figure. At $D = D_j$ the Rayleigh curve is tangent to $\lambda = 1$ and the two critical points merge to form a higher-order singularity having nodal and saddle sectors as shown.

There is a difficulty in starting the integral curves at point O which arises because the simple reaction rate chosen is finite at the state point O and the equations are written to require that the solution be strictly steady. The curve $f_1 = 0$ passes through point O, so the integral curve leaving point O, Fig. 5.32, does so with zero slope, and thus moves off to infinite volume. The difficulty is removed by introducing an ignition temperature $T^* > T_0$ below which the rate is assumed to vanish. Any integral curve from O then moves down $\lambda = 0$ until it reaches the ignition temperature, and then moves off with negative slope.

We now turn to the numerical solutions for the full Navier-Stokes detonation; those for the simpler viscous detonation would be similar. A progression of integral curves for different values of k at fixed $D > D_j$ is shown in the λ-u plane in Fig. 5.34, and in the p-v plane in Fig. 5.35. For very small k the integral curve passes close to point N and terminates at point S. The shock and reaction zone are nearly decoupled, and the pressure profile of the solution is close to that for the ZND model. As k is

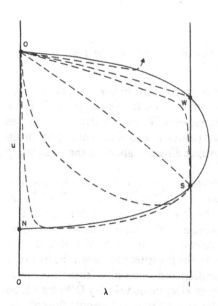

Fig. 5.34. A progression of integral curves for the Navier-Stokes detonation in λ-u space. From Wood (1963). The ignition temperature point is indistinguishable from point O in the figure. The λ-u plane is topologically equivalent to the λ-v/v_0 plane of Fig. 5.32 since v is proportional to u.

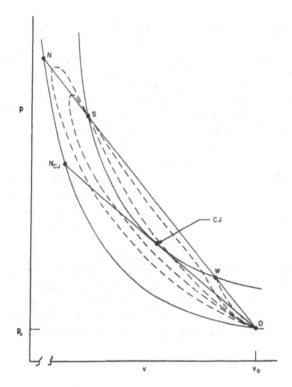

Fig. 5.35. Integral curves of Fig. 5.34 in the p-v plane. From Wood (1963).

increased the integral curve moves to the right in the λ-u plane, eventually passing arbitrarily close to point W on its way to point S. For a particular value of k it terminates at point W, and for still larger values moves off toward $u = \infty$. Note that points W and S are connected by an integral curve, the line $\lambda = 1$, corresponding to a shock in the completely reacted material.

Thus, like other problems discussed in previous sections, this one has, for fixed k, an eigenvalue detonation velocity $\widetilde{D}(k)$, or for fixed D, an eigenvalue rate multiplier $\widetilde{k}(D)$. As D is increased indefinitely at any fixed k, the ZND solution is approached. The eigenvalue $\widetilde{k}(D)$ is found to be a monotone increasing function of D. For values of $k < \widetilde{k}(D_j)$ the piston problem will be "normal," and the pressure profile will be like that for the ZND model with some rounding and smoothing from the viscous effects. For $k > \widetilde{k}(D_j)$ there will be an eigenvalue velocity $\widetilde{D}(k)$ for which the steady reaction zone terminates at a point of supersonic flow on the weak branch of the Hugoniot curve, and the piston problem will be analogous to that for the one-reaction pathological detonation of Sec. 5C. The discussion of the piston problem is given in detail by Wood (1963) for the full Navier-Stokes equations (with certain assumed simplifying relations among the transport coefficients) for an ideal gas with a single forward reaction A → B obeying an Arrhenius rate law.

In gases, at least, the transport effects and the reaction rate are related to the molecular collision rate, and examination of these relations suggests that realistic gaseous reaction rates are too small to give the weak solution (eigenvalue detonation). Measurements of reaction rates confirm this conclusion. The eigenvalue detonation might, of course, occur in a different physical system in which the energy-production process arises from a physical process independent of the molecular collision rate, such as, for example, a nuclear reaction initiated by compression. In condensed-phase explosives the relationships among the transport coefficients are no doubt quite different and are certainly unknown, but it seems unlikely that the eigenvalue detonation occurs in them.

Deviations from the nontransport model in gases can be appreciable. A calculation for ozone by Oppenheim and Rosciszewski (1963) using realistic kinetics and transport coefficients in the Navier-Stokes equations is compared with the nontransport (ZND) solution in Fig. 5.36.

5G. SLIGHTLY DIVERGENT FLOW

This section considers the effect on the steady solution of including the small radial component of the divergence of the flow field (hereafter "divergence" for short) which is always present in a finite-diameter charge. We limit the discussion to the case of small deviations from strictly one-dimensional flow (the "plane" case). That is, the diameter of the charge is well above the failure diameter, and much larger than the reaction-zone length. As a consequence, the amount of lateral expansion in the reaction zone is small. Furthermore, we do not consider the problem of the complete (i.e., including the details of the reaction zone) two-dimensional flow in a cylindrical charge, recently studied by Bdzil (1976). Instead, we focus attention on the flow near the axis, and primarily on the y complexities, but is sufficiently tractable to give useful qualitative results. The flow variables on the axis cannot, of course, be obtained independently of the remainder of the flow field. But a study of the equations in this region yields a rather complete understanding of the nature of the solution and of the way in which the inclusion of the divergence affects the flow.

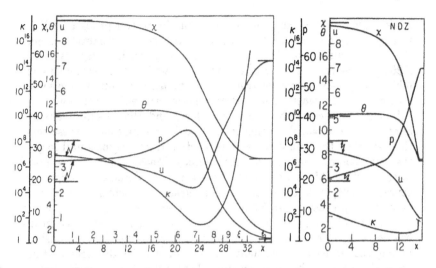

Fig. 5.36. Comparison of steady CJ detonation in ozone calculated (a) with, and (b) without transport properties. The symbols x, p, u, θ, and χ are non-dimensional distance, pressure, material velocity, temperature, and stagnation enthalpy; κ is the average number of intermolecular collisions per molecule of products formed by reaction. The values for unreacted ozone are at the right edge of the graphs, and reaction proceeds toward smaller x. The CJ values are indicated at the left edge of the graphs. From Oppenheim and Rosciszewski (1963).

The divergence arises mainly from lateral expansion of the confinement in condensed explosives, and from boundary-layer effects in gases (see, for example, Fay [1959], and Kurzrock [1966], Chapter 7 and Fig. 16). We consider here the most common experiment in the condensed case. The detonation is started at one end of a long stick (cylinder) of explosive by a planewave lens. As it proceeds down the stick side rarefactions from the lateral expansion enter. At first, there is a central region not yet affected by these rarefactions, but in which the reaction zone has become nearly steady. For the central region at this stage, the strict one-dimensionality assumption used heretofore is appropriate. (Non-one-dimensional effects from imperfections in the initiating lens may be of some importance here, but are too complicated and poorly understood to have received any theoretical treatment.) Eventually the side rarefactions affect the entire flow and ultimately a completely steady (in the frame attached to the shock) two-dimensional (i.e., cylindrically symmetric three-dimensional) flow behind a curved shock is presumably approached. Experimentally, no further change can be detected beyond ten or so diameters. In this steady configuration, of course, all memory of the mode of initiation is lost.

Sec. 5G SLIGHTLY DIVERGENT FLOW

It is this steady-flow problem that we treat here. The front boundary condition is supplied by the conservation relations applied across the curved shock surface in the explosive. The side boundary conditions are similarly obtained from the oblique shocks in the surrounding inert. Because, sufficiently far behind the front, side and rear expansion is limited only by the low ambient pressure of the surrounding medium, the products must eventually expand to this pressure and thus to supersonic velocity; this is the rear boundary condition.

The starting point for the axial flow problem is the full set of partial differential equations for steady two-dimensional reactive flow with cylindrical symmetry. These are reduced to a set of ordinary differential equations by specializing them to the axis. In the original equations the radial velocity ω vanishes on the axis by symmetry, but the partial derivative of ω with respect to the radial coordinate r has in general a finite value there, and appears in the axial equations as an unknown function $\omega_r(x)$, where x is the distance behind the shock. The front and rear boundary conditions are just those of the complete problem. The effect of the side boundaries now appears through the divergence $\omega_r(x)$.

The axial equations have the form of the steady-flow equations for the plane case plus correction terms containing the divergence. Thus they remain incomplete until $\omega_r(x)$ is defined, perhaps through some physical approximation relating it to an assumed shock-front curvature or to the charge geometry. However, it is important to observe that much can be learned by studying the equations with physically reasonable but otherwise arbitrary $\omega_r(x)$. If the complete two-dimensional solution were known, and the function $\omega_r(x)$ on the axis from this solution were inserted into the axial equations, these would, of course, yield the correct solution on the axis. Thus, any general properties of the solution deduced for the axial problem with arbitrary $\omega_r(x)$ are those of the axial flow in the corresponding complete problem. Although appreciable space is given to the discussion of approximate methods of determining $\omega_r(x)$, and the results are used as a guide in picking certain simple forms for use in illustrative examples, it should be remembered that this is not an essential step in reaching the principal conclusions.

The divergence can be related to the rate of change of the area of an infinitesimal streamtube centered on the axis, and when this identification is made, the equations are seen to be identical to those describing nozzle or channel flow in the quasi-one-dimensional approximation. The structure of their solutions is similar to those of some of the other steady-state problems considered earlier in this chapter. The effect of the divergence is in some respects similar to that of the endothermic chemical reaction discussed in Sec. 5B2. The problem is also an approximate analog of the standard convergent-divergent nozzle problem (see, for example, Courant and Friedrichs, 1948, Secs. 144-147) with the effect of the chemical-reaction heat release much like that of the area decrease of the convergent section. Since the flow is subsonic immediately behind the front and supersonic at the rear boundary, there must be a sonic point of transition in between. At the critical point of the equations at which this transition can take place, the rate of energy production by the chemical reaction must be balanced by the rate of withdrawal of energy into radial flow.

Typically this point has saddle character so that the detonation velocity is uniquely determined as the solution of an eigenvalue problem for which the integral curve reaches the sonic point and passes through it into the supersonic regime. The velocity thus depends on the chemical rate law of the particular substance considered. There are two "losses" which make it less than the plane value: (1) the energy in radial motion, and (2) the part of the reaction heat released beyond the sonic point. The first loss is ordinarily several times the second. The second arises from the general property of supersonic flow that no event influences anything farther upstream. It is this property which allows us to confine our attention to the "reaction zone"—the region ending at the sonic point. For if the rear boundary condition for the complete problem is replaced by the requirement that the reaction zone terminate in a sonic point, then the problem of the flow in the reaction zone only is completely defined; its solution includes the determination of the eigenvalue velocity.

In the limit of zero divergence, the reaction-zone portion of the plane steady solution is recovered smoothly. As the divergence decreases, the sonic point moves to the rear, more of the energy is released ahead of it, and the complete one-dimensional steady reaction zone is approached. The rarefaction wave required for the complete unsupported plane detonation solution is, of course, not present in the divergent case as we have formulated it here. To see how it might be included, we must return to the rear boundary condition of the complete flow problem.

We may describe this condition in perhaps more familiar form in terms of the nozzle problem. Specifying the exit pressure is equivalent to terminating the nozzle in a reservoir at this pressure. Alternately we may attach a constant-area section carrying a piston receding at a velocity equal to the calculated particle velocity for the specified exit pressure. For greater piston velocities (in the steady frame) the required exit pressure will be unchanged, but the exit section will contain a receding rarefaction wave which increases the velocity to that of the piston. Now in the limiting process we decrease the divergence by decreasing the nozzle's rate of area expansion with distance, simultaneously lengthening it as required to maintain the large exit area required by the ambient-pressure exit condition. The exit area thus remains nearly constant at a large value for all finite values of the divergence, then drops discontinuously to the value of the much smaller entrance area at the zero-divergence (constant-area channel) limit, at which point the rarefaction wave required for the unsupported plane case must be introduced. To obtain a smooth approach to the limit, we must pose a different problem, in which the area of the exit section decreases continuously to the entrance value as the divergence decreases to zero. In this process, the exit pressure increases, no longer satisfying the ambient pressure exit condition, and, with the piston velocity in the exit section held constant, the rarefaction wave in this section becomes stronger and recedes less rapidly as the divergence decreases, going smoothly into the one-dimensional rarefaction in the limit.

How do our limited considerations relate to the complete problem? Two principal questions are usually asked: (1) how will the detonating stick push the surrounding inert? and (2) what is the "diameter effect," that is, how do the detonation velocity and pressure depend on the charge diameter? We have little to do with the first question. A typical application is the "cylinder test" for explosives performance described by Lee, Hornig, and Kury (1968). The wall motion is described quite well by either a two-dimensional time-dependent calculation, or by a two-dimensional steady calculation by the method of characteristics (see Wilkins [1965]; Richards [1965]; Hoskin et al. [1965]; and Fickett and Scherr [1975]). In these calculations the details of the reaction zone are not resolved. The characteristic calculation, in fact, assumes instantaneous reaction in a plane front moving at plane CJ velocity. The flow field corresponds to our supersonic regime, which we consider only very qualitatively. For the diameter effect, the solution would require a two-dimensional time-dependent numerical calculation with resolution sufficient to calculate details of the reaction zone, or a difficult two-dimensional transonic steady problem somewhat like that of the bow shock ahead of a blunt body but with at least two materials, the explosive and the confining inert. This has not been done in detail, although Bdzil (1976) has done a perturbation treatment. The program usually followed is to first determine in some approximate way a function $\omega_r(x;d)$, with d the charge diameter, and then solve the equations for a sequence of eigenvalue solutions and their associated propagation velocities as a function of d. The trouble with this approach is that finding a reasonable $\omega_r(x;d)$ requires, in effect, guessing part of the solution to the unsolved complete flow problem.

A less ambitious approach makes use of the important property that $\omega_r(x)$ is, at the front, inversely proportional to the radius of curvature R of the shock. Here the function used is $\omega_r(x;R)$, and simple assumptions

STEADY DETONATION   Chap. 5

about the off-axis shape of the shock and the nature of the flow field give some guidance in choosing its form away from the shock. This approach of course retreats to the position of giving properties as a function of shock curvature instead of diameter, but is still of interest since shock curvatures have been measured for a number of explosives. Many calculations of this type have been made. We do not attempt to cover this subject here, as such, but we will have occasion to refer to some of the work for the approximations made to ω_r, and for some of the solutions of the axial flow equations in particular cases.

In our presentation we follow the same plan used in some earlier sections, that is, to precede the discussion of the general problem by some simple examples chosen to illustrate the general properties. We give some attention to the approximations used for ω_r, and use this as a guide to the choice of a simple and reasonable form for use in the examples.

5G1. The Steady Flow Equations

Much of this discussion follows the paper of Wood and Kirkwood (1954). However, all velocities are here referred to the coordinate system in which the flow is steady, i.e., the coordinate system attached to the shock, as shown in Fig. 5.37. The incoming flow is uniform with velocity D. Cylindrical symmetry is assumed. The time-dependent flow equations are

$\dot{\rho} + \rho(u_x + \omega_r) + \rho\omega/r = 0$

$\rho\dot{u} + p_x = 0$

$\rho\dot{\omega} + p_r = 0$

$\dot{E} + p\dot{v} = 0$

$\dot{\lambda} = \hat{r}$

$\dot{} \equiv \partial/\partial t + u\partial/\partial x + \omega\partial/\partial r.$ \hfill (5.22)

The chemical rate is here denoted by $\hat{r}$ to distinguish it from the radial coordinate r.

For steady flow the partial time derivatives vanish and the equations are:

$$u\rho_x + \rho u_x + \omega \rho_r + \rho(\omega_r + \omega/r) = 0$$

$$\rho u u_x + p_x + \rho \omega u_r = 0$$

$$\rho u \omega_x + p_r + \rho \omega \omega_r = 0$$

$$u E_x + \omega E_r + p u v_x + p \omega v_r = 0$$

$$u \lambda_x + \omega \lambda_r = \hat{r}. \tag{5.23}$$

The equations are next specialized to the axis, where, by symmetry, the radial velocity ω vanishes. The result is

$$u\rho_x + \rho u_x = -2\rho \omega_r$$

$$\rho u u_x + p_x = 0$$

$$p_r = 0$$

$$E_x + p v_x = 0$$

$$\lambda_x = \hat{r}/u. \tag{5.24}$$

The starting point for the integration is the shock state determined from the shock conservation conditions.

The dependence of the radial component of the flow divergence ω_r on x remains to be determined by some physical assumption. The equations

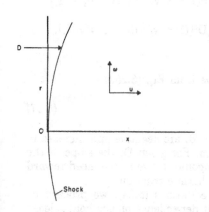

Fig. 5.37. Coordinate system for the curved-front problem. The shock is stationary in this frame. The incoming flow has velocity D and the interior has radial and axial velocity components ω and u.

STEADY DETONATION   Chap. 5

can, however, be integrated formally to yield the one-dimensional conservation equations plus correction terms containing ω_r. If we first define the dimensionless integrals

$$I_1 = D^{-1}\int_0^x \omega_r(x')dx'$$

$$I_2 = (\rho_0 D)^{-1}\int_0^x \rho(x')\,\omega_r(x')\,dx'$$

$$I_3 = (\rho_0 D^2)^{-1}\int_0^x \rho(x')\,u(x')\,\omega_r(x')\,dx'$$

the result is

$$\rho u - \rho_0 D = -2\rho_0 D I_2 \tag{5.25a}$$

$$\rho u^2 + p - (\rho_0 D^2 + p_0) = -2\rho_0 D^2 I_3 \tag{5.25b}$$

$$H + 1/2\, u^2 - (H_0 + 1/2\, D^2) = 0, \tag{5.25c}$$

where we have not yet used I_1. Note that the last equation is the familiar Bernoulli equation, unaffected by the divergence. By the usual transformations, we can also write these as the planewave Rayleigh and Hugoniot functions with divergence corrections. In doing this we neglect p_0 and retain only first-order terms in ω_r; to first order in ω_r we have $I_1 = I_3$, which may be seen by substituting for ρu in the integrand of I_3 from Eq. (5.25a). The result is, in this approximation

$$\mathcal{R} = \rho_0^2 D^2 - p/(v_0 - v) = -2\rho_0^2 D^2 (1 - v/v_0)^{-1}\,[2(v/v_0)I_2 - I_1]$$

$$\mathcal{H} = E - E_0 - 1/2\, p(v_0 - v) = D^2[(1 + v/v_0)I_1 - 2(v/v_0)I_2].$$

For the sonic condition $u = c$ we have from Eq. (5.25a)

$$\rho c = \rho_0 D(1 - 2I_2) \tag{5.27}$$

All of the correction terms in Eq. (5.26) are negative and increase in magnitude with the degree of reaction. For given D, the slope of the Rayleigh line is increased, and the Hugoniot curve is displaced toward the origin as it would be for an endothermic reaction.

Returning now to the differential equations (5.24), we proceed as before, Sec. 4A4, to obtain the explicit dependence on the heat release

Sec. 5G SLIGHTLY DIVERGENT FLOW

term σr and the sonic parameter η. The equations for the axial flow then become

$$u_x = \psi/\eta$$

$$\rho_x = -(\rho/u)(u_x + 2\omega_r)$$

$$E_x + pv_x = 0 \quad \text{or} \quad H + 1/2\, u^2 - (H_0 + 1/2\, D^2) = 0 \quad (5.28)$$

$$\lambda_x = \hat{r}/u,$$

where

$$\eta = 1 - u^2/c^2$$

$$\psi = \sigma\hat{r} - 2\omega_r.$$

These equations differ from those for plane flow in two ways: (1) the relation between ρ and u now depends on ω_r, and (2) the term $-2\omega_r$ is added to ψ. Note that this divergence term in ψ has for positive ω_r the same sign as that for an endothermic reaction. With the addition of an approximation to determine ω_r, which may of course be another differential equation for it, these equations form a determinate set. They have a critical point where ψ and η vanish simultaneously. The vanishing of ψ means that, roughly speaking, the release of energy by the chemical reaction is balanced by its withdrawal by the radial flow; this point is reached before the attainment of chemical equilibrium. As before, it is found that the integral curves cannot cross the sonic locus $\eta = 0$ except at the critical point, so the only path for an integral curve to a supersonic boundary state is through the critical point. These properties are illustrated in the examples.

The flow equations (5.28) are equivalent to the standard equations for reactive flow in a nozzle, under the usual assumption of quasi-one-dimensional flow. Physically the nozzle is to be identified with the bounding surface of a streamtube of arbitrarily small cross-section centered on the axis. To show the equivalence, we must express the rate of area increase of the streamtube with distance in terms of ω_r. For small displacements from the axis the radial velocity is $\omega = \omega_r r$. The radial motion is given by

$$dr/dt = \omega = \omega_r r. \quad (5.29)$$

STEADY DETONATION   Chap. 5

Distance and time are related along the axis by

$$dx = u\, dt, \tag{5.30}$$

so the area $A = \pi r^2$ of the streamtube is given by

$$A^{-1}dA/dx = 2r^{-1}dr/dx = 2(ru)^{-1}dr/dt = 2\omega_r/u. \tag{5.31}$$

The flow equations can then be written

$$(\rho u A)_x = 0$$

$$u_x = (\sigma \hat{r} - uA^{-1}dA/dx)/\eta \tag{5.32}$$

$$\lambda_x = \hat{r}/u$$

$$(H + 1/2\, u^2)_x = 0,$$

which are the nozzle equations used, for example, by Wecken (1965). The momentum-conservation equation is

$$[(\rho u^2 + p)A]_x = pA_x. \tag{5.33}$$

5G2. Approximation for the Radial Derivative

In Secs. 5G3 and 5G4 the assumption

$$\omega_r = \text{constant}$$

is used to illustrate the properties of the equations, and it proves satisfactory for the purpose. Here we show how ω_r is related to some of the quantities which are (or might be) measurable, and give references to work which has provided estimates of

$$\omega_r = \omega_r(x).$$

Two approaches have been used: the first relates ω_r to the radius of curvature at the front, and the second assumes some simplified approximation to the complete flow problem and computes the divergence along the central steamtube.

The relationship between the radius of curvature of the detonation front on the axis and the value of ω_r just behind the shock is easy to derive. The geometrical details are shown, and the notation defined, in

Fig. 5.38. The radial velocity immediately behind the shock front at an off-axis point is obtained by using the requirement that $u_\parallel = D \sin \theta$ is unchanged across the shock. Some simple trigonometry yields

$$\omega = (r/R)(D \cos \theta - u_\perp). \tag{5.34}$$

The velocity perpendicular to the shock surface $u_\perp$ is related to the initial velocity $D \cos \theta$ perpendicular to the shock by the Hugoniot relation, so it can be expressed as a Taylor expansion about the state on the axis at $x = 0$ in powers of the difference between D and $D \cos \theta$. With $u(0)$ denoting the velocity behind the shock on the axis,

$$u_\perp = u(0) - D(du/dD)_{3C}(1 - \cos \theta)$$

$$+ 1/2\, D^2(d^2u/dD^2)_{3C}(1 - \cos \theta)^2 + \ldots, \tag{5.35}$$

with the derivatives evaluated at $u = u(0)$. After substituting this expression into Eq. (5.34), and eliminating $\cos \theta$ with the equivalent

$$\cos \theta = [1 - (r/R)^2]^{1/2} = 1 - 1/2\,(r/R)^2 - (r/R)^4/8 - \ldots,$$

we find

$$\omega = (r/R)[D - u(0)] - 1/2\, D(r/R)^3[1 + (du/dD)_{3C}] - \ldots. \tag{5.36}$$

Thus the radial derivative, correct to terms of order R^{-1}, is

$$\omega_r = [D - u(0)]/R, \tag{5.37}$$

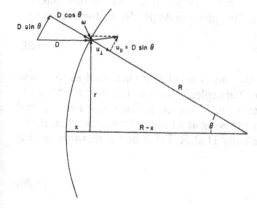

Fig. 5.38. Geometry of the approximation for ω_r. The unlabeled vector behind the shock is the particle velocity. Its component in the radial direction is ω, and its component perpendicular to the shock surface is $u_\perp$.

and this is a good approximation not only on the axis, but for reasonable distances off the axis. Our assumption of ω_r = constant is thus equivalent to assuming that ω_r on the axis at a distance x behind the front is the same as ω_r at the same x but off-axis enough to lie on the shock front. Wood and Kirkwood (1954) employed a similar reasoning but assumed that p_{rr} was the same on the shock front at a given x as on the axis at that x. Because of the large changes in u through the reaction zone, it seems hard to justify this kind of argument. Perhaps it is better to say that Eq. (5.37) relates the curvature of the front to the value of ω_r immediately behind the shock, and that in the absence of knowledge the simplest assumption is that it is constant.

The discussion of Eyring et al. (1949) of the relation between a spherical and a stick detonation is of interest here because it gives some insight into the validity of the approximation ω_r = constant. If we imagine that a large mass of explosive is initiated at an interior point, the detonation wave will be spherically expanding, and after some time will have the radius of curvature R equal to the radius of curvature R of a long small-diameter stick in which the detonation has propagated far enough to become steady. Although the spherical detonation is not a steady flow, we can neglect the time dependence and try to identify terms in the equations describing it with those describing the truly steady flow in the stick, thus identifying ω_r. Because the extra explosive in the sphere confines the expansion somewhat (we can suppose that the stick is made by removing explosive from the sphere), we expect to find ω_r for the complete sphere less than that for the stick. It seems reasonable to think that ω_r for the sphere is a lower limit for ω_r for the stick for both fronts having the same R. Obviously they have to have the same value at the point immediately behind the shock, because the front curvature determines the divergence there, but at a point x farther back in the flow, the value for the stick is larger than that for the sphere.

The mass conservation equation given in Appendix 4C for a sphere is

$$\rho_t + \rho u_z + u\rho_z = -2\rho u/z \tag{5.38}$$

in a coordinate system (z is the radial coordinate) centered at the point of initiation with the undisturbed explosive at rest. Choose t such that at t = 0 the detonation front is at distance R from the center, and transform to a reference frame moving at a constant velocity equal to the instantaneous detonation velocity D at R. Let x be the distance in this frame, related to z by

$$x = R + Dt - z. \tag{5.39}$$

Sec. 5G SLIGHTLY DIVERGENT FLOW

The point x = 0 coincides with the front at t = 0 and nearly so at later times (since the front velocity in general is not exactly constant), and the direction of positive x is opposite to that of positive z to correspond to the steady reference frame that we use for the stick. The transformation equations are

$$(\partial/\partial t)_z = (\partial/\partial t)_x + (\partial x/\partial t)_z (\partial/\partial x)_t \tag{5.40}$$

$$= (\partial/\partial t)_x + D(\partial/\partial x)_t$$

$$(\partial/\partial x)_t = -(\partial/\partial z)_t. \tag{5.41}$$

Applying these to Eq. (5.38) we obtain the corresponding equation in the new frame

$$\rho_t + \rho u_x + u\rho_x = -2\rho(D - u)/(R + Dt - x), \tag{5.42}$$

where u is now the velocity in the new frame and the quantity $(D - u)$, familiar in this type of transformation, appears on the right. The first of Eqs. (5.24), the mass conservation equation for the steady frame

$$\rho u_x + u\rho_x = -2\rho\omega_r$$

is the one we want to compare with Eq. (5.42). Thus we see that when the radius is large enough that ρ_t in the "steady" frame attached to the front is small enough to neglect, and when Dt and x are small, the identification of the remaining terms gives

$$\omega_r = (D - u)/R, \tag{5.43}$$

which is identical with Eq. (5.37) at the front and smaller than the value obtained from the assumption

$$\omega_r = \text{constant} \tag{5.44}$$

(because u increases with x in the steady frame) behind the front.

The approximations to the complete flow problem have been made for many special purposes, and it is difficult to compare or contrast them. About all we can do is give some very brief descriptions and the references. Jones (1947) calculates the flow behind a finite plane disk starting with the Chapman-Jouguet condition, and expanding in a Prandtl-Meyer flow originating at the edge of the disk, and assumes that the area expansion of the central steamtube is the same for a finite reaction zone as he finds for the case of instantaneous reaction. With the axial steamtube area found this way, Eq. (5.31) is used to obtain ω_r. More accurate calculations of the Prandtl-Meyer expansion behind an instantaneous CJ detonation are given by Hill and Pack (1947), Hoskin, et al. (1965), and Richards (1965).

Dabora, et al. (1965) estimate the angle of the interface between explosive and bounding inert by matching the flow in fully reacted products to that behind the oblique shock in the inert; the entire reaction zone is then treated as a conical channel with expansion rate corresponding to this angle. Sichel (1966) uses a perturbation analysis based on the planewave solution, and arrives at the same estimate of ω_r as Eyring, Eq. (5.43). Tsuge et al. (1970) treat the laterally expanding detonation products as a blunt body about which the confining inert material flows. They apply a simple approximation to hypersonic flow theory (the Newton approximation) to relate the cross-sectional area to the pressure. Their work will be referred to again in Sec. 5G6.

Erpenbeck (1969b), in work discussed in Sec. 5G5, uses

$$A(x)/A(0) = (1 + \beta x^2)/(1 + x^2) \tag{5.45}$$

to describe the channel area. The initiating shock stands at the entrance where $x = 0$, and where $dA/dx = 0$, and the channel expands through the reaction-zone region until at large x the area is constant and β times as large as the entrance area. The advantage of this form for his work is that it allows examination of all the possible rear boundary conditions, with fixed velocity pistons imagined in the constant-diameter channel at large x.

5G3. A Simple Example—Irreversible Reaction in an Ideal Gas

A simple example is discussed here to illustrate the structure of the problem. The equation of state is the simple polytropic gas form of Sec. 4B1. The material is initially all species A, and is transformed by the reaction to species B, with no mole change and no change in the specific heat. The reaction $A \rightarrow B$ is assumed irreversible with rate function

$$\hat{r} = k(1 - \lambda). \tag{5.46}$$

The equation of state is

$$H = h(p,v) - \lambda q; \quad h = \gamma pv/(\gamma - 1) = c^2/(\gamma - 1). \tag{5.47}$$

Sec. 5G SLIGHTLY DIVERGENT FLOW

The radial component of the flow divergence is taken in its simplest form, from Eq. (5.37), as

$$\omega_r = \text{constant.} \tag{5.48}$$

From Eqs. (5.28) the determining equations are

$$du/dx = \psi/\eta \tag{5.49}$$

$$d\lambda/dx = \hat{r}/u \tag{5.50}$$

$$H + 1/2\, u^2 = h_0 + 1/2\, D^2, \tag{5.51}$$

where

$$\psi = \sigma\hat{r} - 2\omega_r \tag{5.52}$$

$$\eta = 1 - u^2/c^2. \tag{5.53}$$

The second equation of the set (5.28) is omitted here; for the particular equation of state and rate chosen the remaining equations comprise a complete set for the variables u and λ, and suffice to determine the eigenvalue velocity. To get ρ and p the differential equation for ρ must be added; p can then be calculated from the equation of state. The simplification comes about because with

$$h = c^2/(\gamma - 1) \tag{5.54}$$

substituted in the Bernoulli equation the density does not appear in it, and c^2 can be expressed in terms of u and λ. No additional information is required to calculate the right-hand sides of the first two equations, for σ is given by

$$\sigma = q/h, \tag{5.55}$$

η requires only c^2, and the rate depends only on λ. (Since $T = c^2/\gamma \tilde{R}$, we could equally well let the rate depend also on T.)

The explicit relations for σ and c^2 in terms of u and λ are

$$\sigma = [\lambda - 1/2\, u^2/q + 1/2\, D^2/q]^{-1} \tag{5.56}$$

$$c^2 = (\gamma - 1)(\lambda q - 1/2\, u^2 + 1/2\, D^2), \tag{5.57}$$

where h_0 has been dropped as negligibly small.

It is convenient to work in λ-u^2 space. The distance x is eliminated and u replaced by u^2 by multiplying Eq. (5.49) by 2u and dividing by Eq. (5.50) to get

$$du^2/d\lambda = 2u^2\psi/\eta\hat{r}, \tag{5.58}$$

in which, after substituting from Eqs. (5.46), (5.52), (5.53), (5.56), and (5.57), the only variables are u^2 and λ. The solution to Eq. (5.58) for the planewave detonation, i.e., $\omega_r = 0$, is given by

$$\frac{(\gamma+1)}{(\gamma-1)}u^2 - \frac{2\gamma}{(\gamma-1)}uD + D^2 + 2\lambda q = 0. \tag{5.59}$$

Unfortunately, analytic solutions to Eq. (5.58) for the finite values of ω_r are not known, and the problem has to be discussed by considering qualitatively the path of the integral curve in the λ-u^2 plane.

There are several loci which are required to discuss the path of the integral curve. The integral curve has a vertical tangent on the locus $\eta = 0$ obtained by setting $u = c$ in Eq. (5.57);

$$\eta = 0: u^2 = 2[(\gamma-1)/(\gamma+1)](\lambda q + 1/2\, D^2). \tag{5.60}$$

The integral curve has a horizontal tangent on the locus $\psi = 0$, which can be found by substituting from Eqs. (5.46) for r and (5.56) for σ into Eq. (5.52). It is convenient to define a dimensionless parameter $\epsilon = 2\omega_r/k$ which measures the divergence. The result for $\psi = 0$ is

$$\psi = 0: u^2 = 2q(1+\epsilon^{-1})\lambda + D^2 - 2q/\epsilon; \quad \epsilon = 2\omega_r/k. \tag{5.61}$$

Although the reaction rate chosen for this example, Eq. (5.46), is independent of temperature, the loci of constant temperature are useful for obtaining a qualitative idea of how a temperature-dependent reaction rate would influence the integral curve. For the simple equation of state chosen, constant temperature implies constant sound velocity, so the loci are the straight lines obtained from Eq. (5.57) with c^2 = constant. Note that this equation with $c^2 = 0$ gives the escape velocity, which is a boundary of the accessible part of the λ-u^2 plane.

The λ-u^2 diagram for plane detonation ($\epsilon = 0$) is shown in Fig. 5.39. Its properties are essentially those of the p-λ diagram, Fig. 5.4, except that here the Rayleigh curves cross in the lower portion of the diagram since in the steady frame the particle velocity at point N increases with D.

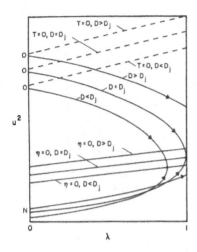

Fig. 5.39. The λ-u^2 diagram for different steady velocities in the plane detonation case ($\epsilon = 0$).

However, the term "sonic locus" here refers to a member of the one-parameter family of lines obtained by solving the Bernoulli equation with D as a parameter and with the condition u = c, i.e., $\eta = 0$, imposed. The sonic locus of Fig. 5.4 is the locus of intersections of the present $\eta = 0$ lines with their respective Rayleigh curves.

With ϵ finite, the only change in the appearance of the λ-u^2 plane is the shifting of the $\psi = 0$ locus from a line coincident with the $\lambda = 1$ line to a straight line of positive slope, intersecting the $\lambda = 1$ line at $u^2 = 2q + D^2$ and the $u^2 = 0$ axis at $\lambda = (1 - \epsilon D^2/2 \, q)/(1 + \epsilon)$. Let us consider how the solution (that is, the integral curve beginning at the von Neumann point N) changes as ϵ increases from zero *at fixed* $D < D_j$ (this is easier to illustrate than the variation of D at fixed ϵ). The case $\epsilon = 0$, Fig. 5.40a, is the same as Fig. 5.39. As ϵ increases from zero, Fig. 5.40b, the $\psi = 0$ locus swings out to the left and the solution curve is displaced to the right of the plane solution, but still terminates on the sonic locus. The arrows indicate the directions of other integral curves. As ϵ increases further, it passes through the special value, Fig. 5.40c, at which the solution becomes the separatrix passing through the saddle critical point at $\psi = 0$, $\eta = 0$ into the supersonic region. For larger values of ϵ, Fig. 5.40d, the solution crosses the $\psi = 0$ locus below $\eta = 0$ and proceeds downward to the nodal critical point $\lambda = 1$, u = 0. The case of interest as the steady (unsupported) detonation is the special value of ϵ which allows the solution to pass through the critical point. The case of larger ϵ might correspond to a detonation supported by a blunt projectile. The case of smaller ϵ with its sonic termination is unsatisfactory for the same reason as in the plane case, Sec. 5A3.

STEADY DETONATION   Chap. 5

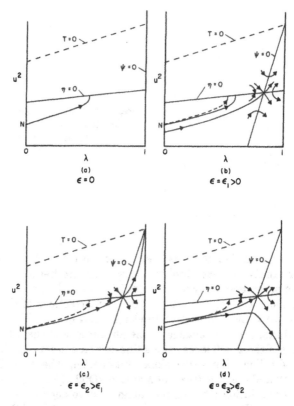

Fig. 5.40. The λ-u^2 diagram for the curved front case with increasing $\epsilon(=2\omega_r/k)$ at fixed $D < D_j$ for constant ω_r and the chemical reaction rate a function only of the degree of reaction. Part (a) is for $\epsilon = 0$; this solution is shown as a dashed curve in (b), (c), and (d).

Denote the eigenvalue D for a given ϵ by $\widetilde{D}(\epsilon)$. This is the desired result. It may be obtained by the above procedure carried out for a sequence of values of D, or equally well by varying D to find $\widetilde{D}$ at each of a sequence of values of ϵ.

This result may be related to the shock radius of curvature R by using the value of ω_r at the shock given by Eq. (5.37):

$$\omega_r = [D - u(0)]/R.$$

Take the eigenvalue D and u for each ϵ, and multiply both sides by 2/k to obtain kR for each D:

Sec. 5G SLIGHTLY DIVERGENT FLOW

$$kR = \epsilon/[\tilde{D} - \tilde{u}(0)].$$

We thus have the product kR with dimensions of reciprocal velocity as a function of D [since $u(0)$ is given in terms of D by the Hugoniot relations]. This result is usually expressed in the form of the velocity deficit as a function of the ratio of R to a characteristic reaction length x^* (roughly proportional to k^{-1}):

$$D_j - \tilde{D} = f(x^*/R).$$

We have done some additional analysis and calculation for this model, too lengthy for inclusion here, to determine the relative size of the divergence and reaction-heat "losses" or corrections to the planewave solutions (by the reaction-heat loss we mean the heat released beyond the sonic point). The result is that the first loss, the divergence, is several times the second.

Let us now consider the probable effect of relaxing some of the simplifications. If the assumption expressed by Eq. (5.48), that the radial component of the flow divergence is constant, had been replaced by one of the other possibilities discussed in Sec. 5G2, then ω_r would become a function of distance

$$\omega_r = \omega_r(x). \tag{5.62}$$

In this case the differential equation (5.58) contains x as a variable in addition to u^2 and λ, so that the problem now requires the solution of the two non-autonomous differential equations (5.49) and (5.50), which may be replaced through use of the streamline relation $dx/dt = u$ by a set of three autonomous equations for λ, u^2, and x with particle time t as the independent variable. The important loci and some integral curves are diagrammed in λ-u^2-x space in Fig. 5.41. The critical point of the original example may be expected to become a line of saddle-type critical points, through which passes a two-dimensional manifold of solutions, one of whose members is, for the eigenvalue D, an integral curve satisfying the initial conditions. The locus $\psi = 0$ is a surface which depends upon x, so the line of critical points is not a straight line, and the manifold of solutions does not intersect the $\lambda = 0$ plane in a straight line. Although the solution becomes more difficult, it is not likely that its general character in the region of interest is changed by allowing ω_r to depend on x in an appropriate way.

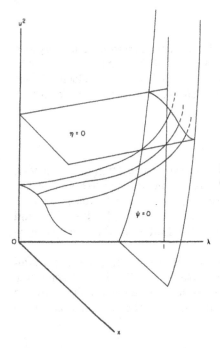

Fig. 5.41. The λ-x-u^2 diagram for the case in which ω_r is a function of x.

In contrast, our assumption of a rate function dependent only on λ, Eq. (5.46), is an extreme oversimplification. Letting it depend on temperature (or other state variables) opens up so many possibilities that no comprehensive discussion is available. We limit ourselves here to some general remarks about certain features of the problem. Examples will appear later in Erpenbeck's comments at the end of Sec. 5G5 about some of his unpublished work, and in Tsuge's calculations of the ozone detonation in Sec. 5G6.

In the λ-u^2 plane, Fig. 5.42, the $\psi = 0$ curve is displaced as shown. For a weak temperature dependence, such as an Arrhenius form with small activation energy, the locus moves only slightly away from its position in our simple example, and the solutions are qualitatively unchanged. As the temperature dependence is increased (as by increasing the activation energy) the $\psi = 0$ curve is displaced further, and its intersection with the $\lambda = 0$ axis moves downward. As this intersection crosses the sonic locus a new critical point (intersection of $\psi = 0$ and $\eta = 0$) appears. Finally the intersection of $\psi = 0$ with the $\lambda = 0$ axis moves below the initial point. Here ψ is initially negative, the divergence effects predominate, and the first part of the flow behind the shock is like that in a (nonreactive) subsonic nozzle, with the pressure and temperature

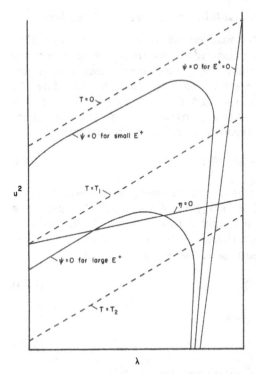

Fig. 5.42. The λ-u^2 diagram for the case of an Arrhenius rate function.

increasing with distance. This continues until the temperature increases sufficiently to get the reaction moving. As the reaction rate increases, the sign of ψ is reversed and the pressure begins to fall. Finally, for still stronger temperature dependence of the rate, the $\psi = 0$ curve will eventually lie entirely below the sonic locus, so that the critical points vanish, only subsonic solutions are allowed, and no steady solutions to the problem as we have posed it are possible. This would presumably represent the point of detonation failure.

In the intermediate temperature range there can be more than one eigenvalue velocity [for a given $\omega_r(x)$ and fluid]; this is perhaps the most interesting result of introducing the temperature dependence of the rate. However, the global phase-portrait remains unclear. Wecken (1965) has shown that, for the nozzle represented by d ln A/d ln x = constant, the right critical point is always a saddle and the left is either a focus or a node. But the physical significance of the left point, if any, remains uncertain. In the ozone detonation described in Sec. 5G6, the critical point reached by both eigenvalue solutions is identified as a saddle point, but no further information about the phase space is given.

5G4. Effect of Chemical Equilibrium (Reversible Reaction)

In the previous section the reaction rate is taken to be irreversible; here a reversible reaction is considered so that the rate becomes zero at a point of chemical equilibrium. The governing equations remain unchanged except Eq. (5.46), the reaction rate expression. As before, the problem is much simplified by using the same simple equation of state and by considering only the approximation that the divergence parameter $\epsilon = 2\omega_r/k$ is independent of x. The Arrhenius rate is

$$\hat{r} = ke^{-E\dagger/RT}(1 - \lambda)(1 - e^{\Delta F/RT}). \tag{5.63}$$

Attention is focused on the case $E^\dagger = 0$. Although we believe that the inclusion of a nonzero activation energy would not change the qualitative conclusions, we have not investigated this in detail. With this simplification, and with the product and reactant species taken to have the same entropy in the standard state, the rate is

$$\hat{r}/k = 1 - \lambda(1 + e^{-q/RT}). \tag{5.64}$$

Chemical equilibrium implies

$$\hat{r}(\lambda_e) = 0, \tag{5.65}$$

which gives for the equilibrium composition

$$\lambda_e = (1 + e^{-q/RT})^{-1}, \tag{5.66}$$

from which

$$\lambda_e = 1 \quad \text{at} \quad T = 0 \quad \text{and} \quad \lambda_e = 1/2 \quad \text{at} \quad T = \infty. \tag{5.67}$$

The $\hat{r} = 0$ curve (equilibrium composition) is a locus of vertical slope for the integral curves, and has the shape sketched in Fig. 5.43. The $\psi = 0$ curve, which is the locus of horizontal slope for the integral curves, lies slightly to the left of $\hat{r} = 0$, and approaches it as $\epsilon \to 0$. The sonic locus where $\eta = 0$ is unchanged from the last section.

The critical point formed by the intersection of $\psi = 0$ and $\eta = 0$ is a saddle point as before. The point of intersection of $\hat{r} = 0$ with $\eta = 0$ is a vertical inflection point as shown. The integral curve of interest, which passes through the saddle point into the supersonic region, is trapped in that region between the $\hat{r} = 0$ and $\psi = 0$ curves. The formal structure of

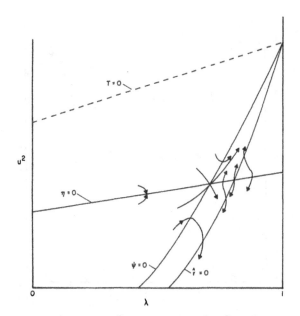

Fig. 5.43. The λ-u^2 diagram for the case of chemical equilibrium (reversible reaction) in the neighborhood of the critical point on the sonic locus.

the problem is thus much the same as for the irreversible reaction, and the results are qualitatively similar.

The behavior of the solution in the plane limit $\epsilon \to 0$ is of particular interest, as discussed in the introduction to this section. An expanded portion of the λ-u^2 plane is shown in Fig. 5.44, and the plane solutions (Rayleigh lines) for the equilibrium and frozen Chapman-Jouguet velocities D_e and D_0 are shown on it. (The slight changes of the positions of the other loci with change in D are neglected.) The Rayleigh lines intersect the $\hat{r} = 0$ locus in the equilibrium and frozen Chapman-Jouguet points C_e and C_0. These are the possible solutions for $\epsilon = 0$. For all finite ϵ, the unique solution satisfying the boundary conditions is the integral curve (separatrix) passing through the critical point at the intersection of $\psi = 0$ and $\eta = 0$. The question is, to which of the plane solutions does it tend in the limit $\epsilon = 0$?

The differential equation can be written in the form

$$du^2/d\lambda = 2u^2\sigma/\eta - 2u^2\epsilon/(\eta\hat{r}/h). \tag{5.68}$$

STEADY DETONATION   Chap. 5

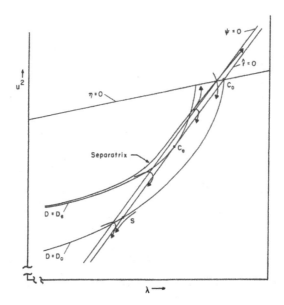

Fig. 5.44. The λ-u^2 diagram for the case of chemical equilibrium, with plane-detonation solutions for the equilibrium CJ and frozen CJ detonation velocities superimposed.

It is evident that for small ϵ the solutions away from $\eta = 0$ must be close to the plane solutions everywhere except in the neighborhood of $\hat{r} = 0$, where $\hat{r}$ is sufficiently small to give the second term an appreciable size. The character of the solutions in this neighborhood is quite different above and below the point C_θ, so the discussion is devoted separately to the portions above and below this point, with the immediate neighborhood of C_θ excluded for the moment.

Consider first a point on $\hat{r} = 0$ well above C_θ, but below C_0. Moving to the left from this point along a line of constant u, an integral curve very near $\hat{r} = 0$ has large negative slope; on $\psi = 0$ an integral curve has zero slope, and sufficiently far to the left, near the plane solution which passes through C_θ, an integral curve can be found with slope greater than that of $\psi = 0$, since the slope of this plane solution is greater than that of $\psi = 0$. Thus, since the function σ/η is well-behaved here, an integral curve can be found, at some point on the line of constant u slightly to the left of $\psi = 0$, with a slope equal to that of $\psi = 0$. The vector field in this region must therefore be approximately as shown, and the presence of the separatrix approximately parallel to $\psi = 0$ is implied.

Sec. 5G   SLIGHTLY DIVERGENT FLOW

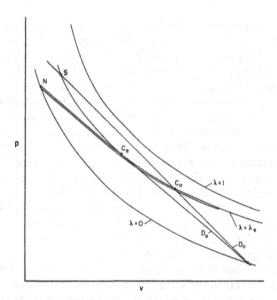

Fig. 5.45. The p-v diagram for the case of chemical equilibrium. The line lying just below the λ_e Hugoniot curve is the solution corresponding to the separatrix passing through the critical point in Fig. 5.44.

Below C_e the plane solutions to the left of $\psi = 0$ have smaller slope than that of the $\psi = 0$ locus, and a similar argument shows that integral curves to the left of this locus must cross it as shown. Thus the separatrix is absent from this region. In the limit, then, it appears that below the point C_e the separatrix approaches the plane equilibrium solution D_e. Above $\eta = 0$ it approaches the equilibrium locus $\hat{r} = 0$. Thus it appears that the separatrix is as shown in Fig. 5.44, passing arbitrarily close to C_e as $\epsilon \to 0$, and that the limiting steady propagation velocity is D_e. It is interesting to note that in the limit the negative-slope separatrix of the saddle point (the intersection of $\eta = 0$ and $\psi = 0$) becomes that portion of the D_0 plane solution lying to the right of $\hat{r} = 0$.

The solution obtained in the limit as $\epsilon \to 0$ propagates at the equilibrium CJ velocity, but the portion of it above C_e has to have its physical significance clarified. It certainly does not approach that part of the plane-solution Rayleigh line which goes on from C_e to reach the sonic locus, but that part has no physical significance anyway. The clarification is obtained by considering the probable behavior in the p-v plane, which is diagrammed, much distorted to exaggerate the interesting region, in Fig. 5.45. In this plane the integral curve for very

223

small ϵ begins near point N and remains very close to the equilibrium CJ Rayleigh line until it has passed point C_0. It then follows close to the equilibrium Hugoniot curve $\hat{r} = 0$, reaching the critical point in the neighborhood of C_0, and then proceeds on a path close to the equilibrium CJ isentrope on to infinite expansion. The portion of the curve between N and C_0 approaches, as $\epsilon \to 0$, the usual planewave reaction zone. The remaining portion of the curve represents nearly equilibrium flow behind the "reaction zone," in which the products expand to match the rear boundary condition. This region of steady (time-independent) expansion replaces the usual nonsteady expansion ordinarily expected to match between the reaction zone and the rear boundary in treatments of plane detonation. The finite deviation from the Rayleigh line below C_0 is allowed, even though the flow is steady, by the corrected Rayleigh line relation, Eq. (5.26), because the expansion region becomes longer as ϵ decreases, so that the correction integrals remain finite.

As in the previous section, we comment briefly on the nature of critical points of less interest and on some aspects of the effects of letting $E^\dagger$ take on a nonzero value and of allowing ω_r to depend on x.

Subsonic integral curves which do not terminate on the sonic locus cross the $\hat{r} = 0$ locus and terminate at $u = 0$, $\hat{r} = 0$. For nonzero activation energy the $\psi = 0$ curve will be altered as described in the previous section, Fig. 5.42. For constant ω_r, supersonic integral curves lying between $\psi = 0$ and $\hat{r} = 0$ above $\eta = 0$ probably proceed to a point of nonequilibrium composition ($\lambda < 1$) on the $T = 0$ line, the reaction being frozen out by the rapid expansion of the nozzle.

If ω_r decreases sufficiently rapidly with x at large distances, then du/dx will approach zero as $x \to \infty$ and the equilibrium locus $\hat{r} = 0$ becomes a locus of critical points at $x = \infty$. Integral curves can then terminate at finite u and T. A subsonic termination of this type is analogous to the final state of an overdriven detonation in the plane case. A detailed discussion may be found in the paper of Erpenbeck (1969b) which is reviewed briefly in the next section.

5G5. The General Case

In Sec. 5E we described Wood and Salsburg's (1960) general analysis of the one-dimensional detonation. Erpenbeck (1968, 1969b) has performed the same task for the quasi-one-dimensional case, that is, for a detonation wave standing in a channel (nozzle) whose cross-sectional area increases with distance behind the front. The equations used are those of the set (5.32), that is, the set we have been considering, with the addition of another rate equation for each additional reaction. His first

Sec. 5G SLIGHTLY DIVERGENT FLOW

paper is a general discussion for an arbitrary number of chemical reactions, an arbitrary well-behaved equation of state, and for nozzles terminating in both finite and infinite area. His second paper contains explicit calculations for an ideal gas with a single reversible reaction in a finite-area nozzle.

For the nozzle terminating in a finite area (at an infinite distance behind the shock), the results are analogous to those of the plane problem. Critical points for the termination of integral curves at $x = \infty$ lie on a locus which is divided into frozen-subsonic and frozen-supersonic portions. The dividing point between these two portions is analogous to the plane C_0 point and goes into it in the plane limit. As in the plane case, the frozen-subsonic portion is further divided into equilibrium-subsonic and equilibrium-supersonic regions by a point analogous to the plane C_e point, which becomes the C_0 point in the plane limit. The manifold of solutions approaching the portion of the locus lying between the analogues of C_0 and C_e has dimension one less than that of the solution space and is approached only by solutions having eigenvalue character. The remainder of the locus is approached by space-filling manifolds. As in the plane case, two-wave solutions are required for certain ranges of the rate constants and piston velocity. The detonation velocity of a solution terminating in the analogue of C_e approaches the plane value D_e in the planewave limit. In the infinite-expansion case, the region between the analogues of C_0 and C_e disappears as both c and c_e go to zero. In both cases, the eigenvalue solution of the type discussed in the preceding section, which passes through the (frozen-) sonic surface at a critical point and terminates at a frozen-supersonic point, can occur. We believe its velocity approaches D_e in the plane limit, but this is not proved.

The results of the second paper illustrate the two main types of solutions. The nozzle used has the area law

$$A(x)/A(0) = (1 + \beta x^2)/(1 + x^2).$$

The initiating shock stands at the entrance $x = 0$ where $dA/dx = 0$. The value $\beta = 2$ is used, so that the nozzle expands to twice the original area. A piston is placed at the end of the nozzle (at $x = \infty$) where again $dA/dx = 0$, and the piston problem is discussed.

For sufficiently slow reaction, there is a nearly nonreactive expansion to an area close to the final value, followed by a nearly plane reaction zone like that for a plane detonation whose von Neumann point is the flow state after expansion (but before reaction). The piston problem discussion is exactly analogous to that for the plane case. The unsupported detonation is a CJ_e-type solution terminating in the C_e point at $x = \infty$, with subsonic flow throughout. Its velocity is about 0.4 times that for the same initial flow without expansion.

For sufficiently fast reaction, a transonic solution of the type we have been describing obtains. The reaction is nearly complete at the sonic point before much area expansion has taken place. The remainder of the flow consists of the supersonic nozzle expansion of the near-equilibrium fluid. The eigenvalue velocity is slightly less than the plane CJ_e value. The piston problem is essentially the same as that for a convergent-divergent nozzle, with the reaction zone acting like the convergent section.

The transition between these two regimes, which we will not discuss here, can be quite complicated. Even for zero activation energy there can be two transonic points, and for nonzero activation energy the eigenvalue velocity function analogous to our $\tilde{D}(\epsilon)$, Sec. 5G3, can be triple-valued over an appreciable range.

5G6. Applications and Results

We list here some numerical and approximate solutions to the axial flow problem as described above. We remark that since, in condensed explosives, estimates of the reaction rates are probably not correct to within an order of magnitude, any measurement which can yield some information about them is of considerable interest. The central radius of curvature of the front can be measured (at least in liquids) as a function of detonation velocity by varying the charge diameter. A solution of the axial flow problem with the approximate ω_r related to the shock front curvature by the exact relationship for its value at the front, Eq. (5.37), then allows an estimate of the reaction-zone length.

Wood and Kirkwood (1954) have obtained an approximate analytic solution by solving the algebraic divergence-corrected Rayleigh, Hugoniot, and sonic conditions, Eqs. (5.26, 5.27), for the eigenvalue velocity. They estimate the divergence integrals from the approximation to ω_r described in Sec. 5G2, and from first-order corrections to the planewave solution (for the state variables in the integrands). The value of λ at the sonic point is taken to be that of the plane solution, thus neglecting the effective reduction in exothermicity caused by some of the reaction taking place beyond the sonic point. The planewave solution itself is approximated by a square wave (the limit of large activation energy), in which the state is constant from the front back to the plane in which the reaction takes place instantaneously. (This correction is probably considerably smaller than that for lateral motion.) With an approximate equation of state for nitromethane, they obtain the result

Sec. 5G SLIGHTLY DIVERGENT FLOW

$$(D_\infty - D)/D_\infty = 3.5 \, x^*/R,$$

where D_∞ is the planewave velocity, R the front radius of curvature, and x^* the reaction-zone length, that is, the distance from the front to the sonic point.

Erpenbeck's calculations are described in the previous section. He avoided a curved shock by taking $d\mathring{A}/dx = 0$ at the channel entrance. We give one numerical result with reference to our discussions in Secs. 5G3 and 5G4. This is for the case $\gamma = 1.2$, $q = 50 \, RT_0$, $E^\dagger = 0$, the equilibrium relation of Sec. 5G4, $k = 10 \, c_0/\hat{x}$ (with $\hat{x}$ the distance from the shock at which the channel area has the mean of its values at $x = 0$ and $x = \infty$), and $\beta = 2$. The eigenvalue velocity at this k (the largest k used) is 0.85 times the plane wave CJ value, and the reaction is half complete at $x = 0.9 \, \hat{x}$.

Wecken (1965) has discussed the phase plane and integrated the nozzle equations numerically for the case of an ideal gas equation of state and a reaction rate similar to the Arrhenius form. For the rate of streamtube expansion he assumes that the derivative of the logarithm of the area is constant at its value immediately behind the shock

$$d \ln A/dx = R^{-1}[D - u(0)]/u(0),$$

which is a more rapid expansion with distance than the considerations of Sec. 5G2 suggest since, in Eq. (5.31), u increases with x. He does not cast his results in the same form as do Wood and Kirkwood, but by reading from his graphs for a particular case (his $\gamma = 3$, $h'_* = 2$, $q'_0 = 1.2$, $h'_0 = 0.4$) we find $(D_\infty - D)/D_\infty = 3.9 \, x^*/R$. The objective of his work is somewhat different from that of our discussion here, with his numerical results serving largely as a basis for a qualitative discussion of the effects of large divergence and the possibility of failure. It would be of interest to present a comparison of a calculated pressure profile of the reaction zone with that for the plane-front solution. Unfortunately, the manner of presentation of Wecken's results does not easily lend itself to such a comparison.

Cowperthwaite (1971) has used a different type of assumption. Given an equation of state, he assumes that u is a monotone function of x with adjustable slope, leaving the rate function to be determined. Under these assumptions, choosing a pair of values for D and R determines the rate function and the solution, which is presumably forced to be an eigenvalue solution by the requirement of monotone u(x). The choice of values of D and R is not completely free; some result in physically impossible solutions. This form for u(x) gives in the plane case a simple rate function complete in finite time. The rate functions for finite divergence are not given. It seems likely that they can not be expressed as functions of the thermodynamic state alone, so that the results may be physically unrealistic to this extent.

Tsuge et al. (1970) and Fujiwara and Tsuge (1972) have integrated the nozzle equations numerically with realistic kinetics for a detonation in gaseous hydrogen/oxygen confined by nitrogen. They obtain the rate of area expansion from a simple approximation to hypersonic flow theory, treating the laterally expanding detonation products as a blunt body about which the inert confining gas flows. This estimate of the expansion rate contains the cross-sectional area of the body (and thus the charge diameter) so that the calculation gives the velocity defect as a function of diameter. For sufficiently small diameter the eigenvalue problem has no solution. For larger diameters there are two eigenvalues whose separation increases with diameter. The upper one approaches the plane CJ velocity, while the lower one, not calculated at large diameters, appears to be approaching a shock Mach number of one, that is, a sound wave.

Tsuge (1971) has also treated the case in which the losses arise from a growing boundary layer in a rigid tube. The results are similar. The dependence of the eigenvalue velocities on diameter and two solution profiles for ozone are shown in Figs. 5.46 and 5.47. The lower-velocity solution has a reaction zone about five times as long as the higher-velocity solution. Fig. 5.48 shows the same solutions in the u-λ plane. Here u is used instead of u^2 to better display the minima. The curves shown terminate just below the sonic locus and correspond to the upper branches of the M-curves in Fig. 5.47.

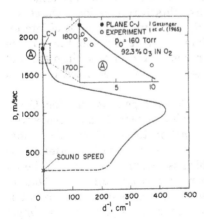

Fig. 5.46. Eigenvalue propagation velocity vs. reciprocal tube diameter d^{-1} for a detonation in 92.3% ozone/oxygen mixture at p_0 = 21.3 kPa, T_0 = 300 K. The results are compared with the experimental data (available only at the relatively large diameters shown) in the inset. Profiles of the two solutions marked by open circles are shown in Fig. 5.47. From Tsuge (1971).

Sec. 5G SLIGHTLY DIVERGENT FLOW

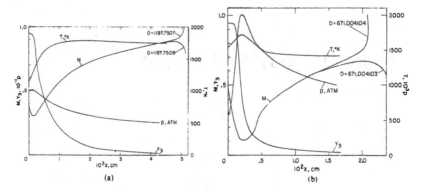

Fig. 5.47. Profiles (Y_0 is the mass fraction of ozone) for the (a) faster, $M_0 = 4.6$, and (b) slower, $M_0 = 2.58$, steady solutions at diameter 0.00333 cm, marked with open circles on Fig. 5.46. The branches at the end of the M curves in (b) show solutions on either side of the saddle-point separatrix. From Tsuge (1971).

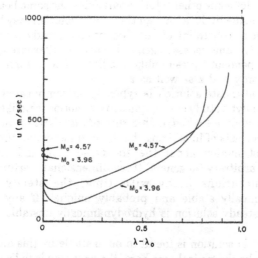

Fig. 5.48. Particle velocity u vs. degree of reaction $\lambda - \lambda_0$ for the solutions of Fig. 5.47. From Tsuge (private communication, 1972). Here $\lambda = 1 - m_{o_3}$ is the mass fraction of ozone. The circles at the left show the approximate intersections of the sonic loci with $\lambda = \lambda_0$.

6

NONSTEADY SOLUTION

Having completed our survey of the various types of one-dimensional steady solutions, we next ask whether, for a given rear boundary condition, the steady solution is the unique long-time limit. If it is not, we have to look for some other solution satisfying the same boundary conditions in the same limit. Although the steady solution may be unique under the special constraint of one-dimensional steady flow, we want to know about the uniqueness in general, where an alternate solution would be time dependent (presumably oscillatory) and three-dimensional, depending on y and z as well as x.

Such an alternate solution is typically so complicated that its discovery by analytic study of the equations of motion is unlikely. The usual first step is to do a *hydrodynamic stability analysis* of the steady solution. This consists of linearizing the three-dimensional, time-dependent equations of motion about the steady solution and studying their response to arbitrary (in general three-dimensional) perturbations. If all such perturbations decay with time, the steady solution is hydrodynamically stable and probably unique[1]. If any perturbation grows, the steady solution is hydrodynamically unstable and thus not unique.

If the steady solution is found to be unstable by this analysis, which turns out to be the typical case here, the next step is to look for an alternate solution. The results of the linear analysis are a great help here, for they define the region of instability in the parameter space, thus telling us where to look. They also give us a good estimate of the growth rate of

---

[1] We say "probably unique" because the usual situation is like that of a marble either at the bottom of a bowl or the top of a dome. It could of course be sitting in a dimple at the top of a dome, from which it would not be displaced by an infinitesimal perturbation. This is the exceptional case in problems of fluid flow.

Chap. 6   NONSTEADY SOLUTION

perturbations on the steady solution, and an estimate of the temporal and spatial period of the solution we seek.

Sec. 6A covers the question of hydrodynamic stability. The formal analysis has been carried to the point where stability can be determined by locating the zeroes of a complex function whose evaluation at each point requires the numerical integration of a set of 14 or more coupled ordinary differential equations. Numerical results have been obtained for an idealized (one-reaction, polytropic-gas) system. The result is that these detonations are unstable to perturbations (over a range of transverse wavelengths) down to quite small heats of reaction, and are unstable even when the reaction rate is "decoupled" from the rest of the system by being made independent of all state variables except the composition. Moreover, if the reaction rate depends strongly enough on temperature, the interesting special case of *longitudinal instability* appears. By this we mean that the steady solution is unstable under the constraint of longitudinal motion only, presumably going over into an *oscillatory galloping detonation*.

The formal treatment, unfortunately, offers little insight into the *mechanism* of instability. Some progress in this direction comes from qualitative intuitive treatments most conveniently applied to a particular idealization which we call the *square-wave model*. This consists of a shock, followed by an *induction zone* in which no heat is released, followed by the fire, a plane of instantaneous complete reaction in which all the heat is released. (In general we use the term "fire" to denote a region of rapid but not instantaneous heat release.) The induction time of each particle is allowed to depend on its state history. Interestingly enough, the discontinuities in this model have so far prevented an exact treatment of its hydrodynamic stability. However, its simplicity facilitates qualitative discussion and has inspired some intuitive derivations of stability criteria. The most useful result of this work is the insight it gives into the mechanism of instability. For this reason, and because it will be used later, we discuss at some length the local instability of the fire to breakup into a multiwave system.

Sec. 6B covers various theoretical approaches to the time-dependent replacement of the steady solution. At the cost of considerable increase in complexity, some nonlinear terms can be retained in the standard stability analysis, and the problem can be reduced to a complicated set of ordinary differential equations. This approach gives quite good results in one dimension (the galloping detonation) but is only qualitatively successful in two. A second and quite different approach is to study, in the geometrical acoustic approximation, the transverse propagation and amplification by chemical reaction of acoustic wave fronts in the reaction zone. This approach also offers some useful insights into the mechanism. The laterally propagating rays bounce back and forth between the shock and the fire (here a rapid but not instantaneous reaction) and the wave front becomes convoluted, eventually producing a series of regularly spaced contacts of the acoustic wave and the shock. The regular spacing is, presumably, equal to a preferred wavelength for transverse waves. Finally, there is a quite successful approximate treatment of galloping detonation for the square-wave model. It begins with the fire instability discussed in Sec. 6A, and is completed by conventional wave-matching techniques, applied under reasonable thermodynamic and hydrodynamic approximations, to the longitudinal waves bouncing back and forth between the fire and the shock.

The last section 6C deals with the full numerical treatment via time-dependent finite-difference calculations. The existence of the longitudinal instability is the saving feature here, since detailed calculations in one space variable can be made quite accurately. These have been done for several systems, including condensed explosives, and have produced a number of interesting examples of galloping detonation. The oscillation of the shock pressure is typically very nonlinear and of large amplitude, the peak pressure being as high as twice that of the steady solution. There is one very preliminary two-dimensional calculation, which taxed the capabilities of the fastest computer then available. Although this was carried to only a short time, the results did show definite growth of a point disturbance. Advances in computer capability since this work was done warrant a new attempt.

Many of the numerical results are given in units characteristic of the problem. These are specified wherever they occur, with one exception: the most common choice, for which the units are often omitted in the text. These units are

energy — $p_0 v_0$

length — $x_{1/2}$

time — $t_{1/2}$,

where subscript o denotes the initial state, and $x_{1/2}$ and $t_{1/2}$ are the half-reaction length and time for the steady solution. Precisely, $x_{1/2}$ is the distance from the shock to the point at which $\lambda = 1/2$, and $t_{1/2}$ is the time for a particle to react from $\lambda = 0$ to $\lambda = 1/2$ after passing through the shock. Two commonly used symbols are

$\epsilon = 2\pi/$(transverse wavelength)

$f = D^2/D_J^2$, the degree of overdrive.

Sec. 6A   STABILITY THEORY

6A. STABILITY THEORY

The most careful and complete treatment of the hydrodynamic stability of the steady solution is that of Erpenbeck (1962a, 1964b, 1965, 1966, and 1969a, a summary). We present mainly his treatment in Sec. 6A1, discuss briefly some of the qualitative work of others in Sec. 6A2, and review the conclusions and results, again mostly Erpenbeck's, in Sec. 6A3.

6A1. General Theory

The question of the hydrodynamic stability of the one-dimensional steady solution is posed by Erpenbeck (1962a) as an initial-value problem for the partial differential equations of the (three-dimensional) flow, together with the conservation conditions at the shock front. The hypothetical steady detonation is assumed to take place in an infinite medium, with a reaction-rate law which gives a reaction zone of formally infinite length. Only over-driven detonations are explicitly considered, with the CJ case as a lower limit. Thus a constant-velocity piston may be imagined to follow at an infinite distance behind the front, where chemical equilibrium has been attained. The problem studied is then a "steady one-dimensional detonation upon which is superimposed at time $t = 0$ a perturbation in the variables which characterize the flow. The subsequent growth or decay of the perturbation (for arbitrarily small initial magnitude) determines the stability of the basic flow to any given disturbance. If the flow is stable to *every* disturbance, it is said to be stable. Should any disturbance grow in time, the flow is regarded as unstable."

The equations of motion are those of chapter 4, in which transport processes are neglected, written for three dimensions with independent variables S, v, λ, u:

$$\dot{\rho} + \rho \, \text{div} \cdot \mathbf{u} = 0$$

$$\dot{\mathbf{u}} + v \, \text{grad} \, p = 0$$

$$\dot{S} = -(\Delta F/T) \cdot \mathbf{r} = \phi$$

$$\dot{\lambda} = \mathbf{r}.$$

These are supplemented by the conservation conditions for the shock, with the usual additional statement that all velocities appearing there are the components normal to the shock surface; velocity components parallel to the surface are the same on both sides of the shock.

The analysis proceeds by the conventional route of studying the behavior of infinitesimal perturbations in the linearized equations. However, here the linearization is preceded by the unusual step of transforming to an accelerated coordinate system attached to the perturbed shock. The unperturbed shock is the plane surface x = Dt proceeding in the x-direction. The y and z coordinates are unchanged by the transformation, but the new x is the distance from the perturbed-shock surface at each y and z. In addition to complicating the analysis by adding additional terms to the equations, this transformation introduces a novel state of affairs: As expected, the complete solution is the sum of the unperturbed steady solution and a time-dependent perturbation term, but, in the original reference frame, the unperturbed part depends on y, z, and t as well as on x, because at each (y,z,t) the unperturbed steady solution is displaced in the x-direction so that its front coincides with the perturbed-shock position. The linearized equations are more complicated because the transformation to the perturbed-shock frame introduces additional inhomogeneous terms containing derivatives of the shock position.

It is tempting to think that the transformation will always be necessary, since otherwise unshocked points near the shock are approximated by shocked ones, and vice versa. The real problem is more subtle. The transformation is actually required only when there is a nonzero gradient in the unperturbed solution, for the terms introduced by the transformation all vanish when the gradient vanishes. When the gradient is zero, the result is the same whether the transformation is used or not. A case in point is the shock stability problem described in Sec. 6A4. Here the unperturbed solution is just the flat-topped shock, and Erpenbeck, who used the transformation, and Fowles, who did not, both obtained the same result. We are not aware of this transformation having been used by anyone else, but the usual situation is that the gradient is absent. An example is the perturbation treatment by Whitham (1974), p. 270, of the change in strength of a shock traveling down a channel of slowly varying area.

The equations are linearized about the steady solution. The resulting coefficients are thus functions of x. They contain the first derivatives of p, ϕ, and r with respect to S, v, and λ, and derivatives with respect to x of these derivatives. The shock conservation conditions are linearized about the shock state at the front of the steady solution, so that the resulting coefficients are constants. An error in one of these coefficients, h_t, in Erpenbeck (1964b) is corrected in Erpenbeck (1967).

The linearized equations are studied by transforming them into a set of ordinary differential equations in x through Fourier transformation in y and z with wave number parameters α and β, which combine into the single transverse wave number ϵ

Sec. 6A STABILITY THEORY

$$F(x, t, \epsilon) = \int_{-\infty}^{\infty} \int_{-\infty}^{\infty} f(x,y,z,t) e^{-i(\alpha y + \beta z)} \, dy \, dz$$

$$\epsilon^2 = \alpha^2 + \beta^2,$$

followed by Laplace transformation in t with parameter τ, corresponding to a complex frequency $\tau/2\pi$,

$$G(x, \tau, \epsilon) = \int_0^{\infty} F(x, t, \epsilon) e^{-\tau t} \, dt.$$

The result is a set of inhomogeneous coupled linear ordinary differential equations for the transforms of S, v, λ, and u, plus an algebraic equation for ξ, the transform of the displacement of the shock from its unperturbed position. In the differential equations, dependence on ξ and the initial data is confined to the inhomogeneous part. To determine stability, it turns out that only the homogeneous part needs to be solved.

After application of appropriate boundedness conditions, it is found that ξ can be written in the form

$$\xi(\tau,\epsilon) = U(\tau,\epsilon)/V(\tau,\epsilon),$$

where both U and V are complicated functions requiring for their evaluation x-integrals over the solution of the set of homogeneous differential equations for the transforms of S, v, λ, and u. The appropriate solution of these equations is selected by applying a boundedness condition at x = ∞. The initial data appear in U but not in V. Furthermore, U is regular in the right half (Re $\tau > 0$) of the complex τ-plane.

The location of the poles of ξ determines the hydrodynamic stability. A pole of ξ located in the right half of the complex τ-plane, where the real part of τ is positive, implies exponential growth of the original disturbance, with a time frequency corresponding to the imaginary part of τ. Since U is regular in the right half-plane, all such poles are produced by zeroes of the complex function $V(\tau,\epsilon)$, and the stability is independent of the initial data, as expected.

The theory has been applied only to cases in which there is a single reaction. This offers a simplification in that the reaction variable λ can be used to replace the distance variable x, thus obviating the necessity of determining the explicit spatial dependence of the steady solution. When this is done the function V can be expressed in terms of the values of the solutions of a set of fourteen nonlinear differential equations at $\lambda = 1$. These are integrated numerically, starting from the point of complete reaction $\lambda = 1$ and proceeding to $\lambda = 0$. This is a singular point, so the integration is started by a power-series expansion. With this calculation of V in hand, the next step is to determine if V has any zeroes in the right half of the complex τ-plane. Now the number of zeroes less the number of poles of a complex function f(z) within a closed contour is equal to $1/2\pi$ times the increase in its argument as z makes one complete traversal of the contour in the counterclockwise direction. The procedure is to calculate $V(\tau)$ as τ traverses the D-shaped contour shown in Fig. 6.1. It can be shown that V has no poles in the right half-plane, so the change in the argument gives the number of zeroes. The radius of the circle is taken large enough so that no roots of V lie outside it, and so that an asymptotic expansion of V for large τ may be used over the portion of the contour indicated by the crosshatching. V must be calculated on the imaginary axis to a value of τ sufficiently large that the asymptotic expansion can be employed from there on. Conjugate symmetry of V allows the calculation to be done for positive $\text{Im}(\tau)$ only. The roots can of course be approximately located by calculating V at interior points.

Erpenbeck states that the possibility of distinctly atypical behavior at the CJ detonation velocity should not be ignored. The calculation is not performed at this velocity, for in this case the singularity at $\lambda = 1$ becomes higher-order and the radius of convergence of the power series about it vanishes.

In a later paper, Erpenbeck (1966) treated the stability problem in the short (transverse spatial) wavelength limit $2\pi/\epsilon \to 0$. The analysis is

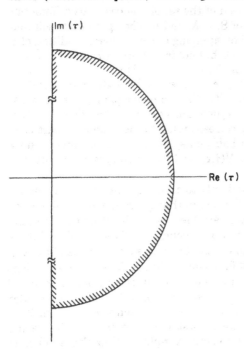

Fig. 6.1. Contour for the determination of the number of roots of V in the right half of the complex τ-plane.

rather involved, but a relatively simple stability condition emerges. The quantity $c^2 - u^2$ through the steady solution (with u the particle velocity in the steady frame) plays a crucial role. Three types of profiles are shown in Fig. 6.2. For profile D, the solution is stable. For profiles M and I the steady solution is unstable if for any value of the (real) variable z in the range

$$(c^2 - u^2)_+ \leq z \leq (c^2 - u^2)_{max},$$

as inequality

$$\alpha(z;y^+) \, e^{\beta(z;y)} > 1$$

Here the presence of $y = y(x)$ in the function indicates the appearance in it (not as a vector) of some of the variables of the steady solution, including thermodynamic first derivatives, and the subscript + denotes values immediately behind the shock. The function α is algebraic. The evaluation of the function β requires a quadrature over the steady solution:

$$\beta = z \int_0^{x^*(z)} |z^2 - (c^2 - u^2)|^{-1/2} \, [f(y) + g(y)/(u^2 + z^2)] dx,$$

where f and g are algebraic functions, and the upper limit is the smallest x for which $z = c^2 - u^2$, i.e., x^* is the solution of

$$[c(x^*)]^2 - [u(x^*)]^2 = z.$$

Since $\beta(z)$ becomes infinite as $z \to (c^2 - u^2)_{max}$, a *sufficient* algebraic condition for instability is that the square bracket be positive at the value of x for which $c^2 - u^2$ is a maximum.

Pukhnachev (1963) carried out an analytic treatment along the same lines as Erpenbeck (1962a) but used a round tube of finite diameter at the outset and employed the "normal modes" approach instead of

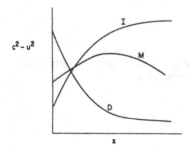

Fig. 6.2. Profiles of $c^2 - u^2$ in the steady solution; x is the distance from the shock. The letters D, M, and I stand for "decreasing," "maximum" and "increasing."

treating the initial value problem explicitly. However, he omitted the complication of transforming to the coordinate frame attached to the perturbed shock.

To sum up, the theoretical analysis of the hydrodynamic stability problem for the overdriven one-dimensional steady solution is substantially complete. For finite spatial wavelength $2\pi/\epsilon$ of transverse disturbances, the steady solution is unstable if the function $V(\tau,\epsilon)$ of the general theory has any zeroes in the right half complex τ-plane. The calculations required to apply the theory are rather complex even for the reported case of a single reaction in a perfect gas. For more complicated equations of state, or for more than one reaction, a still more complicated calculation would be required. The asymptotic analysis for the short wavelength limit $\epsilon \to \infty$, although quite involved, yields a relatively simple result requiring for its evaluation the calculation of a simple quadrature over the steady solution for a range of end points. There is also an algebraic sufficient condition for instability which is evaluated at only one point of the steady solution.

The unsupported case, with the approximately steady CJ reaction zone followed by a time-dependent rarefaction wave, has not been carefully analyzed. Of course, the long run limit of this case is simply the supported CJ detonation, with the piston moving at CJ particle velocity. Mathematical difficulties peculiar to this case prevent its explicit inclusion in the general theory.

6A2. The Square-Wave Detonation

There has been much discussion in the literature of the stability of the "square-wave" model of a detonation. In this model instantaneous complete reaction follows a state-dependent induction time, so that the steady profile is that of Fig. 6.3.

There have been several more or less intuitive considerations of this problem, and one attempt at a formal analysis, but no complete and correct treatment. Erpenbeck (1963, 1969a) reviews this work in some

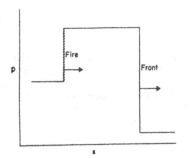

Fig. 6.3. Steady square-wave detonation.

detail, and discusses the difficulties inherent in the model. We limit ourselves here to a brief review of what has been done. As background, we first present the basic idea behind the intuitive approach to one-dimensional instability. This gives some physical insight into what is going on, and forms the basis for Toong's model, Sec. 6B3.

We may represent this reaction mechanism by two reactions with progress variables λ_1 and λ_2

$$\dot{\lambda}_1 = r_1 \; ; \quad q_1 = 0$$

$$\left. \begin{array}{l} \lambda_2 = 0 \text{ for } \lambda_1 < 1 \\ \lambda_2 = 1 \text{ for } \lambda_1 = 1 \end{array} \right\} \quad q_2 \neq 0.$$

In this scheme the first reaction is a thermally neutral timer for the second. The induction time for any particle is the time for the first reaction to reach completion under the conventional state-dependent reaction rate r_1; the second reaction then goes instantaneously and releases all the heat. We call the plane of instantaneous heat release the "fire."

6A2.1. Longitudinal Fire Instability

In addition to the steady square-wave solution of Fig. 6.3, there are other simple self-similar solutions which may exist for a finite time. We call these solutions perturbations, positive perturbations if they contain shocks, and negative perturbations if they contain rarefactions. Before considering them, we give some background.

Figure 6.4 shows a complete Hugoniot curve for $\lambda = 1$, including both the detonation and deflagration branches. Elsewhere in this book only

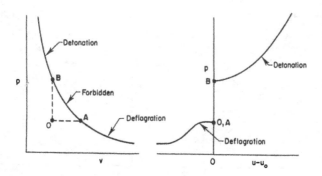

Fig. 6.4. Complete Hugoniot curves in the p-v and p-u planes for exothermic reaction.

NONSTEADY SOLUTION Chap. 6

the detonation branch, with $v < v_o$, is of interest, but here the deflagration branch, with $v > v_o$, also enters. The initial state (p_o, v_o, u_o) is arbitrary, except that we require the degree of reaction to be zero. The particle velocity jump $u - u_o$, given by

$$(u - u_o)^2 = (p - p_o)(v_o - v),$$

is imaginary in the forbidden region between points A and B, and is zero at A and B. By differentiation we find

$$(dp/du)_{\mathcal{H}} = 2(u - u_o)/[(v_o - v) - (p - p_o)(dv/dp)_{\mathcal{H}}],$$

where the derivatives are along the Hugoniot curve, from which we see that $(dp/du)_{\mathcal{H}}$ is zero at both points A and B. Hence the Hugoniot curve in the p-u plane is as shown.

Now we consider in Fig. 6.5 the complete Hugoniot diagram for a detonation. The light lines are the usual diagram showing the Hugoniot curves for no reaction and complete reaction, both centered at the initial state O, and the Rayleigh line, for an overdriven detonation, through O. The heavy line, not usually shown, is the deflagration Hugoniot curve for complete reaction, centered at N, the shocked, unreacted, induction-zone state. This deflagration Hugoniot curve is, of course, the deflagration branch of the complete Hugoniot curve of Fig. 6.4, centered at point N. It must pass through the final state points S and W, the intersections of the Rayleigh line with the usual complete-reaction Hugoniot curve centered at the initial point O. This requirement is similar to property 7, Sec. 5A1. It also has its own deflagration CJ point at J, where a Rayleigh

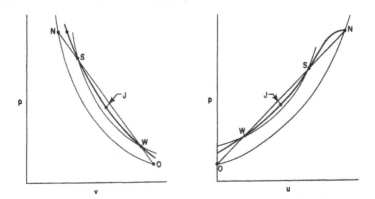

Fig. 6.5. Conventional p-v and p-u diagrams for a steady detonation with the deflagration Hugoniot curve for the fire (heavy-line) added.

line through N, if it has the right velocity, is tangent. As the velocity of the Rayleigh line through O is decreased, S and W move closer together and finally coalesce at the CJ point on the detonation Hugoniot curve centered at O. The CJ point of the deflagration Hugoniot curve then coincides with the detonation CJ point.

The usual steady detonation solution, treated in Sec. 2C without any reference to the deflagration Hugoniot curve, can be described as a shock to point N, followed by a deflagration. Application of the conservation relations gives the deflagration Hugoniot curve and deflagration Rayleigh line centered at point N. The requirement that the flow be steady fixes the deflagration as the one which connects point N to point S, and integration of the rate law gives the progress variable as a function of time or distance from the shock. It is easy to show that the deflagration and detonation Rayleigh lines coincide in this case. Although the point of view and the words are different, both treatments gives the same result.

Now we are ready to consider the perturbation solutions. The two types are shown in t-x (steady frame) and p-x in Fig. 6.6. (To agree with

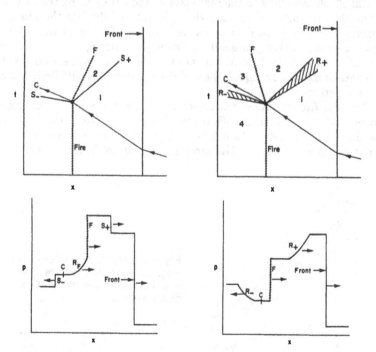

Fig. 6.6. Positive and negative perturbations; t-x (steady frame) and p-x diagrams.

NONSTEADY SOLUTION Chap. 6

the published work on this subject, we abandon here our usual convention of choosing our steady frame with all velocities positive, as in Fig. 4.2, and let the velocities be negative.) A right-facing shock S_+ or centered rarefaction R_+ runs ahead of the fire F, altering both the particle velocity and the rate r_1 of the timer reaction, and thus the velocity of the fire. The right-facing rarefaction R_F (similar to the Taylor wave in an ordinary detonation), whose head coincides with the fire, and which is shown in the p-x diagram but not the t-x diagram, may or may not be present. A left-facing shock S_- or rarefaction R_- completes the match to the original complete-reaction state 4. The contact discontinuity C separates the particles passing through the old and new fire.

In contrast to many matching problems, the solution of this one requires the use of the rate equation $\dot\lambda = r$ in addition to the conservation relations. For background discussion see Courant and Friedrichs (1948), pp. 218-227, Shchelkin and Troshin (1963), Oppenheim, Urtiew, and Laderman (1964) and Chernyi (1975). If we choose the strength of the leading shock S_+, we can calculate the corresponding fire velocity in two different ways: (1) from the conservation relations alone, by conventional shock matching to the rear state 4, and (2) by using the rate law (plus the streamline deflection through S_+) to calculate the changed reaction locus. In general, these two calculated values of the fire velocity are not equal, implying no solution for the given strength of S_+, because not all of the required conditions are satisfied. If for some value of the strength of S_+ they are equal, then all the equations are satisfied, and we have a perturbation solution.

Consider first the conservation relations. Fig. 6.7 is the p-u diagram, showing the unperturbed deflagration Hugoniot curve as the line connecting points 1 and 4. Point 1 is the same as point N in Fig. 6.5, and point 4 is the same as S. The jump across the fire in the unperturbed

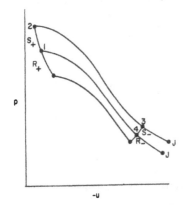

Fig. 6.7. Perturbation-solution p-u diagram. Note that the abscissa is $-u$, because of the choice of coordinate system.

solution is from point 1 to point 4. In a positive perturbation the front shock S_+ raises the state from 1 to 2 (the line connecting them is a shock Hugoniot curve centered at 1). The jump through the fire drops the state to point 3 on the deflagration Hugoniot curve centered at 2. There it is met by the rear shock S_- coming up from point 4 (on the shock Hugoniot curve centered at 4) to close the cycle. If the detonation is not sufficiently overdriven, or the shock S_+ is not sufficiently weak, the deflagration CJ point J may lie above the S_- shock locus, in which case the deflagration jump stops at J, and the rarefaction R_F (shown in the p-u diagram of Fig. 6.6) connects the state to the shock locus. The treatment of the negative perturbation is similar, with the shocks replaced by the rarefactions R_+ and R_-; the corresponding curves are shown in the figure, lying below the unperturbed deflagration Hugoniot curve. Thus for every shock strength S_+ (or rarefaction R_+) measured, say, by $u_2 - u_1$, the conservation relations give a corresponding value of U_c, the increase (or decrease) of the fire velocity from the unperturbed value. A typical U_c vs. u_2 curve is given in Fig. 6.8; the point where $U_c = 0$ corresponds to $u_2 = u_1$.

Any point on the U_c vs. u_2 curve satisfies the conservation relations and could be a solution. The rate equation, not yet used, allows us to select a single point as a solution. With it we make an independent calculation of the fire velocity. For a given (constant) strength of S_+, the state behind it is known; we need the particle velocity and the reaction rate. The calculation is relatively simple because the states between the waves are known and constant. A particle passes through the lead shock and proceeds at a constant velocity and constant timer-reaction rate, until it encounters S_+. There it has achieved a degree of reaction proportional to the time of passage. Behind S_+ it proceeds at a different constant velocity and increased timer rate, until it reacts, which point is by definition on the fire locus. This construction, carried through for all particles, and for a perturbing rarefaction R_+ as well as for a shock S_+,

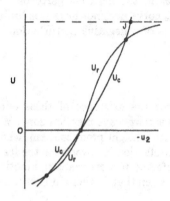

Fig. 6.8. Fire velocities U_c and U_r from the conservation and rate relations vs. shock strength $-(u_2 - u_1)$.

determines the (constant) fire velocity implied by the rate equation. We call this increase (or decrease) of the fire velocity from the unperturbed value U_r (r stands for rate, and c for conservation).

A typical U_r curve, along with that for U_c, is shown in Fig. 6.8. There is, of course, an intersection of the two curves at $u_2 = u_1$, the unperturbed steady solution. If they intersect at another point above or below, as shown in the figure, then a perturbation solution exists, and u_2 for that solution is given by the intersection, where both the conservation relations and the rate law are satisfied. By the following simple argument we can determine the condition for the existence of perturbation solutions. Looking at Fig. 6.6, we see that there is an upper limit on the fire velocity following the shock: that is, the fire can go no faster than the shock. When the fire coincides with the shock, the detonation (shock and fire) inside the unperturbed reaction zone is steady and unsupported, so it is a CJ detonation in the shocked material, and it has a defined and known velocity. This condition could be satisfied only if the rate of the timer reaction were infinite, so there would be no delay between shock and fire. In our system this can occur only in the limit of infinite shock strength, so the U_r curve bends over as shown in Fig. 6.8 in the positive perturbation region. The U_c curve has no such limit, and it extends across the CJ velocity limit for finite u_2. Therefore the condition for a positive perturbation solution to exist is that, in the unperturbed state,

$$|dU_c/du_2| < |dU_r/du_2|,$$

which ensures that the two curves intersect, as shown in the figure.

For the negative perturbation the limit on the fire velocity occurs when the fire coincides with the contact surface C in Fig. 6.6, and this corresponds to the rate of the timer reaction going to zero. This zero rate can occur, in our case, only for the limit of an infinitely strong rarefaction. Therefore the U_r curve also bends over in the negative perturbation region as shown in Fig. 6.8, while the U_c curve does not have this limit. Therefore the condition for the existence of a negative perturbation is also that, in the unperturbed state,

$$|dU_c/du_2| < |dU_r/du_2|.$$

We have now found the condition for the existence of these one-dimensional perturbation solutions in a square-wave reaction zone. We do not know what sort of initial perturbation might produce them, what happens after the leading shock S_+ or rarefaction R_+ overtakes the lead shock, or how the criterion might be affected if more realistic kinetics were used. All we really have is the suggestion that further study of these

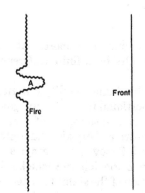

Fig. 6.9. Two dimensional perturbation of the fire in Shchelkin's (1959) discussion of fire stability.

perturbations is warranted. Alpert and Toong (1972) have constructed a semiquantitative picture of the mechanism of one-dimensional oscillation in a detonation, based on the ideas presented in this section. Their results are discussed in Sec. 6B3.

6A2.2 Treatments of Square-Wave Stability

Shchelkin (1959), considered the stability of the fire to a two-dimensional perturbation, such as the dimple of Fig. 6.9. He arrived at a simple stability criterion by qualitative consideration of the way rarefactions move into the lower pressure dimple A from its sides. Zaidel' and Zeldovich (1963), and Shchelkin (1965) both treat the longitudinal stability of the fire by what appears to be essentially calculation of the initial slopes of the U_c and U_r curves of Fig. 6.8. The reasoning is that if a state perturbation produces an increased fire velocity U_r owing to increased rate greater than the increased fire velocity U_c calculated from the conservation conditions, then the fire is unstable and will break up into the multiple-wave positive perturbation described above. However, the two papers give different answers, and no one has resolved the difference.

Zaidel' (1961) earlier carried out a conventional stability analysis of the complete square-wave detonation; his result was later obtained by Zaidel' and Zeldovich (1963) in their more intuitive treatment. Zaidel's approach is similar to that of Erpenbeck (1962a), but Erpenbeck (1963) showed that his assumed Laplace transform fails to exist.

All of the above are covered in Erpenbeck's (1963, 1969a) reviews. There are some other papers not covered there: Il'kaeva and Popov (1965) have a good discussion of the problem of longitudinal fire stability. Their conclusion is the same as that given above, that the fire is unstable if

$|dU_c/du_2| < |dU_r/du_2|$.

They applied it to a system with a simple rate function proportional to a power of the density, and confirmed the results by a finite-difference calculation.

Chernyi and Medvedev (1970) evaluated this same stability criterion for a polytropic gas with induction time proportional to $e^{-E^{\dagger}/RT}$ and gave numerical results locating the stability boundary in the space of heat of reaction, activation energy, and degree of overdrive. They also did finite-difference calculations. Neither Il'kaeva and Popov nor Chernyi and Medvedev comment on whether Zaidel' and Zeldovich or Shchelkin is correct, or give enough details of their evaluation of the stability criterion to allow the reader to make a judgment. Finally, Aslanov et al. (1968) approximated a more realistic reaction zone by a sequence of steps in each of which the square-wave model applies, that is, each step is assigned a constant state terminated by a small deflagration discontinuity, in which the appropriate portion of reaction takes place instantaneously. They make no reference to Erpenbeck, and give no results other than the qualitative statement that "a detonation wave with a steep drop in pressure (density) behind the shock ... turns out to be considerably more stable than would follow from the use of the one-step (square-wave) approach."

6A3. Results

Erpenbeck has applied his theory to a polytropic gas with reaction A → B (no back-reaction) obeying in Arrhenius rate law, with no change in the number of moles or the heat capacity with reaction. Thus the equation of state and rate are

$$E = pv/(\gamma - 1) - \lambda q$$

$$\dot{\lambda} = k(1 - \lambda)e^{-E^{\dagger}/RT}.$$

We use the term "finite theory" for the general theory, Erpenbeck (1962a), and the term "asymptotic theory" for the limiting case of short transverse wavelength, Erpenbeck (1966). The results described here are selected from Erpenbeck's calculations: those with the finite theory (1964b), including zero activation energy (1965), and those for the asymptotic theory (1966). His complete results are much more extensive. Concerning the range of applicability of the asymptotic theory, a rigorous statement cannot be made in the absence of an explicit investigation of transport properties; presumably it applies for

STABILITY THEORY

wavelengths much smaller than the reaction-zone length but much larger than a mean free path.

As described in Sec. 6A1, the determination of stability requires only that the existence of roots of the complex function $V(\tau,\epsilon)$ in the right half of the complex τ plane be demonstrated: the roots themselves need not be found. The demonstration is effected by evaluating V on a contour. The roots themselves are not usually located, since this would require much more calculation. The growth rates and oscillation frequencies of perturbations are known only if the roots are found; some typical cases will appear in later sections.

The parameters of the calculations are

(1) fluid properties: $\gamma, q, E^\dagger$

(2) drive: $f = (D/D_j)^2$

(3) transverse wave length: $2\pi/\epsilon$.

The rate multiplier k enters in a trivial way, serving only to determine the time scale. The propagation velocity D may be regarded as determined by a given constant velocity assigned to a piston an infinite distance behind the front. The ranges of the parameters covered by the investigation are

$\gamma = 1.2$

$0.1 \leq q \leq 50$

$0 \leq E^\dagger \leq 50$

$0 \leq \epsilon \leq 16$, finite theory

$15 \leq \epsilon \leq \infty$, asymptotic theory.

A few larger values of $E^\dagger$ appear for the asymptotic theory. The values $q = 50$ and $E^\dagger = 50$ are typical of real gaseous systems.

Fig. 6.10 shows, for $q = 50$, the stability boundaries in the f-ϵ plane for $E^\dagger = 10$ and $E^\dagger = 50$. The boundaries shown are interpolated from a set of calculated points; those for $E^\dagger = 50$ are indicated by dots. These results show several interesting features. We take them up in turn for increasing ϵ.

For $\epsilon = 0$, i.e., infinite transverse wavelength or longitudinal instability (one-dimensional flow), we see that, for large enough $E^\dagger$, the

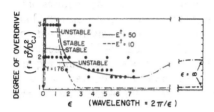

Fig. 6.10. Stability boundaries in f-ϵ for q = 50 and two activation energies. Recall that length and energy are in units of the steady half-reaction length and $p_o v_o$, respectively. The dots are the calculated points from which the stability boundaries for $E^\dagger = 50$ are interpolated. The chain lines indicate possible connections between the finite-wavelength and zero-wavelength-limit results.

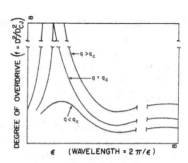

Fig. 6.11 Probable q-dependence of the stability boundary of Fig. 6.10.

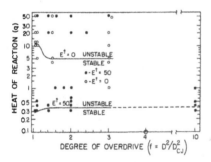

Fig. 6.12. Stability boundaries in q-f for two activation energies. From Erpenbeck (1965).

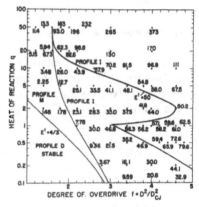

Fig. 6.13. Asymptotic short wavelength ($\epsilon \to \infty$) stability in q-f. The light lines divide the plane into regions of steady-solution profiles types D, I, and M (Fig. 6.2). The neutral stability value of $E^\dagger$ (larger values unstable) is given at each point, as well as contours for $E^\dagger = 50$ and $E^\dagger = 4/3$ (coincides with the D-M boundary). From Erpenbeck (1966).

steady solution is unstable if not overdriven too strongly: for $E^\dagger = 50$ the critical value of f is 1.76. Since numerical time-dependent calculations can be done quite accurately in one dimension, this offers a good opportunity to study one case of the long-time unsteady solution in some detail. Calculations of this type are described in Sec. 6C1.

For ϵ in the approximate range 0.2 to 2, that is, for transverse wavelengths 3 to 30 times the steady reaction-zone length, the steady solution is unstable even for strong overdrive. Erpenbeck (1964b) believes that at least for the case q = 50, $E^\dagger = 50$ a finite range of unstable wavelengths persists to $f = \infty$. It appears that as q is reduced the top of the instability boundary finally closes up at some critical q = q_c, as shown in Fig. 6.11, and that for smaller values of q all wavelengths in this range become stable for sufficiently large f. The value of q_c appears to be remarkably small, approximately 0.3.

For $\epsilon \to \infty$, we have the asymptotic theory, giving the limiting values shown at the right of Fig. 6.10. A minimum in the curve for $E^\dagger = 50$ of Fig. 6.10 is clearly implied for some $\epsilon > 7$, and similarly, a probable minimum is suggested for $E^\dagger = 10$ for some $\epsilon > 2.5$.

Stability in the q-f plane with $E^\dagger$ as parameter is shown in Fig. 6.12. Here the ϵ-dependence is condensed to a single attribute: the steady solution is stable if perturbations of all wavelengths considered (here roughly $0 < \epsilon < 7$) decay; otherwise it is unstable. Neutral stability boundaries are shown for $E^\dagger = 50$ and $E^\dagger = 0$. The instability persists to quite small values of q: to $q \cong 0.3$ for $E^\dagger = 50$ and to q = 5 to 10 even for $E^\dagger = 0$. The results for $E^\dagger = 0$ are remarkable in showing that, although the sensitivity of the rate to temperature is obviously important, it is not a necessary condition for instability. Taking $E^\dagger = 0$ removes the "feedback" by removing the state-dependence from the rate equation, decoupling it from the others.

The $E^\dagger = 50$ curve, from Erpenbeck (1964b), includes a long interpolation (dashed line) between calculations at f = 2 and f = 10. The large open circle indicates a later calculation, Erpenbeck (1966), at q = 0.1, $E^\dagger = 50$, $\epsilon = 15.8$, which turns out to be unstable, showing that the original interpolation is wrong and that the curve has a more complicated shape. (The new calculation was made for the purpose of checking the asymptotic theory, with which it agrees.)

The asymptotic-theory results are shown in Fig. 6.13, a q-f diagram showing the neutral stability values of the activation energy (with larger values unstable) at a number of points. For the particular system chosen, all three types of steady-solution profiles shown in Fig. 6.2 are obtained. The two light curves divide the plane into three regions, one for each profile type. Recall that profile D is unconditionally stable. The boundary between regions D and M is also the contour line for $E^\dagger = 4/3$. The heavy curve at the right is the contour line for $E^\dagger = 50$. The cusp in this curve arises from the nature of the stability condition, in which the inequality may be satisfied for any value of z in the range of values of $c^2 - u^2$ occurring in the steady solution. The cusp occurs when the region in which the inequality is satisfied switches from the neighborhood of the shock to that of the end of the reaction zone.

Pukhnachev (1963) gives numerical results for the case $\gamma = 1.2$, $q = 47$, $E^\dagger = 97$, $f = 1$, for a round tube of radius r_o. His results are given as a function of the ratio d/r_o, where d is the reaction-zone length. For the lowest mode, with wavelength twice the tube diameter, Erpenbeck's wave number ϵ is about $1/2 \; \pi d/r_o$. For the lowest mode (Bessel function J_1) there are two distinct roots with positive real part. One has positive real part for $0.475 < d/r_o < 1.35$, and the other for $0.557 < d/r_o < 2.15$. Higher modes exhibit similar behavior. Thus the longest unstable wavelength is at $d/r_o = 0.475$ or $\epsilon = 0.75$. This result is in apparent contradiction to that of Erpenbeck, Fig. 6.10, which for the much lower activation energy of $E^\dagger = 50$ has the instability extending to infinite transverse wavelength ($\epsilon = 0$). But Pukhnachev's analysis is incomplete: his normal-mode functions have radial dependence $J_n(r)$ (with J_n the Bessel function of order n and $n = 1$ the lowest mode), so that he does not have a term with no radial dependence, which presumably would have to be included to discover longitudinal instability. In addition, as pointed out by Erpenbeck (1969a), he did not transform to the shock frame, and thus would have missed the one-dimensional instability in any case.

Finally there is the square-wave detonation. For a polytropic gas with Arrhenius temperature dependence of the ignition time the criteria for instability from the works discussed in Sec. 6A2 are

Shchelkin (1959): $E^\dagger/RT_i > (1 - T/T_i)$

Zaidel' and Zeldovich (1963): $E^\dagger/RT_i > f(\gamma)$

Shchelkin (1965): $E^\dagger/RT_i > (1 + c/c_i)/[M_i(q_i/c_o^2)]$; $M_i = u_i/c_i$.

Here subscript i denotes the induction-zone state and no subscript the complete-reaction state. These are all approximate treatments of the fire stability, the first being two-dimensional and the second and third one-dimensional. The function $f(\gamma)$ in the second varies from 5.4 to 7.7 for γ from 7/5 to 11/9.

Erpenbeck (1963, 1969a) criticizes these as approximations to the (non-square) model with first-order Arrhenius rate. He points out that the second is obviously erroneous, since it is independent of the heat release q and thus predicts instability for any degree of exothermicity, in

Sec. 6A   STABILITY THEORY

particular for a shock (q = 0). The first and third overestimate the stability, predicting that detonations which are actually unstable are stable.

Chernyi and Medvedev (1970) evaluated the (longitudinal fire) instability criterion of Sec. 6A2.1

$$|dU_r/du_2| > |dU_c/du_2|$$

for infinitesimal displacements in a polytropic gas with Arrhenius temperature dependence of the induction time. Unfortunately they did not write down completely their analytic result, but presumably it is different from any of those above. They present some numerical results for $\gamma = 1.4$ and varying heat of reaction q and degree of overdrive f. Reading from their figure we have converted their results, with some extrapolation, to Fig. 6.14, a q-f diagram like Fig. 6.12. Il'kaeva and Popov (1965) have had some success with this same criterion applied to a density-dependent induction time. Their work is discussed in Sec. 6C1.

All these results present a confusing picture, but the main point is that all the cases relevant for detonation problems are unstable for some value of ϵ. We speculate that the situation can be summarized as shown in Fig. 6.15, where q-f diagrams for three activation energies and three

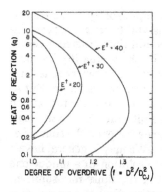

Fig. 6.14. Local *one-dimensional* instability for a square-wave detonation with $\gamma = 1.4$ according to Chernyi and Medvedev (1970). The curves are neutral-stability boundaries for the activation energies shown, with the stable region on the right.

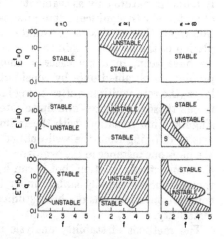

Fig. 6.15. Summary of detonation stability results.

values of ϵ are divided into stable and unstable regions by the neutral stability lines. With $E^\dagger = 0$, the detonation is stable in the longwave ($\epsilon = 0$) and shortwave ($\epsilon \to \infty$) limits, but unstable for medium ($\epsilon = 1$) wavelengths, as long as q is large enough. For $E^\dagger = 10$, a little "nose" of instability appears for small f and a range of q values in the longwave limit, and a diagonal band shows up in the shortwave limit, with the medium wavelength limiting q decreasing. When $E^\dagger = 50$, the longwave instability region grows larger. The shortwave instability region also grows larger and develops a "nose" sticking out from it, and the limiting q in the medium wave region decreases even more, to a very low value. One may wonder whether the "nose" of the shortwave limit, which corresponds to instability near the fire while the band corresponds to instability near the shock, and the similarly-shaped region of the longwave limit, are in any way connected, perhaps indicating the same feedback mechanism.

6A4. Shock Stability

By "shock stability" we mean the hydrodynamic stability of a step shock in an inert fluid. With the shock regarded as a jump discontinuity, and the flow away from the shock locally stable, the stability depends solely on the shape of the Hugoniot curve. Certain shapes of a segment of the curve give rise to instability.

There is some question as to the physical applicability of the results, because the Hugoniot shapes giving rise to instability are almost certainly found in nature only as segments of the equilibrium Hugoniot curves of systems with significant rate processes. The shock-stability idealization treats these with strictly infinite rate, assuming instantaneous completion of the process within the discontinuous shock transition. But as we have seen, in many cases a unique solution to the propagation problem is obtained *only* by considering a finite rate, no matter how large (the rate multiplier often being just a scale factor on the time). Let us call these the infinite- and finite-rate cases. We may in general have a singular limit situation, with the infinite-rate solution (if a unique one exists) differing from the well-defined limit of the finite-rate solution. We have one instance in the overdriven detonation. Both treatments give the same steady solution, but it can be unstable under finite rate even though unconditionally stable under infinite rate. A more interesting comparison, which has not been studied would be a case of infinite-rate instability.

The methods of stability analysis used are (1) conventional linear stability analyses done in two different ways, (2) stability to longitudinal perturbations in the presence of a reflecting rear boundary via a physical argument, and (3) stability to longitudinal perturbations by consideration of alternate two-wave solutions. Method (1) is presumably rigorous, but does not apply to all of the cases of interest. Method (2), while not rigorous, is convincing, and gives some useful insights. Method (3) is least convincing, but gives additional insight, and is supported by observation in one case. Although the three methods do not give identical results, the main conclusions are fairly clear. But the disagreements, and difficulties of detailed interpretation, support our suspicion that the infinite-rate model generally leaves out some of the essential physics. We remark in passing that Morduchow and Paullay (1971) have shown by the conventional analysis that weak shocks of finite thickness (by inclusion of viscosity and heat conduction) are stable.

The main conclusion is that the two shapes of Hugoniot curves shown in Fig. 6.16 have unstable segments AB. The first case is typical of various real systems, the one shown being for a material with a first-order phase transition. In such systems it is observed experimentally that single shocks to points between A and B do not exist, but two-wave structures are observed instead. The second case requires that the Hugoniot curve have a region of positive slope in which the magnitude of the slope becomes sufficiently small. Although this Hugoniot curve has the general shape of that for a dissociating material, the small magnitude of the slope in AB required for instability is probably not realized in nature. Unpublished calculations by Fickett have shown that for a dissociating ideal gas the slope is unlikely to be small enough in this region.

We limit our discussion to the case of (unperturbed) one-dimensional flow without lateral boundaries. Even so, there are a large number of papers. The ones we have found most helpful are

Method (1):
  Swan and Fowles (1975) — normal modes
  Erpenbeck (1962b) — initial-value problem by transforms (see Sec. 6A1)

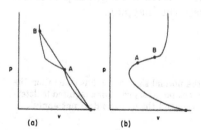

Fig. 6.16. Two Hugoniot-curve shapes with unstable segments AB.

Method (2):
Fowles (1976)

Method (3):
Fowles (1976), Swan and Fowles (1975)
Cowperthwaite (1968)
Duvall (1962).

The two treatments of the problem under method (1), the conventional linear stability analysis (to three-dimensional perturbations), give the same result. The shock is unstable when[2]

$(dp/dv)_{\mathcal{K}} > 0$ and $(dp/dv)_{\mathcal{K}} < [(p - p_o)/(v - v_o)]/(2M + 1)$.

Equivalent forms are

$(dp/du)_{\mathcal{K}} < 0$ and $|(dp/du)_{\mathcal{K}}| < \rho c$

$1 + M - M^2(v_o - v)(\partial p/\partial S)_v/T > 0$

$(M + 1)\gamma/\Gamma < 1 - p_o/p$

$(M + 1)/(\Gamma M^2) < v_o/v - 1$,

where M is the Mach number of the flow behind the shock, and Γ is the Gruneisen coefficient

$M = (D - u)/c$, and $\Gamma = v(\partial p/\partial E)_v$.

Figure 6.17 shows the unstable segment AB of the Hugoniot curve in the p-v and p-u planes. In p-v the slope of the Hugoniot curve must have the unusual positive sign and sufficiently small magnitude. At the end points A and B in p-u the isentropes (acoustic loci) $dp/du = -\rho c$ are tangent to the Hugoniot curve. Points C (vertical tangent in p-v) and D (same in p-u) are included to illustrate the mapping:

[2] A second part of the formal result is

$(dp/dv)_H < 0$ and $|dp/dv|_H < |(p - p_o)/(v - v_o)|$.

that is, the slope of the Hugoniot curve has the normal sign, but is less steep than the Rayleigh line. Thus the flow is supersonic, the shock boundary condition required to determine the solution in the shocked material is lost, and the analysis does not apply.

Sec. 6A   STABILITY THEORY

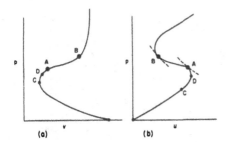

Fig. 6.17. Unstable Hugoniot segment AB, with tangent isentropes in p-u.

$$(dp/du)_{\mathcal{K}} = 2\rho_0 D/[1 - \rho_0^2 D^2/(dp/dv)_{\mathcal{K}}]$$

$$\rho_0^2 D^2 = -(p - p_0)/(v - v_0), \quad \rho_0 D = (p - p_0)/u.$$

The normal-modes treatment can be specialized in straightforward fashion to the case of longitudinal perturbations, that is, to infinite transverse wavelength. When this is done, the result is the incorrect one of neutral stability, independent of the shape of the Hugoniot curve.

Method (2) is a discussion, with reasonable approximations, of longitudinal stability in the presence of a rear boundary. Let a step shock be supported by a constant-velocity piston. Consider first the stable case, Fig. 6.18a. Increase the piston velocity an infinitesimal amount, from u to u + δu, generating the weak shock 1-2 overtaking the lead shock. The overtake raises the lead shock state to point 3, and sends the shock 2-3 back to the piston. This reflects from the piston as shock 4-5, and the process repeats, converging to Hugoniot point P where the velocity is u + δu. The contact surfaces generated at the overtakes are ignored in this discussion.

If the Hugoniot curve has negative slope, the process diverges. We distinguish two cases: (1) points of negative slope outside segment AB of Fig. 6.17, and (2) points within AB. In case 1, the process diverges, as shown (for the region to the left of B) in Fig. 6.18b, until the Hugoniot curve turns back. In case 2, Fig. 6.18c, the process diverges more strongly: the isentropes are now steeper than the Hugoniot curve and there would be no solution for the first overtake if the Hugoniot curve did not turn back.

Method (3) in its usual form looks for an alternate two-wave solution which satisfies the given rear boundary condition and whose lead shock moves faster than the single shock in question. If such a two-wave solution exists, it is assumed to be the preferred solution, so that the single shock is unstable. For our discussion we will take a constant-velocity piston as the rear boundary condition.

NONSTEADY SOLUTION   Chap. 6

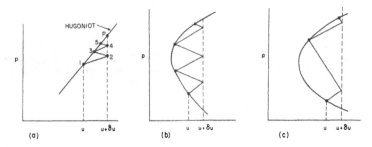

Fig. 6.18. The physical argument, method (2), for longitudinal instability with a rear boundary.

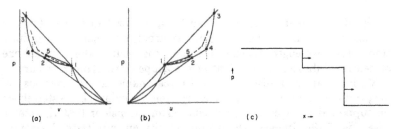

Fig. 6.19. Hugoniot curve for a material with a first-order phase transition with unstable segment 1-3. The dashed curve is the Hugoniot curve centered at point 1, and the dotted lines are the boundaries of the phase-transition region.

Fig. 6.19 shows an Hugoniot curve for a material with a first-order phase transition. The unstable segment is 1-3, point 3 being the end of the Rayleigh line through the cusp at point 1. We remark in passing that segment 1-4, spanning the transition region, is supersonic. A point in segment 1-3 can be reached either by a single shock or by a two-wave structure fulfilling the conditions given above. The two-wave structure, shown in the figure, has a weaker but faster leading shock to point 1, followed by a slower shock 1-5 to the point on the Hugoniot curve centered at point 1 having the given particle velocity. Its final state is subsonic.

For the case of Fig. 6.17, there is no two-wave solution satisfying the required conditions, but a different discussion indicates instability. The first step is to inquire into the possibility of single-shock *breakup*; that is, can the single shock, once established, break up into a two-wave structure having the same p-u state in the rear? The answer is yes, but the second wave moves backward instead of forward. Fig. 6.20 shows the possible modes of breakup. To facilitate illustration, we suppose that ρc, the slope of the p-u isentrope, indicated by the short segments crossing the

STABILITY THEORY

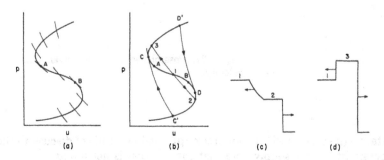

Fig. 6.20. Range and modes of possible shock breakup.

Hugoniot curve in Fig. 6.20a, varies as shown. Points A and B are the boundaries of the unstable region according to the linear analysis; we suppose that the isentrope is less steep in this range than outside it. Recall that the isentrope is steeper than the Hugoniot curve within this range and tangent to it at the end points. Now consider a shock to point 1, Fig. 6.20b. It could break up in two ways: (1) a weaker lead shock to state 2 with a back-facing rarefaction 1-2, Fig. 6.20c, or (2) a stronger lead shock to state 3 with a back-facing shock 1-3, Fig. 6.20d. Without considering all the cases in detail, we see that the range of possible shock breakup in at least one of these ways is much larger than AB. Tracing along the Hugoniot curve in the direction of increasing pressure, the first point at which a single shock could break up is point D, where the Hugoniot curve for the back-facing shock first intersects the original Hugoniot curve again at the higher point D'. At point D and for some way beyond it the shock can break up, but only in the second mode, Fig. 6.20d. At least one mode of breakup remains possible until we encounter point C', the last point at which the expansion isentrope intersects the original Hugoniot at a lower point C. Over some range before point C', the shock can break up, but only in the first mode, Fig. 6.20c. (In looking at the figure one must remember that isentropes may cross in the p-u plane; the value of ρc at a p-u point depends on the origin of the curve on which it lies.)

Method (3) equates instability with the possibility of shock breakup. It thus predicts the entire range CD as the unstable range, considerably larger than the unstable range AB given by the linear analysis.

We now ask: what is the solution for a given piston velocity between the extrema F and G, Fig. 6.21? There are three possible single-shock solutions 1, 2, and 3. Recalling the piston-reflection analysis discussed earlier, we see that if the center solution 2 were first established, breakup in the rarefaction mode, Fig. 6.20c, would eventually produce state 1, while breakup in the shock mode, Fig. 6.20d, would eventually produce

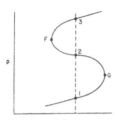

Fig. 6.21. Possible solutions for given piston velocity.

state 3 (if state 3 itself did not break up). Again we will not discuss all the possibilities, but simply note that the solution is not unique.

6B. APPROXIMATE THEORIES

The linear stability theories of the previous section determine whether perturbations on the one-dimensional steady solution will grow or decay, but give no information about the flow pattern which ultimately replaces the steady solution, other than the initial growth rate and oscillation frequency of small disturbances. What happens, as in other problems of this general type, is that an unstable steady solution is replaced by an oscillatory solution. The linear theory of course gives no information about the amplitude, but its predictions of frequency turn out to be surprisingly good.

In this section we present three approximate theoretical treatments of the oscillatory solution. The first is an extension by Erpenbeck of his linear theory to include some nonlinear terms. The analysis is lengthy and complicated. Application is limited to systems near the stability boundary having a single frequency of oscillation. In one dimension the results are quite good (as judged by comparison with numerical finite-difference calculations, Sec. 6C). The two-dimensional treatment suffers from the assumption at the outset of sinusoidal transverse spatial dependence, which excludes transverse shocks, but still gives results with some qualitative resemblance to those seen in real systems.

The second approach, by Strehlow and his coworkers, uses geometrical acoustics to examine the propagation of acoustic wave fronts through the reaction zone. This provides some interesting physical insights and a system of transverse acoustic waves with well-defined spacing again having some qualitative resemblance to the real thing.

The third treatment, by Toong and his coworkers, is based on the square-wave model, and considers only longitudinal instability, which is observed to occur ahead of blunt bodies propelled through detonable gas mixtures. This work goes beyond the study of the local stability criterion described earlier, providing an approximate calculation of repeated cycles of the oscillation, even in the case where two frequencies are present. The predicted frequencies are in reasonable agreement with experiment.

Sec. 6B APPROXIMATE THEORIES

6B1. Nonlinear Perturbation Theory

Erpenbeck (1967, 1970) extended his linear theory to include nonlinear terms, thus obtaining equations with amplitude limiting. Both the one- and two-dimensional cases are treated. A limitation applying to both is that the application is confined to cases close to the stability boundary having only a single root (actually a complex-conjugate pair) of the stability function $V(\tau,\epsilon)$ and thus a single frequency of oscillation. In the two-dimensional case there are the additional limitations that (1) transverse shocks are excluded at the outset by the assumption of sinusoidal transverse spatial dependence, and (2) consideration is limited to a suitably narrow range of finite transverse wavelengths such that the arbitrary selection of one of these, by imposition of a periodic boundary condition, makes it the only transverse wavelength present.

The analysis consists of expansion as in the linear theory but with nonlinear terms retained on the right side of the equations, followed by an iterative perturbation treatment. There are complications having to do mainly with ensuring boundedness at each order of perturbation and separating out terms, representing stable solutions, which decay with time. In the two-dimensional case the Fourier transform is replaced by a Fourier series over the finite range of the transverse coordinate y established by the assumption of a periodic boundary condition.

To obtain a tractable result, it is necessary to express the solution associated with the unstable (complex conjugate) pair of roots as the sum of two separate solutions, each given by a set of differential equations associated with only one of the two roots. Making this separation requires an arbitrary choice of how to split certain terms between the two sets of equations. The numerical results for one dimension presented below have been calculated for two of the many possible choices; those for two dimensions are for the choice which gives the best results in one dimension.

Carrying the perturbation to second order suffices to give the desired oscillatory perturbation of the shock front. It turns out, however, that the x-dependent perturbations of the state variables can be expressed in terms of these shock perturbations and the solutions $\varphi(x)$ of the first order (linearized) transformed equations, that is, just those of the linear stability theory.

The one-dimensional results have the form

$$D(t) = D_1(t) + D_2(t)$$

$$q(x,t) = D_1(t)\varphi_1(x) + D_2(t)\varphi_2(x).$$

NONSTEADY SOLUTION Chap. 6

Here D is the shock-velocity perturbation, the components of **q** are the perturbations in the state variables v, u, T, and λ, and φ is the (Fourier-Laplace) transform of **q** as given by the first-order (linearized) equations. The subscripts 1 and 2 refer to the two conjugate roots of τ. The quantities on the left are real and those on the right are complex. The φ's are solution of the (first perturbation order) linear differential equation and the D's of the (second perturbation order) nonlinear ones, both of whose coefficients are functions of x. The conjugate relation between the roots of τ allows the solution to be written as

$$D(t) = 2 \, \text{Re}[D_1(t)]$$

$$\mathbf{q}(x,t) = 2\text{Re}[D_1(t)\boldsymbol{\phi}_1(x)].$$

The differential equation for $D_1(t)$ is found to have a stable limit cycle in the complex D_1-plane, representing the desired oscillatory solution approached at long times.

In two dimensions the usual problem arises: to select from the continuous range of transverse wavelengths given by the linear theory a particular one. Generally some relatively arbitrary selection is made. Here this is done by first narrowing the field by considering only those systems sufficiently close to the stability boundary that the upper and lower bound on the range of unstable transverse wavelengths differ by less than a factor of two. Imposition of periodic boundary conditions and expansion in Fourier series then picks out only a single unstable wavelength. In general, the solutions found under these boundary conditions are traveling-wave solutions. Standing-wave solutions satisfying rigid wall boundary conditions can also be obtained by choosing suitably symmetric initial data, but, as we shall see, the fundamental and all even harmonics are in this case omitted from consideration.

The analysis starts by expanding the perturbations ψ in the shock position and **q** in the state variables (now including w, the transverse velocity component) in Fourier series in the transverse coordinate y over a strip of width l:

$$\psi(y,t) = \sum_{n=-\infty}^{\infty} \hat{\psi}(n,t) \, e^{i(2n\pi y/l)}$$

$$\mathbf{q}(x,y,t) = \sum_{n=-\infty}^{\infty} \hat{\mathbf{q}}(x,n,t) \, e^{i(2n\pi y/l)},$$

where the caret denotes a Fourier coefficient. The periodic boundary conditions are

Sec. 6B APPROXIMATE THEORIES

$\psi(y = 0,t) = \psi(y = 1,t)$

$q(x, y = 0, t) = q(x, y = 1, t).$

The nth term of the series represents a transverse wavelength l/n. The distance scale (steady half-reaction-zone length) is then chosen so that only wavelength l is unstable, that is wavelengths l/n are

stable for $|n| = 0, 2, 3, 4, \ldots$

unstable for $|n| = 1$.

(Here n = 0 of course means infinite transverse wavelength or longitudinal instability.) It is then found, as expected, that the Fourier coefficients for $|n| = 1$ grow with time, while all others decay. The parameter ϵ is defined as before, the unstable wavelength being

$l = 2\pi/\epsilon$.

The fundamental and first overtone in a rigid tube of width l, Fig. 6.22, correspond to n = 1/2 and n = 1 in this framework. Thus the fundamental and all odd harmonics (3/2, 5/2, ...) are excluded from consideration by the periodic boundary conditions imposed. The treatment is thus deficient in this respect. However, as we shall see in Chapter 7, the fundamental is difficult to generate in the narrow-channel geometry in which two-dimensional flow is most closely approached.

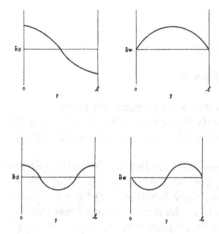

Fig. 6.22. Fundamental and first overtone pressure (δp) and transverse velocity (δw) perturbations in a rigid-walled tube of width l. The first overtone is the same as the n = 1 term of the two-dimensional theory, the longest wavelength allowed by periodic boundary conditions.

NONSTEADY SOLUTION   Chap. 6

The analysis is like that for the one-dimensional case, except that there are many more nonlinear terms. The general result has a similar form

$$\psi(y,t) = 2\text{Re}\{[\psi_1(t) + \psi_2(t)]e^{i\epsilon y}\}$$

$$q(x,y,t) = 2\text{Re}\{\psi_1(t)\phi_1(x) + \psi_2(t)\phi_2(x)]e^{i\epsilon y}\},$$

where ψ is the perturbation in the shock position and the subscripts 1 and 2 refer to the two (conjugate) roots τ_1 and τ_2 as before. Everything comes from the n = 1 term of the Fourier series, with $\psi_1(t)$ and $\psi_2(t)$ coming from ψ (n = ±1,t) and with $\epsilon = 2\pi/l$. Again $\psi_1(t)$ and $\psi_2(t)$ satisfy nonlinear differential equations arising from the second-order terms of the perturbation and $\phi_1(x)$ and $\phi_2(x)$ are given by the linear theory.

Analysis of these equations shows that there are limit cycles corresponding to both traveling- and standing-wave solutions. The traveling-wave solutions have the form

$$\psi(y,t) = 2\text{Re}[f_1(t) + f_2(t)]e^{i\epsilon(y-\omega t)}$$

$$q(x,y,t) = 2\text{Re}[f_1(t)\phi_1(x) + f_2(t)\phi_2(x)]e^{i\epsilon(y-\omega t)}$$

$$f_1(t) = a_1 e^{-(1/2)i\epsilon\omega t}, \quad f_2(t) = a_2 e^{-(1/2)i\epsilon\omega t},$$

with phase velocity ω. The coefficients a_1 and a_2 are (complex) constants, so that $f_1(t)$ and $f_2(t)$ have sinusoidal time dependence. The standing-wave solutions have the form

$$\psi(y,t) = 4\text{Re}[\hat{\psi}_1(t)]\cos \epsilon y$$

$$q'(x,y,t) = 4\text{Re}[\hat{\psi}_1(t)\phi'(x)]\cos \epsilon y$$

$$\delta w(x,y,t) = -4\text{Im}[\hat{\psi}_1(t)\phi(x)] \sin \epsilon y,$$

where the prime on q and ϕ indicates all components except δw, the transverse component of the velocity perturbation, which has a slightly different form and is given separately. [For this case $\hat{\psi}_2(t)$ is the complex conjugate of $\hat{\psi}_1(t)$.]

The one-dimensional theory has been applied to the polytropic-gas system of Sec. 6A, with $\gamma = 1.2$, q = 50, f = 1.6, and $E^\dagger$ = 45, 46, 50. Table 6.1 gives the roots of the stability function $V(\tau)$ from the linear theory and the oscillation periods from the linear and nonlinear theories (all rounded to three places). The linear-theory period is just $2\pi/\text{Im}(\tau)$.

Sec. 6B
APPROXIMATE THEORIES

Table 6.1 PERIODS OF THE ONE-DIMENSIONAL OSCILLATORY SOLUTION.

$E^\dagger$	$\tau\, t_{1/2}$	Period/$t_{1/2}$		
		Linear	Nonlinear	
			Split "a"	Split "b"
44.8	$0 \pm 0.798i$	7.88	---	---
45	$0.005 \pm 0.798i$	7.88	7.86	7.85
46	$0.027 \pm 0.797i$	7.88	7.77	7.75
50	$0.121 \pm 0.789i$	7.97	7.49	7.41

The two values from the nonlinear theory correspond to two choices for the arbitrary splitting of the nonlinear terms, which we here designate as splittings "a" and "b." The neutral stability point at $E^\dagger = 44.8$ is also included. The linear theory is seen to overestimate the period, with the error increasing as we move away from the stability boundary. Figure 6.23 shows the v and u components of ϕ for $E^\dagger = 50$, and the shock pressures over a cycle for $E^\dagger = 45, 46,$ and 50. The bar over λ denotes the unperturbed composition. Splitting "a" gives better results as judged by comparing the shock pressure with the finite-difference calculation of Sec. 6C1, which it reproduces reasonably well.

The two-dimensional theory has been applied to the polytropic gas system with $\gamma = 1.2$, $E^\dagger = 50$, $f = 1.1$, and the values of the heat release q listed in Table 6.2. The unstable range of ϵ is shown in the third column, and the values of ϵ used in the calculation in the fourth column. The traveling- and standing-wave solutions are given for each case. The traveling-wave solutions are exactly sinusoidal and the standing-wave ones are within one per cent of being so. Here p_{max} is the maximum pressure and ω is the phase velocity; p_+ and c_+ are pressure and sound speed immediately behind the shock in the steady solution.

Although the stability properties of the two types of solutions have not been completely established, the situation under periodic boundary conditions appears to be as follows: omitting for the moment the last two lines of the table, each case has both a traveling and a standing solution, but only one of these is stable; presumably a system set up in the unstable solution would revert to the stable one. Starting at the top of the table, the traveling solution is the stable one until we come to q = 2, $\epsilon = 1.3$, where the situation is reversed. For the last two lines, the traveling solution remains unstable but there is no standing solution. (Values of δp and ω are not given in the original paper for the unstable traveling-wave solutions). This last result is suspect since the amplitudes here are so large and the small-perturbation assumption is violated.

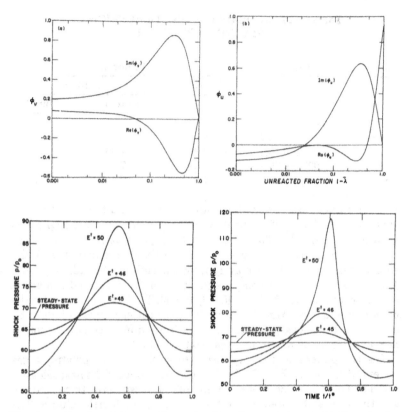

Fig. 6.23. Results of the one-dimensional nonlinear theory. (a,b) The v and u components of the transformed state perturbations φ vs. $(1-\bar{\lambda})$ [distance from the shock is roughly proportional to log $(1-\bar{\lambda})$]. (c,d) One cycle of shock pressure vs. time (in units of the period t*) for splittings "a" and "b." From Erpenbeck (1967).

Figure 6.24 shows one traveling-wave solution. The reference frame is traveling with the transverse wave at its velocity across the steady frame; it moves in the x-direction at the steady detonation velocity, and in the y-direction at the traveling wave velocity. A streamline of the flow in the unreacted explosive therefore makes an angle with the reference frame, as shown in Fig. 6.24a. The very slightly curved, and not quite parallel, streamlines in the reaction zone are also shown. The width of the diagram is one wavelength, and the full solution, satisfying the periodic boundary conditions, is an infinite repeating array. One might expect to find behavior something like the full solution in a very large tube, persisting until it reached a boundary, or to find reflection, with waves

Table 6.2 RESULTS FROM THE TWO-DIMENSIONAL THEORY

Parameters				Travelling-Wave Solution			Standing-Wave Solution	
$q/p_o v_o$	$Et/p_o v_o$	Unstable ε-range	ε	$p_{max}/\bar{p}_+$	ω/c_+	Stability	$p_{max}/\bar{p}_+$	Period/$t_{1/2}$	Stability
0.3	50	0.51 - 0.92	0.7	1.094	0.624	s	1.11	11.39	u
			0.9	1.065	0.608	s	1.07	9.12	u
0.5	40	0.31 - 1.35	0.7	1.208	0.703	s	1.23	9.63	u
2.0	20	0.18 - 1.60	0.9	1.565	0.897	s	1.56	3.73	u
			1.1	1.647	0.874	s	1.82	4.08	u
			1.3			u	2.11	3.41	s
			1.4			u			
			1.5			u			

traveling in both directions. Fig. 6.24b shows the pressure profiles along lines of constant y at an instant of time, spaced uniformly from y = 1 to y = 0.11. The abscissa is $(1 - \bar{\lambda})$ instead of x, as in Fig. 6.23 a, b, and the shock is at the right-hand edge of the diagram. The wave front itself is symmetric about 1/2, but the pressure curves are not because the wave is moving in the y-direction, and the streamlines of the unreacted explosive intersect the shock at different angles over the full wavelength.

What about the applicability of these solutions to rigid-wall boundary conditions? We can make a few precise statements, plus some guesses based on the nature of superposition in the linear case and (it must be admitted) on what is observed experimentally. The standing-wave solution satisfies rigid boundary conditions for channels of width

k(1/2), k = 1, 2, 3, 4, ...

With rigid boundary conditions imposed, all the standing-wave solutions become stable (with the *caveat* that the theory as formulated excludes the fundamental mode). A possible solution for a channel whose width is a half-integral multiple of any wavelength in the unstable range is the appropriate number of laterally-stacked standing-wave solutions (with half a solution at one edge for k odd). The traveling-wave solution does not, of course, satisfy rigid-boundary conditions. But for systems with a range of unstable wavelengths covering many terms of the Fourier series,

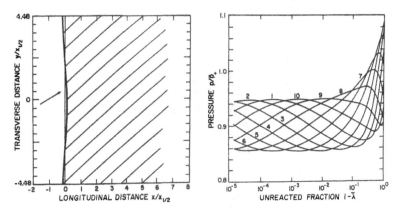

Fig. 6.24. Traveling-wave solution from the two-dimensional theory for q = 0.3, ϵ = 0.7. (a) Shock and streamlines in the steady frame. The light vertical line is the unperturbed shock; the arrow gives the direction of the unperturbed flow. (b) Pressure profiles (shock at right). Distance from the shock is roughly proportional to log $(1-\bar{\lambda})$. The labels 1 through 10 denote equally spaced values of y from y = 1 to y = 0.11. From Erpenbeck (1970).

Sec. 6B APPROXIMATE THEORIES

one could have, in wide tubes, traveling-wave packets which move back and forth, reflecting from the walls. One might expect to detect semblances of such packets as components of the standing-wave solution.

6B2. Geometrical Acoustics

A study by Barthel and Strehlow (1966) and Strehlow and Fernandes (1965) of the lateral propagation of acoustic wave fronts in the reaction zone suggests a way in which regularly spaced transverse waves could originate, and shows what properties of the steady solution are important to the process. This work does not provide a complete solution of even the linearized equations of motion. It treats only the wave fronts, following them by ray-tracing methods analagous to those of geometrical optics. The treatment is two-dimensional — the wave fronts considered are initiated by line sources parallel to the shock. We take the shock normal to the x-axis and the source parallel to the z-axis and describe the wave front by its cross section in the x-y plane.

There is a large change in the acoustic index of refraction through the region of rapid reaction — the "fire" — which acts like an inversion layer in the atmosphere, trapping some rays between itself and the shock. The trapped rays, bouncing back and forth between the fire and shock, lose energy at each reflection from the shock, and require amplification at the fire reflection to retain their strength. If sufficient amplification is available the propagation is sustained, and the initially circular wave front eventually convolutes near the shock into a set of fronts with periodic transverse spacing.

There are some interesting connections with Erpenbeck's (1966) short wavelength analysis: the most obvious is that the plane at which $c^2 - u^2$ has an extremum in the steady solution (see Fig. 6.2) plays an important part in the ray trapping.

6B2.1 Ray Theory

We present here just the main ideas. For details, the reader is referred to the original papers, or to the brief review by Engelke (1974).

We switch on our source instantaneously and follow the resulting wave front. The amplitude must of course be small enough that the acoustic approximation holds. Thus we are concerned with the propagation of acoustic waves in the high frequency limit, and wave fronts follow the characteristic surfaces. For a medium in which the frozen sound speed c and particle velocity **u** are functions of position **r**, the rays propagate according to

$$d\mathbf{r}/dt = \mathbf{w}(\mathbf{r}) = c(\mathbf{r})\mathbf{n} + \mathbf{u}(\mathbf{r}),$$

where **r** is the position vector of a point on a ray, and **n** is the unit outward normal to the wave front at **r**, Fig. 6.25. The problem to be solved begins with a wave front surface of a given shape at a given time, and current ray points distributed on this surface in such a way that their density is everywhere proportional to the intensity. In a time dt, each ray moves out a distance d**r**, and the new wave front is obtained by connecting all the new ray positions. In the application at hand, with x the distance behind the shock, we have c = c(x), u = u(x) from the steady solution (with u the x-component of **u** and the other components zero). Then

$$d\mathbf{r}/dt = \mathbf{w}(x) = c(x)\mathbf{n} + u(x)\mathbf{i}.$$

As stated earlier, we treat the two-dimensional case, with all **n** lying in the x-y plane and the wave fronts cylinders perpendicular to this plane.

As a simple example, take the case c = constant, u = constant ≠ 0, Fig. 6.26. The wave front from a point source (projection of a line source) is the circle of radius ct with its center displaced downstream a distance ut from the source. The rays are straight lines, but are not, as in the geometrical optics of an isotropic medium, normal to the wave front. In general, the rays are curved.

Fig. 6.27 is another example showing how a convolution arises in the wave front. A sound-speed gradient in the atmosphere acts as a reflecting layer. The convolution is formed essentially by the arrival of different rays at the same point via different paths. The cusps are points of increased intensity.

A ray may be said to describe the propagation of an element of the wave front with the group velocity **w**. Energy also moves with the group velocity and moves along the rays. The intensity at any point on the wave front is the density of rays there. The application of this property to the geometrical optics of a uniform medium is just the familiar inverse-square law. These statements are precisely true only for a nonreactive medium. With reaction, the propagation of the wave front, which moves along characteristics, is unchanged, since the characteristics move with

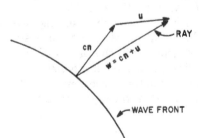

Fig. 6.25. Ray propagation in geometrical acoustics.

APPROXIMATE THEORIES

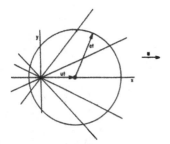

Fig. 6.26. Wave front and rays from a line source at $x = y = 0$ in a medium with constant c and u (in the x direction). The wave front is a circle of radius ct with center displaced from the origin by ut. The rays are straight lines through the origin.

Fig. 6.27. Convoluted wave front produced by a sound-speed gradient in the atmosphere acting as a reflecting layer. (a) Sound speed as a function of height. (b) Ray paths for a source at $x = y = 0$. (c) Convoluted wave front.

frozen sound speed. But the group velocity is more like $c_e \mathbf{n} + \mathbf{u}$, with c_e the equilibrium sound speed, so the intensity properties are probably somewhat modified. We are not aware of any careful study of the geometrical acoustics approximation in a reactive medium. However in the application at hand we are more concerned with the position of the wave front than with the intensity.

6B2.2 Ray Trapping and Amplification

In the first paper Strehlow and Fernandes (1965) take up two questions: (1) Omitting the question of reflection at the shock, how do acoustic rays propagate in the reaction zone? (2) Under what conditions does amplification of the acoustic signals take place?

Examination of the equations reveals a critical plane at the point where $c^2 - u^2$ has an extremum in x: a ray moving in this plane remains in it forever. The propagation is shown in Fig. 6.28 for the CJ detonation

$$\gamma = 1.4, \quad q = 24.8 \ p_0 v_0 \ (D = 6 \ c_0)$$

$$r = k\rho e^{-E^\dagger/RT}, \quad E^\dagger = 24.8 \ p_0 v_0.$$

For a source between the shock and the critical plane, Fig. 6.28a, there is a critical angle; rays with smaller angles return to the shock and those with larger angles pass through the critical plane and leave the reaction zone. The ray starting at the critical angle approaches the critical plane

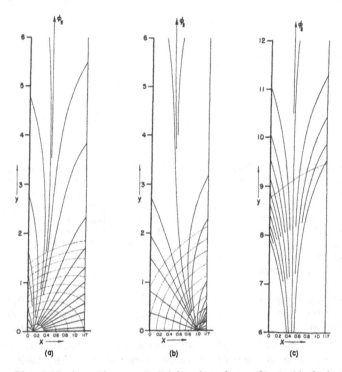

Fig. 6.28. Acoustic rays (solid lines) and wavefronts (dashed lines) in a steady CJ reaction zone from a line source perpendicular to the page. The distance unit is arbitrary but x and y are plotted to the same scale. (a) Source at $\lambda = 0.1$, between the shock and the critical plane. (b) Source at $\lambda = 0.9$, behind the critical plane. (c) Late behavior of the ray bundle (from any source) around the critical plane. From Strehlow and Fernandes (1965).

asymptotically. The situation for a source behind the critical plane, Fig. 6.28b, is similar but reversed. A slightly divergent ray bundle centered on the critical plane, Fig. 6.28c, continues to diverge. Note that the region behind the critical plane has no effect on what goes on in front of it.

There is some reason to believe that most of the amplification takes place near the critical plane, where the reaction is most rapid. What is studied is the amplification in a narrow bundle of rays centered on the critical plane. By means of an intuitive physical argument, this ray bundle is approximated as a flow in a hypothetical diverging channel with a fixed reaction rate corresponding to that at the critical plane. The

Sec. 6B APPROXIMATE THEORIES

growth rate of acoustic waves propagating down this hypothetical channel is investigated. Two reaction rate forms are considered

(1) $r = k\rho^{n-1} e^{-E^\dagger/RT}$

(2) $r = k\rho^{n-1} T^b$,

the second being typical of recombination reactions. The acoustic amplitude is found to be e^{At}, with

$A = 1/2 \, (\gamma-1)(qr/c^2)[(\gamma-1)z + n - (2+\gamma)^{1/2}]$

$z = E^\dagger/RT$ for reaction (1)

$z = b$ for reaction (2),

where r and c are the steady-solution values at the critical plane. Curves of the neutral stability value of z vs. γ for n = 1, 2, and 3 are given. For positive qr and γ = 1.4, the critical z for n = 1 is 2.11. For n = 2 it is −0.390. Thus not much sensitivity to temperature is required for amplification; for k = 2 there is amplification even for slightly negative b or $E^\dagger$.

6B2.3 Formation of Transverse Waves

Barthel and Strehlow (1966) extended the above calculation to long times, taking into account the reflection of the acoustic waves from the shock. They used the same CJ detonation, but with a specified $\lambda(t)$ with an induction zone

$\lambda = 0, \; 0 < t < t_i$

$\lambda = 1/16 \, (\cos 3\pi\theta - 9 \cos \pi\theta + 8), \; t_i < t < t_r$

$\theta = (t - t_i)/(t_r - t_i)$.

Here t_i is the induction time and $t_r - t_i$ the reaction time, and the trigonometric representation of λ in the reaction zone is chosen to make the first three derivatives of λ vanish at its boundaries. The corresponding $\lambda(x)$ is diagrammed in Fig. 6.29.

A source at x = y = 0 (that is, at the shock) is pulsed at t = 0. The resulting wave fronts at a sequence of times are shown in Fig. 6.30. (The picture is of course symmetric about y = 0; only the upper half is shown.)

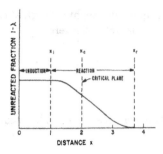

Fig. 6.29. Induction and reaction zones of the steady solution for the assumed λ(t).

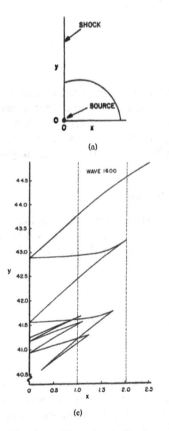

Fig. 6.30. Wave fronts from a source at $x = y = 0$. The distance unit is the same as Fig. 6.29, where x_i and x_c are defined. (a) Magnified view at infinitesimal time. (b), (c) Snapshots at later times; times in arbitrary units. From Barthel and Strehlow (1966).

Sec. 6B   APPROXIMATE THEORIES

Fig. 6.30a shows an expanded view at infinitesimal time. In Fig. 6.30b we see a convolution forming near the end of the induction zone. The mechanism is similar to that shown in Fig. 6.27. The convolution grows and moves toward the shock; at the next time shown its left edge has reflected from the shock. As time proceeds, Fig. 6.30c, the wave front separates into two branches. The lagging branch consists of a number of closely bunched reflections. In this region each reflection leaves the amplitude unchanged but changes the sign, so that this branch dies out by cancellation. The leading branch consists of a set of waves periodic in y and of the same asymptotic shape. In Fig. 6.30c, the second wave of this set is emerging, not yet quite fully formed, from the lagging branch. On the leading branch the reflection at the shock produces a loss in amplitude but leaves the sign unchanged. If this reflection loss is made up by amplification in the reaction zone, this branch will grow. We have then a growing set of equally spaced transverse waves.

Although they present detailed results only for this case, the authors discuss the dependence on the sign and magnitude of heat release and (for the exothermic case) the degree of overdrive. For a slightly endothermic reaction there are no trapped rays. As the endothermicity is increased, rays are trapped, but for reasonable reaction rates there is no amplification. Thus endothermic systems are expected to be stable.

For exothermic systems, consider first the CJ case. For small enough heat release, in the range corresponding to profile D, Fig. 6.2, there is no critical plane in the reaction zone, and there are no trapped rays. Hence this case is stable. For larger heat release at CJ we have profile I, with a critical plane in the reaction zone and trapped rays. Instability then requires sufficient amplification in the reaction zone to offset the reflection loss at the shock. A quantitive analysis of amplification has not been made, but the qualitative analysis for rays near the critical plane shows that a modest temperature sensitivity of the reaction rate will produce amplification. Thus we may expect instability here for typical systems.

With increasing overdrive, the critical plane moves to the rear and finally disappears as the profile changes from type M to type I. In this case, there are still some trapped rays, but both the trapping and the amplification are less effective, so that sufficient overdrive is expected to stabilize the steady solution.

Barthel (1974) proposed a different criterion for obtaining a transverse-wave spacing from the same type of calculation. The wave front from a source between the shock and the critical plane is symmetric about the x axis, so either half may be considered. Two convolutions appear at separate points on the wave front after it has begun to reflect from the shock. The upstream one (nearest the shock) comes from rays which were initially directed upstream and have reflected from the shock. The downstream one, which forms later, comes from rays which were initially directed downstream and have reflected from the fire. These two convolutions move toward each other and eventually collide. The point at which their edges first collide will be a hot spot because of the energy concentration there. This hot spot turns out to be at the same x (distance from the shock) as the source, and both this property and its transverse distance y from the source are independent of the position of the source so long as it is in the interior of the interval. If the hot spot is at y then it has a mirror image at $-y$. The transverse wave spacing is taken to be 2y, the distance between the hot spots.

6B2.4 Results

The transverse wavelength (distance between shock contacts) calculated by Barthel and Strehlow (1966) is shown in Fig. 6.31 as a function of the ratio of reaction-zone to induction-zone length for three values of γ and two values of the detonation Mach number D/c_0, related to q by

$$q/c_0^2 = (D^2/c_0^2 - 1)^2/[2(\gamma^2 - 1)D^2/c_0^2].$$

There are several points of contact between this work and Erpenbeck's asymptotic short wavelength theory. The wavelength is the same as the longest unstable Erpenbeck wavelength. The condition for amplification at the critical plane depends on the steady solution only at this point and has this feature in common with Erpenbeck's sufficient condition for stability. But the critical value of $E^\dagger$ is, in one case which has been compared, an order of magnitude larger than Erpenbeck's value. Finally, Erpenbeck has pointed out to us that the integral for the contact spacing Δy (see Strehlow, Maurer, and Rajan [1969])

$$\Delta y = 2 \int_0^{x^*} [\eta^{-1}(\eta^*/\eta - 1)]^{1/2} \, dx$$

$$\eta = 1 - M^2; \quad M = u/c,$$

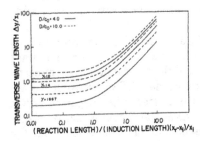

Fig. 6.31. Transverse wavelength (shock-contact spacing) vs. (reaction-zone)/(induction-zone) length. From Barthel and Strehlow (1966).

Sec. 6B　　　　　　　　　　　　　　　APPROXIMATE THEORIES

where u is the steady-frame velocity and an asterisk denotes the value at the critical plane where η is a maximum, also appears as one term of the Erpenbeck stability function $V(\tau, \epsilon)$.

Barthel (1974) applied his hot spot method to $H_2/O_2/Ar$ with a realistic representation of the steady reaction zone. He obtained spacings considerably larger than those from the shock contacts of Barthel and Strehlow. With increasing overdrive the calculated spacing goes through a minimum at $D/D_j < 2$ and then rapidly increases to infinity as the critical plane is removed and the ray-trapping property is lost. Chiu and Lee (1976) repeated this calculation for simplified models of the reaction zone, including the square-wave limit approached as the ratio of recombination to induction-zone length approaches zero with increasing pressure. In this approximation the calculated spacings are somewhat smaller.

6B3. One-Dimensional Oscillation in the Square-Wave Detonation

In attempting to explain the beautiful and complicated flow patterns around blunt projectiles fired into gaseous explosives (see Chapter 7), Alpert and Toong (1972) have worked out a complete semiquantitative picture of the mechanism of longitudinal oscillation in a square-wave detonation. They have calculated two frequencies of oscillation which are in reasonable agreement with the observed ones for hydrogen/oxygen, for which the square-wave model is a fairly reasonably approximation.

The simpler higher-frequency mode of oscillation is shown in the t-x diagram of Fig. 6.32. Referring first to Fig. 6.32b, we may take a cycle to

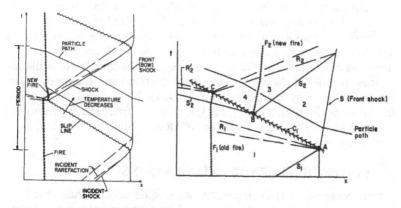

Fig. 6.32. Wave diagram for the simple (higher frequency) oscillation. (a) Complete cycle. (b) Details of the interactions. From Alpert and Toong (1972).

begin with the interior shock S overtaking the front shock at point A. The overtake produces a weak rear-facing rarefaction R_1, whose effects can be neglected, an acceleration and strengthening of the front shock, and a contact discontinuity above which the fluid is hotter from passing through the strengthened front shock. The first hotter particle reacts at point B, initiating the new fire (reaction surface) F_2, and the front- and rear-facing shocks S_2 and S_2' required for matching. The p-u diagram with state 2 as origin is shown in Fig. 6.33. The matching is similar to that described in Sec. 6A2. Here, however, the fire is a supported detonation with no following rarefaction, for the CJ point on the deflagration Hugoniot lies well below its intersection with the S_2' shock Hugoniot curve at point 4. The original fire terminates at point C, Fig. 6.32b, with the reaction of the last particle passing through the undisturbed front shock. Its termination generates the front- and rear-facing rarefaction waves R_2 and R_2'. The overtaking of the front shock by S_2 initiates the next cycle. The timing of rarefaction R_2 is important. If points B and C are close together, R_2 can overtake and effectively destroy S_2 before S_2 reaches the front. Thus the steady solution is stable with small enough temperature sensitivity of the rate. With the timing shown in Fig. 6.32b, the main effect of R_2 is to decelerate the front shock back toward its original strength after S_1 has overtaken it. The repeated cycle is seen in Fig. 6.32a.

The lower frequency mode of oscillation, Fig. 6.34, is more complex, and actually contains two frequencies: one approximately that of the simple mode, and one several times lower, with the ratio probably depending on both the kinetics and the sound speed. Here the first interior shock S_1 is so strong that the second fire F_2 starts quite early, so that the next interior shock S_3 overtakes R_2 from behind. Several of the basic cycles are required to get back to the original state.

The frequencies and approximate amplitudes of these oscillations are determined through approximate calculations of all the wave interactions and reaction times. For hydrogen/oxygen, the calculated oscillation periods in units of the steady solution induction time are about 1.6 for the high-frequency oscillation, and between 3.8 and 5.4 for the low-frequency oscillation. The typical jumps in the front shock pressure at overtake are on the order of 4% and 18% for the high- and low-frequency modes, respectively.

6C. FINITE-DIFFERENCE CALCULATIONS

One-dimensional time-dependent finite-difference calculations have been made for a variety of equations of state and reaction rates. In two dimensions there has been only a single preliminary attempt.

Sec. 6C FINITE-DIFFERENCE CALCULATIONS

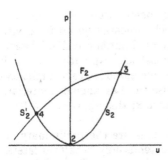

Fig. 6.33. The p-u diagram for the formation of the new fire F_2 and associated shock S_2 and S_2' of Fig. 6.32. The curves S_2 and S_2' are the Hugoniot curves centered at 2, for forward and backward shocks. Curve F_2 is the deflagration Hugoniot curve centered at 3.

Fig. 6.34. Wave diagram for the low-frequency oscillation. From Alpert and Toong (1972).

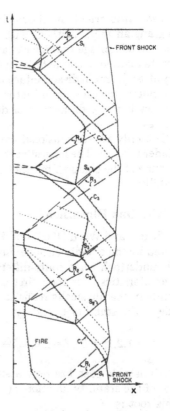

The one-dimensional calculations have correlated nicely with the results of the linear stability analysis, provided a detailed picture of the oscillatory solution, furnished a testing ground for approximate treatments such as Erpenbeck's nonlinear analysis and Toong's square-wave treatment, and located portions of stability boundaries.

Fickett and Wood (1966) calculated the system treated by Erpenbeck (γ = 1.2 polytropic gas, first-order Arrhenius rate), and were guided by his results, Sec. 6A3. Fickett, Jacobson, and Schott (1972) performed similar calculations with a simple rate law intended to mock up branching-chain kinetics and thus produce the induction zone typical of hydrogen/oxygen systems. Chernyi and Medvedev (1970) made calculations for a γ = 1.4 polytropic gas with a half-order Arrhenius rate. Howe, Fyre, and Melani (1976) extended the work of Fickett and Wood to unsupported detonations and to γ = 1.4 and an induction-zone rate. Il'kaeva and Popov (1965) used a square-wave rate law, and compared their results with the simple stability criterion of Sec. 6A2. Finally, Mader (1965) calculated oscillatory solutions for condensed explosives.

Two numerical methods have been used. The conventional Lagrangian mesh method, with shocks smeared by artificial viscosity, has the advantage that shocks are treated automatically. The more accurate method of characteristics treats shocks explicitly as jump discontinuities. Its main disadvantage is the increased logical complexity of the computer program. Both methods use the Lagrangian (material) coordinates h = ∫ ρ dx instead of distance x; this appears in a few of the figures.

The only two-dimensional calculations are the preliminary ones by Mader (1967b). The longest of these extends only to half the lateral transit time of the channel, but they do appear to show the initial onset of instability.

6C1. One Dimension

Fickett and Wood (1966) and Fickett, Jacobson, and Wood (1970), inspired by the one-dimensional instability discovered by Erpenbeck, Sec. 6A, undertook some finite-difference calculations using the method of characteristics. For their first calculation they chose two neighboring points in the physically realistic region of parameter space, one on each side of the stability boundary:

$$\gamma = 1.2, q = 50, f = 1.6, E^\dagger = 40 \text{ (stable)}, E^\dagger = 50 \text{ (unstable)}.$$

The unstable case has only a single frequency of oscillation, only one root of the stability function $V(\tau)$ having entered the right half τ-plane. This root is

$$\tau = 0.113 + 0.789i,$$

with a growth-rate e-folding time of $[\text{Re}(\tau)]^{-1} = 8.85$ and a period $[2\pi/\text{Im}(\tau)] = 7.97$. Other roots "waiting in the wings" are shown in Fig. 6.35. For the above parameter values with $E^\dagger = 40$, all roots are in the left half-plane with this one [root (1)] about to cross over. As the parameters are changed to move the system into the unstable regime the roots move as shown by the arrows, with the lower-frequency roots crossing the imaginary axis first. As the degree of overdrive f is lowered at constant $E^\dagger = 50$, for example, roots of higher frequency [larger $\text{Im}(\tau)$] cross over into the right half-plane and these frequencies appear in the oscillatory solution. Fig. 6.36 shows all the points in the $E^\dagger$-f plane for $\gamma = 1.2$, q = 50 for the calculations described here, together with the stability boundary and three other points calculated by Erpenbeck.

Fig. 6.37 shows the partial-reaction Hugoniot curves for this system, with Rayleigh lines for f = 1.0, 1.6, and 2.0, and also the steady solution

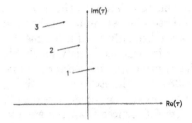

Fig. 6.35. Schematic diagram of the movement of the roots in the τ-plane as the parameters are varied so as to make the system more unstable. The roots of lowest frequency [smallest Im(τ)] cross over first. The lowermost root (1) is the main contributor to the first pair of calculations, although the frequency of (2) also appears as a decaying mode.

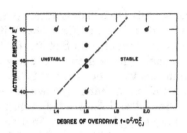

Fig. 6.36. The stability boundary (dashed line) in the $E^\dagger$ - f plane for $\gamma = 1.2$, $q = 50$ (recall that the energy unit is p_0v_0). Erpenbeck, Sec. 6A3, has determined the stability for all of these points; the time-dependent calculations described here have been done for the flagged points. From Fickett and Wood (1966).

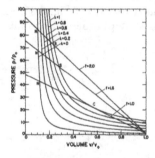

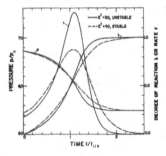

Fig. 6.37. (a) Partial-reaction Hugoniot curves and Rayleigh lines for three degrees of overdrive f (= D^2/D_j^2) for $\gamma = 1.2$, $q = 50$. (b) Steady-solution particle histories for $E^\dagger = 40$ and $E^\dagger = 50$ at f = 1.6. From Fickett and Wood (1966).

for $E^\dagger = 40$ and 50 (at f = 1.6). Notice that this change in $E^\dagger$ makes only a modest change in the steady solution.

For this and all of the following calculations by these workers, the initial condition is the explosive in the initial (spatially uniform, unreacted) state. The (rear) boundary condition is a piston whose velocity is a specified function of time; its motion generates the detonation. If this specified piston velocity function is made identical with that of a particle passing through the steady solution, then the steady solution is generated. The steady-solution particle-velocity history for the two cases at hand is shown in Fig. 6.38a. With a first-order rate function the final velocity (that of point S of Fig. 6.37a) is reached only at infinite time; for reasonably large activation energies it is in practice indistinguishable from this value beyond t = 2. The piston prescription used for the two calculations at hand is shown in Fig. 6.38a; it is the straight line joining the initial and final steady-solution values over the range t = 0 to 2, followed by the constant final value. Its differences from the steady-solution prescriptions constitute fairly small perturbations on the steady solution.

Fig. 6.38b and c show the results. The $E^\dagger$ = 40 case is found to be stable as predicted, with the perturbation decaying and the motion settling down into the steady solution at about t = 20. The $E^\dagger$ = 50 case is dramatically unstable, with the perturbation growing rapidly into a final peak-to-peak amplitude about three-fourths of the steady-solution pressure. The final amplitude is reached in about the same time, t = 20. In one time unit (half-reaction time) the shock traverses about 4 half-reaction distances. The main period of oscillation is about 8 time units, essentially the same as that given by the linear theory. The growth rate is also about the same. The early high frequency oscillation, which quickly damps out, corresponds to the next root (2) of Fig. 6.35. An interesting result is that, in spite of the large amplitude and nonlinearity

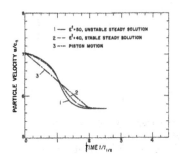

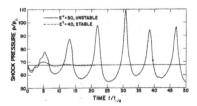

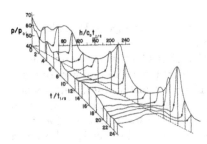

Fig. 6.38. Time-dependent finite-difference calculation for γ = 1.2, q = 50, f = 1.6, $E^\dagger$ = 40 (stable) and $E^\dagger$ = 50 (unstable). (a) Piston velocity prescription, with the two steady-solution particle histories for comparison. (b) Shock-pressure histories. The horizontal line is the steady-solution value. (c) Pressure profiles in Lagrangian (material) coordinate h at fixed times for $E^\dagger$ = 50. The circles mark the points where the reaction is 99% complete. From Fickett and Wood (1966).

Sec. 6C FINITE-DIFFERENCE CALCULATIONS

of the oscillation, the time-average shock pressure is within one percent of the steady-solution value.

In another calculation, Fig. 6.39, a stable solution was first established and then caused to go over into an unstable one by reducing the piston support. The $E^\dagger = 50$ solution is stable at $f = 2$. It was first established by programming the piston to follow its particle history. The piston velocity was then decreased rapidly but smoothly to that for the $f = 1.6$ steady solution. The characteristic oscillation appears shortly after the rarefaction produced by the decrease in piston velocity overtakes the shock.

The irregularity of the oscillation in Fig. 6.38 is due to the presence of the driving piston. The large oscillations at the front emit small waves to the rear which reflect off the piston and return to overtake the shock at different places in its natural cycle. A regular oscillation, Fig. 6.40, is obtained if the piston is in effect removed. This was done by using a rear-facing characteristic as the rear boundary, along which the steady-solution particle velocity (appropriate to a characteristic path) was prescribed. Since signals traveling to the rear move at characteristic velocity there is then nothing for them to reflect from. It turns out that numerical noise—truncation errors in the finite-difference method—furnishes a sufficient perturbation to produce the oscillatory solution in a reasonable time.

The internal structure is shown in Fig. 6.40b. Starting at the trough, with relatively long reaction zone and low pressure, a compression wave forms at the rear of the reaction zone and moves forward, growing

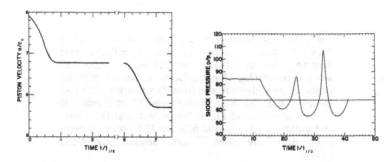

Fig. 6.39. Stable to unstable transition via reduced piston support. (a) Piston velocity: (1) $t < 6$: follows the supported steady-solution for $f = 2.0$, establishing this stable steady solution, (2) $t \geq 6$: decreases smoothly through a quarter sine wave to the support value for $f = 1.6$, generating a rarefaction wave which triggers the transition to the unstable solution for this drive. The time unit $t_{1/2}$ is that for $f = 1.6$. (b) Shock-pressure history. From Fickett and Wood (1966).

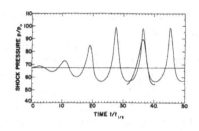

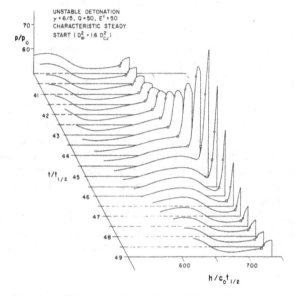

Fig. 6.40. The "natural" regular oscillation for $\gamma = 1.2$, $q = 50$, $f = 1.6$, $E^{\dagger} = 50$ obtained with a nonreflecting rear boundary (steady-solution particle velocity specified along a rear-facing characteristic). (a) Shock-pressure history. The added curve is from Erpenbeck's nonlinear theory, Fig. 6.23d. From Fickett and Wood (1966). (b) Profiles like those of Fig. 6.38b, except that the circles and squares mark points of 50% and 99% complete reaction, respectively. From Fickett, Jacobson, and Wood (1970).

rapidly in amplitude as it is fed by the reaction accelerated by compression heating, and carrying the fire forward with it. By the time it has reached the front, the pressure and temperature are high, the reaction zone is short, and the configuration is approaching that of an unsupported detonation running forward through the partially reacted material ahead of it, and followed, of course, by a rarefaction (like the Taylor wave). Since it is growing so rapidly, the reaction-zone profile becomes steeper than it would be in the steady state. The shock pressure peaks as the shock is overtaken by this interior detonation, and is then rapidly reduced by the steep reaction-zone gradient and following rarefaction. The entire configuration then drops back into the trough and is ready to start again. The compression wave emitted to the rear appears to form at the end of the reaction zone at about the time of overtake. Its peak-to-trough pressure range is 52 to 58 (symmetric around the steady complete-reaction value of 55). The corresponding range at the shock is 57 to 102 (around the steady value of 67), and T/T_o there varies from about 7 to 11, changing the reaction rate by a factor of 20. Additional details may be found in Fickett, Jacobson, and Wood (1970).

Fig. 6.40a also compares the calculated pressure with Erpenbeck's nonlinear theory, Sec. 6B1. The two agree fairly well.

Fickett, Jacobson, and Wood (1970) tried to obtain an accurate value of the period by extrapolating to zero computation-mesh size. Even though four meshes covering a factor of 8 in size were used, it was not clear that the functional form of the mesh-size dependence had been adequately determined. They obtained a period of about 8.2, compared with the linear theory value of 7.97 and the nonlinear theory value of 7.49.

Two cases of multiple unstable frequencies were also calculated. The first is the $E^\dagger = 50$ system above, but with the degree of overdrive f reduced to 1.4, bringing the second root, Fig. 6.35, into the right half-plane. The calculation, carried through the first peak, is shown in Fig. 6.41. (The computer program had to stop at the last time shown because of the appearance of a backward-facing shock, which it was not equipped to handle.) The high-frequency component which decayed in Fig. 6.38 is now seen to grow. The period is that given by the linear theory. The second case is a less violent example with very small heat of reaction

$$\gamma = 1.2, q = 1, E^\dagger = 50, f = 1.6.$$

It has four unstable roots. The result, Fig. 6.42, is complicated as expected.

Having in mind systems like hydrogen/oxygen or hydrocarbon/oxygen, Fickett, Jacobson, and Schott (1972) did additional method-of-characteristics calculations for a simple mockup of branching-chain kinetics, and a polytropic-gas equation of state with variable heat capacity. They used two reaction rates, one (rate 1) with much longer reaction (recombination) zone than induction zone, intended to represent the real system $2H_2 + O_2 + 9Ar$ at $p_o = 10.13$ kPa, $T_o = 300$ K, and the other (rate 2) with reaction and induction zones of about equal

NONSTEADY SOLUTION Chap. 6

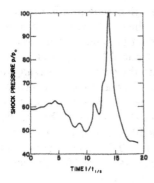

Fig. 6.41. The case $\gamma = 1.2$, $q = 50$, $f = 1.4$, $E^\dagger = 50$ with two unstable roots, started by a linear piston-velocity prescription like that of Fig. 6.38a. The lowest frequency root (1) of Fig. 6.35 has moved farther into the right half of the τ-plane and now grows very rapidly. The higher frequency root (2), which has now moved into the right half of the τ-plane and thus become unstable, can be seen growing in the calculation. From Fickett and Wood (1966).

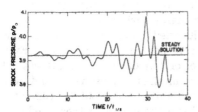

Fig. 6.42. Shock pressure history for $\gamma = 1.2$, $q = 1$, $f = 1.6$, $E^\dagger = 50$, a detonation with very small heat release and four unstable roots, started by a steady-solution piston. From Fickett, Jacobson, and Wood (1970).

length, more like what would be expected for hydrogen/oxygen or hydrocarbon/oxygen systems at higher initial pressures. Both rates have the same activation energy, $E^\dagger = 30$. The steady Chapman-Jouguet solutions for these two rates are shown in Fig. 6.43a. Also shown for comparison is the realistic (ten reactions, experimental rate constants) steady solution for $2H_2 + O_2 + 9Ar$ described in Sec. 5A8.

For rate 1, the $f = 1$ (CJ) steady solution is slightly unstable, with (peak-to-peak shock pressure) amplitude 2.0 and (time) period 3.1. Presumably it would be stabilized by mild overdrive. The results for the more interesting rates are shown in Figs. 6.43b and c. Above $f = 1.5$, the steady solution is stable. Coming down in f, we find first the low-amplitude (4.5), high-frequency (period 1.7) oscillation of Fig. 6.43b. At $f = 1.225$, a high amplitude (3), low-frequency (period 7.0) mode appears, with the high-frequency one still in evidence, as shown in Fig. 6.43c. A shorter calculation at $f = 1$ (to $t = 28$), suggested a pattern similar to Fig. 6.43b but with slightly larger amplitude and longer period. For all of these, the spatial period (in units of the half-reaction length) is between 3 and 4 times the time period.

Referring back to Fig. 6.35, we see that the order of entry (with decreasing f) of the roots (1) and (2) into the right half-plane is reversed from that for the first-order Arrhenus rate; here the higher frequency root (2) crosses over first.

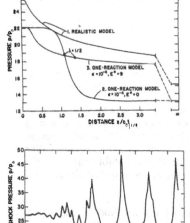
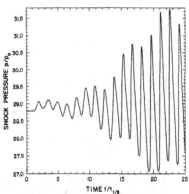

Fig. 6.43. Oscillating detonation for branching-chain (induction zone) kinetics. (a) CJ steady solutions: (1) realistic calculation for $2H_2 + O_2 + 9Ar$ at $p_0 = 10.13$ kPa, $T_0 = 300$ K, (2) rate 1 mocking up this system, and (3) rate 2 representing hydrogen/oxygen or hydrocarbon/oxygen at higher initial pressure. One distance unit for rates 1 and 2 is about 1 cm. (b), (c) Oscillations for rate 2: Low amplitude, high frequency at $f = 1.3$, and high amplitude, low frequency at $f = 1.225$. (Note change in vertical scale.) From Fickett, Jacobson, and Schott (1972).

Chernyi and Medvedev (1970) have done some method-of-characteristics calculations for a polytropic gas with $\gamma = 1.4$ and half-order Arrhenius rate

$$r = k(1-\lambda)^{1/2} e^{-E^\dagger/RT}.$$

These are similar to the calculation of Fig. 6.39. The starting state is a steady $f \cong 1.1$ (apparently stable?) detonation. The rarefaction is a strong centered rarefaction started after five complete-reaction times, so that the final state is an unsupported CJ detonation. Shock pressure histories are shown in Fig. 6.44a and an $E^\dagger$-q stability diagram for $f = 1$ in Fig. 6.44b. The calculations extend far enough in time to determine stability but not far enough to determine the final pattern of oscillation.

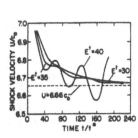 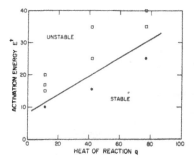

Fig. 6.44. Oscillating CJ detonation in $\gamma = 1.4$ polytropic gas with half-order Arrhenius rate. (a) Shock velocity histories for $q = 34.3$ and three activation energies. The time unit t^* is the steady-solution complete-reaction time. The energy unit is $p_0 v_0$. (b) $E^\dagger$ - q stability diagram for $f = 1$; the dots and squares mark stable and unstable calculated points and the line the neutral stability boundary. From Chernyi and Medvedev (1970).

Howe, Frey, and Melani (1976) have done method-of-characteristics calculations for a polytropic gas similar to those of Fickett and Wood (1966) except that they concentrated on the case of the unsupported detonation. Their computer program inserted (+) characteristics as needed to keep the net fine enough. They calculated for $\gamma = 1.4$ as well as for $\gamma = 1.2$, and used a rate mocking up a branching-chain process with an induction zone as well as the first-order Arrhenius form. They were able to make the interesting generalization that the parameter values which postpone the heat release toward the end of the reaction zone—large activation energy, induction delay, and small γ—make the steady solution less stable. A second interesting result is another case, like that of Fickett, Jacobson, and Schott (1972) in which a higher-frequency root is the first to appear as the stability boundary crossed. The stable system $\gamma = 1.4$, $q = 33$, $E^\dagger = 50$ with first-order Arrhenius rate was made unstable by increasing $E^\dagger$. A single unstable root is present at $E^\dagger = 55$; at $E^\dagger = 58$ there is a second one of about twice the period and twice the amplitude.

Il'kaeva and Popov (1965) did a short Lagrangian mesh calculation for a square-wave detonation. They used a very simple equation of state representing a condensed explosive

$$p = 1/3 \; (\rho^3 + \lambda - 1), \quad \rho_0 = 1.$$

Sec. 6C   FINITE-DIFFERENCE CALCULATIONS

The "square-wave" rate function is zero in the induction period below an ignition density ρ_1, and proportional to a high power of the density above ρ_1:

$d\lambda^*/dt = 0, \quad \rho < \rho_1$

$d\lambda^*/dt = k\rho^\alpha, \quad \rho \geq \rho_1$

$\lambda = 0$ for $\lambda^* < 1$

$\lambda = 1$ for $\lambda^* = 1$.

All the heat is released instantaneously at the end of the induction period. The ignition density ρ_1 is taken to be the induction-zone density in the steady solution. The square-wave stability criterion of Sec. 6C2 predicts instability for $\alpha > 7.77$. For $\alpha = 0$, the steady solution is stable, as expected. Fig. 6.45 shows a calculation for $\alpha = 15$ for which a small positive perturbation (see Sec. 6C2) is introduced in an unspecified manner via the initial conditions. The behavior is as described in Sec. 6C2, but the detonation fails after the shock from the perturbation reaches the front and reduces the density below the ignition value. It would be interesting to run this calculation longer with a lower ignition density, so as to see the oscillatory state.

Mader (1965) has done calculations for supported detonations in liquid nitromethane and liquid TNT with a Lagrangian mesh code, using a first-order Arrhenius rate function. We describe the results for nitromethane; those for TNT are similar. The activation energy is taken to be $E^\dagger/R = 27,000$ K ($E^\dagger = 53.6$ kcal/mol), a value consistent with lower-pressure shock-initiation experiments. The range of temperature through the calculated steady CJ reaction zone, 2600-3200 K, changes the exponential $e^{-E^\dagger/RT}$ by a factor of only 6.4, giving a less square profile than that for the ideal gas calculations of Fig. 6.37. The oscillatory solution for $f = 1.06$ is shown in Fig. 6.46. The steady half-reaction time and distance are about 0.017 ns and 63 nm; the oscillation period is about 17 half-reaction times or 36 half-reaction distances. The peak-to-peak amplitude at the shock is 14% of the steady-solution pressure. The calculations at lower f cover less than one period of the oscillation, so precise values are not available, but, as f is decreased, the period (in units of the half-reaction time) appears not to change too much, while the amplitude

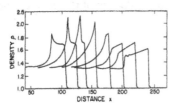
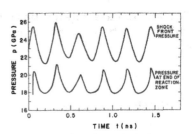

Fig. 6.45. Lagrangian mesh calculation for an idealized condensed explosive with square-wave rate (instantaneous heat release after an induction time) proportional to a high power of density. The pressure profiles shown are snapshots at a sequence of times. The distance scale is such that the steady reaction zone length is about 30. From Il'kaeva and Popov (1965).

Fig. 6.46. Lagrangian mesh calculation of the pressure history at the shock and at the end of the reaction zone for an oscillatory detonation in liquid nitromethane at f = 1.06 with an (assumed) activation energy $E^\dagger/R$ = 27,000 K ($E^\dagger$ = 53.6 kcal/mol). The half-reaction time is 0.017 ns. From Mader (1965).

grows rapidly. At f = 1.03 the amplitude is at least 60% of the steady-solution pressure. As f is increased the steady solution is stabilized somewhere between f = 1.06 and f = 1.10. The f = 1 steady solution is stabilized by reducing the activation energy; a calculation at f = 1, $E^\dagger$ = 40 kcal/mole was stable. The stability and the nature of the oscillation were found to be rather insensitive to reasonable variations in the equation of state.

6C2. Two Dimensions

Mader (1967b) has done some preliminary two-dimensional finite-difference calculations. He prepared two programs for the IBM 7030 (STRETCH) computer, one using the Lagrangian scheme, in which each computation cell represents a particle of the fluid, and the other using the Eulerian scheme, in which each cell is fixed in space. Both used artificial viscosity to handle the shock. His most extensive results are for calculations with cylindrical symmetry, each cell representing a ring of material. A rigid plane piston at one end of the cylinder initiates the detonation. Its velocity is programmed to reproduce the motion of a particle passing through the steady one-dimensional solution, so that it generates this solution in the stable case. At the beginning of the calculation, a perturbation is introduced halfway between the axis and the outer wall by increasing the mass of a single cell at the piston by 10 to 30%.

Sec. 6C   FINITE-DIFFERENCE CALCUALTIONS

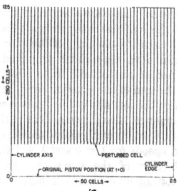

Fig. 6.47. Snapshot of the cell corners at t = 1.2 from the two-dimensional Lagrangian calculation of ideal gas detonation in a cylinder for $\gamma = 1.2$, q = 0.2, $E^\dagger$ = 50, f = D^2/D_j^2 = 2, having a stable steady solution. The perturbation, introduced by a heavy high-pressure cell at t = 0, r = 1.25, decays with time. Note that the horizontal and vertical distance scales are different. As before, the length unit is the steady half-reaction length; the unit of energy is $p_0 v_0$. From Mader (1967b).

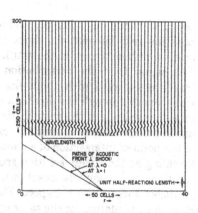

Fig. 6.48. Snapshot of the cell corners at t = 8.4 from the two-dimensional Lagrangian calculation for the unstable case q = 50, with other parameters the same as in Fig. 6.47. Note that the horizontal and vertical scales are again different. From Mader (1967b).

The volume and specific energy of this cell are kept constant in the perturbation so that its pressure and density are increased. The calculation is stopped when signals from the perturbation reach the edge of the cylinder. We show the results for two of Erpenbeck's (Sec. 6A) ideal-gas systems. Both have $\gamma = 1.2$, $E^\dagger = 50$, and f = 2, differing only in the value of q. The first, with q = 0.2, is stable to all perturbations. The second, with q = 50, is shown in Fig. 6.10. It is unstable for wavelengths in the range $3.5 < 2\pi/\epsilon < 63$ or $0.1 < \epsilon < 1.8$.

Fig. 6.47 shows a snapshot of the cell-corner positions from the Lagrangian calculation for the stable case q = 0.2, with 250 cells along the axis and 50 cells along the radius. Each cell is a square of side 0.05 in the undisturbed material. The signal generated by the perturbation is seen to decay rapidly, as expected. Fig. 6.48 is from a similar calculation for the unstable case q = 50. Here each undisturbed cell is a square of side 0.8. The average compression factor in the reaction zone is about seven, so that along a line parallel to the axis there are about nine computation cells between the shock and the half-reaction point, a rather small number in view of the artificial-viscosity smearing of the shock. The signal from the perturbation retains an appreciable amplitude, traveling outward with about the speed of an acoustic front moving transversely through the products at the end of the reaction zone. The larger magnitude of the perturbations on the left side of the figure is no doubt attributable to cylindrical convergence. Also shown are the paths of the point of intersection of an acoustic wave front normal to the shock traveling at the steady-solution sound speeds for $\lambda = 0$ and $\lambda = 1$ (no reaction and complete reaction). A part of the signal, at least, propagates with the $\lambda = 0$ speed. Fig. 6.49 shows pressure contours of an Eulerian calculation for the same system. Here there are 100 cells along both the symmetry axis and the radius, with each cell a (fixed) square of side 0.04. This gives better spatial resolution, with about 25 cells in the half-reaction zone on a line parallel to the axis, at the cost of shortening the distance of run to about three half-reaction zone lengths.

We estimate the transverse wavelength from Fig. 6.48 as the distance between the first maximum leftward particle displacements behind the shock, as shown. This gives a value of 10.4, lying within the predicted unstable range of 3.5 to 63. The calculation of Fig. 6.49 has probably not run long enough to give a wavelength estimate. We note that the distance between the pressure peaks is much smaller, about 0.28, and that the transverse pressure variation is about one percent of the shock pressure.

Two-dimensional calculations were also performed for nitromethane and liquid TNT. The one-dimensional stability results described earlier were reproduced, but no evidence of two-dimensional instability was seen.

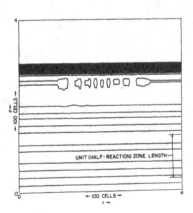

Fig. 6.49. Snapshot of the pressure contours at $t = 0.3$ from the Eulerian calculation for the detonation of Fig. 6.48. Adjacent contours differ by 1 in p/p_0. From Mader (1967b).

7

STRUCTURE OF THE FRONT

Our knowledge of the actual structure of the front rests almost entirely on observation, interpreted with the help of after-the-fact calculations based on standard shock matching and blast wave theory. The theoretical work described in Chapters 5 and 6, even the treatments of galloping detonation and transverse waves, do little more than hint at the structure.

For gases there has been a large experimental effort, resulting in quite detailed knowledge of the structure and identification of the important parameters. The relatively modest effort with the more difficult solids and liquids has been less successful, but some structural features have been identified, and it is clear that the structure is of the same general type as that in gases.

The key feature of the structure is the transverse wave, an interior shock joined to the leading shock in the conventional three-shock configuration. The Mach stem and the incident shock are part of the leading shock, and the transverse wave is the reflected shock. The transverse waves move back and forth across the front. Groups of them moving in the same direction take up a preferred spacing on the order of 100 reaction-zone lengths. They are not steady waves, but are continually decaying, and stay alive only by periodic rejuvenation through collision with other transverse waves moving in the opposite direction.

Secs. 7A through D deal with gases, and Sec. 7E with liquids and solids. Section 7A surveys the main features of the structure, defines the notation, and briefly reviews the experimental methods and the shock-matching calculations. Sec. 7B looks at the general structure of the front and the behavior of the transverse waves—how they couple with the acoustic modes of the tube and how they behave in transient situations. Sec. 7C moves in for a closer look at the details of the structure, of necessity considering only the simplest cases—single spin in a round

STRUCTURE OF THE FRONT   Chap. 7

tube and low-mode-number detonations in narrow rectangular tubes. Sec. 7D compares the theoretical and semi-empirical predictions of transverse-wave spacing with experiment.

Definitions and Symbols
- Wave designations
  M — Mach stem
  I — incident shock
  R — reflected shock
  S — shock strength: pressure ratio minus one
  slip — slip line
  fire — region of rapid energy release

- Cell and tube geometry (Sec. 7A3)
  x,y,z — space coordinates, Fig. 7.5
  L,Z — cell length and height, Fig. 7.5

- Mixture composition
Percentages in mixture composition are mole percent. Thus

$$30\% \ (2H_2 + O_2) + 70\%Ar = 2H_2 + O_2 + 7Ar.$$

In the literature the first percentage is sometimes omitted:

$$2H_2 + O_2 + 70\%Ar.$$

7A. OVERVIEW

A detonation typically has a cellular structure. Fig. 7.1 shows the traces left by a well-established expanding spherical detonation in acetylene-oxygen on a soot-coated flat plate mounted normal to a radius.

The cellular structure is self-sustaining, with the chemical reaction furnishing the driving energy. The cell boundaries are *transverse waves* propagating in the reaction zone at approximately acoustic velocity, so the pattern changes with time. Since they intersect the main shock, it is wrinkled. The transverse-wave spacing is approximately proportional to the (steady-solution) reaction-zone length, with the constant of proportionality on the order of 100. The soot-film technique of Fig. 7.1 is extensively used, and capable of recording great detail. As suggested by this record, there are appreciable variations in pressure and velocity within the structure—sufficient to record the structural details by scrubbing off more soot in some spots than in others.

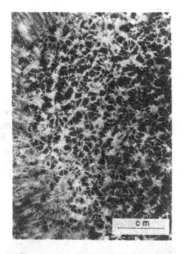

Fig. 7.1. Soot imprint (on a flat plate normal to a radius) from an expanding spherical detonation in $C_2H_2 + 3O_2$, $p_o = 6.7$ kPa. From Duff and Finger (1965).

When the detonation is confined in a tube whose diameter is not too much larger than the characteristic cell size, many materials yield very regular patterns. These structures have been extensively studied; the flow field and its dependence on the system properties is now quite well understood.

7A1. An Intuitive Picture

A good way to visualize the cellular structure is to imagine how it might arise from a grid of regularly spaced perturbations introduced into a square-wave detonation.

Suppose we have a one-dimensional square-wave detonation, and somehow introduce a hot spot into the reaction zone just ahead of the fire. This initiates a hemispherical detonation moving into the induction zone and eventually overtaking the front. The observed cellular structure can be pictured as that resulting from the collisions of a number of such microdetonations, as they have been called by Dremin (1968). With a grid of regularly spaced hot spots, we would have the sequence of events shown in Fig. 7.2. A similar picture would of course apply to an array of line sources generating cylindrical waves. Beginning at the bottom of the figure, we first have the microdetonations starting to collide with each other at their lateral edges and penetrating the leading shock at the top. The details are complicated, but the main features are sketched in the succeeding lines of the figure. The shock reflection at the collisions of the adjacent microdetonations is initially regular, but as the angle of collision increases, a Mach stem (see below) is formed. In the second line of the figure, this Mach stem has overtaken the original

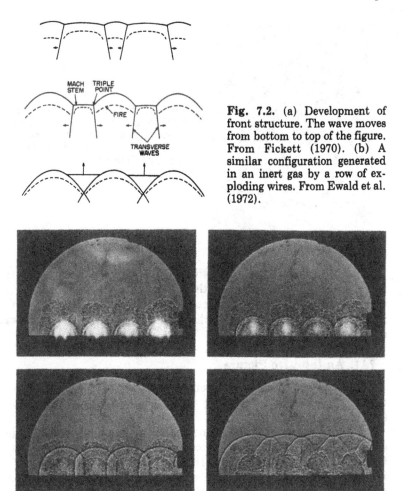

Fig. 7.2. (a) Development of front structure. The wave moves from bottom to top of the figure. From Fickett (1970). (b) A similar configuration generated in an inert gas by a row of exploding wires. From Ewald et al. (1972).

front. Behind the Mach stem the pressure and temperature are high and the induction zone is short. Where the decaying spherical front of the microdetonation has overtaken the original front, and passed into unshocked material, the induction zone is long. The reflected shocks, moving laterally into the unreacted material of this long induction zone, become something like transverse detonations. As the transverse waves continue to approach each other, the Mach stem is being degraded by rarefaction from its sides and rear. If conditions are right, as the

Sec. 7A
OVERVIEW

(a) REGULAR REFLECTION

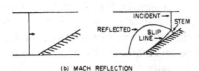

(b) MACH REFLECTION

Fig. 7.3. Regular and Mach shock reflection in a nonreactive fluid.

transverse waves reverse direction through collision they again encounter unreacted material in which to propagate—the now lengthened induction zone behind the weakened Mach stem—and the process can repeat.

Rough criteria for the transverse-wave spacing are evident from this simple picture: It must be large enough to allow an appreciable induction zone to form between two transverse waves receding from each other, so that when their direction is reversed by collision with their outer neighbors they will encounter enough unreacted material to sustain their propagation through the next collision. It must be small enough to preclude the formation and appreciable growth of a new microdetonation between two transverse waves approaching each other.

7A2. The Triple Point

Although we postpone discussion of the details of the structure until Sec. 7C, we will take a first look at the key feature here: the transverse wave and its intersection with the front at the triple point. The configuration is similar to Mach reflection in nonreactive materials, which we first briefly review. The reader interested in more detail may consult a standard reference such as Courant and Friedrichs (1948).

Consider a plane shock moving down a channel and encountering half of a thick wedge, Fig. 7.3a. The shock reflects from the surface of the wedge in straightforward fashion; the process is termed regular reflection. As the wedge is made thinner, a point is eventually reached at which the reflected shock is unable to turn the flow parallel to the wedge surface and a third shock, the Mach stem, appears, Fig. 7.3b. This process is called Mach reflection. The original incident and reflected shocks now join the stem at the *triple point*. The *slip line* beginning at the triple point separates material passing through the stem from that passing through the incident and reflected shocks. Roughly, nature inserts the stem to form a thicker wedge (whose upper surface is the slip line) on which regular reflection can occur. An important point is that the flow in the neighborhood of the triple point is almost steady—so nearly so that analysis under the steady assumption is widely used with considerable success.

Mach reflection of a detonation wave is similar. Fig. 7.4 depicts such an experiment. For practical reasons, the channel geometry, Fig. 7.4a, is different from that of Fig. 7.3, but the reflection which generates the Mach stem is essentially the same. The first requirement is to generate a momentarily stable one-dimensional detonation wave without the transverse structure which would normally be present. This is accomplished by passing a CJ detonation wave through a convergent-divergent section. The strong overdrive in the convergent section is sufficient to make the one-dimensional solution stable and remove the transverse structure. The following divergent section returns the wave to the CJ state quickly enough so that the transverse structure does not have time to reappear.

As the wave emerges from the divergent section, its upper end encounters the change in angle where the upper edge of the divergent section rejoins the channel wall. Here we have the situation of Fig. 7.3b, with the upper edge of the divergent section replacing the channel wall and the upper channel wall replacing the wedge. The important feature is of course just the abrupt change in wall angle seen by the wave. As indicated by the triple-point track, the triple point is allowed to proceed most of the way across the channel before the interferogram is taken.

The sketches show the main features of the interferogram. The fire is seen to have an observably finite width; in the magnified view of the triple point, Fig. 7.4d, it is indicated by just a dotted line. The incident portion of the front shock is about like a normal plane detonation. Particles entering it near enough to the triple point A encounter the reflected shock AC before burning. The shock heating shortens their induction time, producing the inclined fire CD. The stem portion of the front shock is much stronger with correspondingly shorter induction zone. As described in Secs. 6A2.1 and 6B3, the start-up of the fire at E should produce forward- and rear-facing shocks, and the termination at D a similar pair of rarefactions. There is probably insufficient resolution at E to show the shocks; PD may be the rear-facing rarefaction. QD is probably the diffuse extension of the slip line. It is not clear what PQ is. Note that the reflected shock AC bends sharply downward as it encounters the higher sound speed of the burned gases behind the fire.

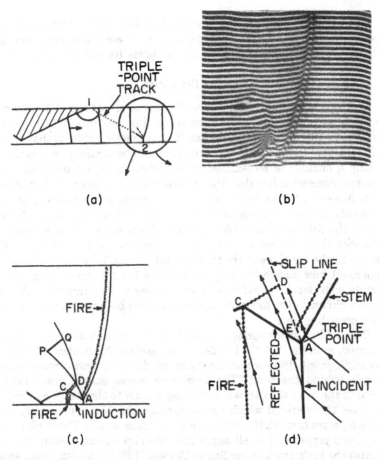

Fig. 7.4. Interferogram of Mach reflection in a reacting gas, $2H_2 + 2CO + O_2$, $p_0 = 2.0$ kPa. (a) The detonation, stabilized by passage through a constriction, undergoes Mach reflection at point 1. The triple point has almost reached the opposite wall at point 2 where the picture is taken. (b) Interferogram. (c) Sketch of the interferogram. (d) Magnified view in the quasi-steady frame attached to the triple point. From White and Cary (1963).

7A3. The Simplest Regular Structure

This triple-point structure has the main features of those observed in steadily propagating detonations, but there are some important differences. It is, however, sufficiently accurate for present purposes.

7A3. The Simplest Regular Structure

We are now ready to describe the simplest cellular detonation in somewhat idealized form. At the same time, we introduce some terminology that will be needed later. The case we consider is that of two-dimensional flow. A good approximation can be achieved in practice by using a channel of rectangular cross section of high aspect ratio with narrow dimension less than the natural transverse-wave spacing. Figure 7.5 shows soot tracks from a laboratory example, and a diagram. Ideally, the only transverse waves present are those moving back and forth between the narrower walls. (Actually Fig. 7.5 is something of a fluke; ordinarily there is some evidence of an orthogonal family of waves). The main tracks are those of the triple points. Figure 7.6 shows (for a wider channel) how successive snapshots of the front superimposed on the triple-point tracks would look to an observer looking into the wide wall of the channel. The following discussion refers both to this figure and to the preceding one.

The leading shock surface, or *front*, is composed of alternate sections termed, as in Fig. 7.4, the *incident shock* and the *Mach stem*. These join at *triple points* from which the *reflected shock*, or *transverse wave* extends behind the front. The transverse waves extend much farther behind the front than shown, becoming weaker to the rear. The *slip line* divides the material which passes through the Mach stem from that which passes through the incident and reflected shocks. The *triple point*, or, more properly, a small segment of the slip line emanating from it, writes the main track: according to Crooker (1969), "the slipstream sheet removes most of the soot from the Mach-wave side of the triple-point trajectory, while the relatively smooth flow on the incident side removes considerably less." The triple "point" is of course actually the projection onto the side wall of the line of intersection of the three shocks, which we occasionally refer to as a *confluence line*. The word *wave* usually refers to a shock, but is also used for more extended configurations such as rarefactions. We usually describe the *reaction zone* in terms of the *square-wave* model: a constant-state *induction zone* followed by a zone of instantaneous or very rapid heat release—the *fire*. In the literature the fire is often called the *recombination zone*, since most of the heat comes from recombination of the dissociation products of the thermally neutral reactions of the induction zone.

At each transverse-wave collision the Mach stem and incident shock are in effect interchanged. Two transverse waves receding from each

Sec. 7A   OVERVIEW

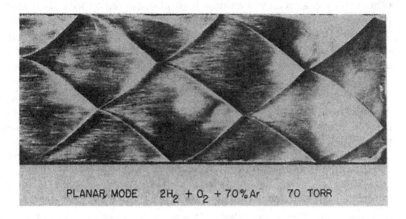

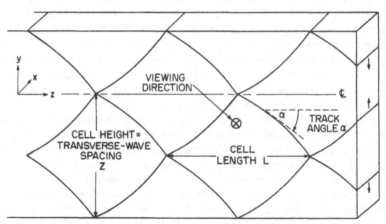

Fig. 7.5. Soot track and sketch of a two-dimensional (planar mode) cellular detonation in $2H_2 + O_2 + 7Ar$, $p_o = 9.3$ kPa. Soot track from Strehlow (1968b).

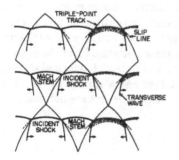

Fig. 7.6. Snapshots of detonation front structure superimposed on triple-point tracks.

299

other after collision leave behind a high-pressure region with short reaction zone—a new Mach stem (see the collision calculation of Sec. 7A5). They move into the now weakened old Mach stems which become the incident shocks for the new triple points coming out of the collision. Along a line in the direction of propagation passing through points of collision, the history of the front (shock) state parallels that for the one-dimensional instability of Chapter 6, with the shock pressure jumping well above CJ at collision, and then decaying to a value below. Unless the cell size is very much smaller than the tube diameter, the average propagation velocity is somewhat below the CJ value.

We now return to Fig. 7.5 to define some terms for the track geometry. The direction of propagation in all figures is from *left* to *right*, or from *bottom* to *top*. The repeated four-sided figure is called a *cell*. A *centerline* joins successive collision points lying on a line in the direction of propagation. The *axial velocity* D is the velocity of the point of intersection of the main front with the centerline. The *cell length* L and *transverse-wave spacing* (cell height) Z are shown in the figure. The transverse-wave spacing is of course just the distance between two transverse waves moving in the same direction. In the figure there is only one family of transverse waves, whose propagation directions lie along the positive or negative y-axis. The flow is two-dimensional, with flow variables independent of x. Such a detonation is called a *planar detonation*, and the associated family of transverse waves the *principal waves*. The detonation is characterized by the *number of transverse waves* or *mode number*, which is twice the number of cells across the tube. In the figure there are 1 1/2 cells and three transverse waves.

If the tube is not too narrow in the x direction, there will also be a well-developed orthogonal family of transverse waves moving back and forth between the wider walls. We then have a *rectangular detonation* with *principal waves* moving up and down in the y direction and leaving tracks on the side walls and *slapping waves* moving back and forth in the x direction and leaving tracks on the top and bottom walls. A rectangular detonation is characterized by two numbers, such as 3×2, the second number being the number of transverse waves moving back and forth in the x direction. Such a detonation imprinting on a soot-coated end plate produces a pattern with approximately square cells (the word *cell* also being used to describe an end-plate pattern). In a tube of square cross section the two families of waves are of course of identical strength, but we retain the terminology of principal and slapping waves. A few more terms having to do with details of the structure are defined at the beginning of Sec. 7C.

7A4. Experimental Methods

Detonations with cell size much smaller than the cross section of the container have an irregular structure. Details of the flow in such a system are of course difficult to measure, and most of the experimental work has been done in simpler if slightly atypical systems in which the degree of dilution and initial pressure are chosen to make the natural cell size comparable to the cross section of a square or circular tube. This simplifies the observational problem at the cost of introducing wall effects (boundary layer, heat conduction, catalysis) which would be of negligible importance in the large-diameter limit. Even so, except possibly in very narrow channels, the principal effect of the boundaries is just their presence as rigid walls and the resultant determination of the cell size and shape and direction of propagation of the transverse waves.

Probably the most important consideration in choosing a gaseous detonation mixture to study is that it be one which gives a regular pattern of cells of convenient size. Other considerations are that the kinetics be well understood, so that the observed structures can be correlated with calculated induction times, and perhaps that the system have some special property which facilitates observation. The mixtures most used are H_2/O_2, C_2H_2/O_2, and CO/O_2, often diluted with inert gases. These have kinetic schemes which have the common feature of a rapid exothermic reaction preceded by a relatively long thermally neutral induction zone. This is produced by a branching-chain mechanism in which atom and radical concentrations build up exponentially with time from very small initial concentrations. The induction time has been measured under a wide range of conditions (White, 1967) and can be reliably calculated from the initial concentrations of reactants. All of these give quite regular patterns, and convenient cell size is easily obtained by proper choice of dilution at reasonable initial pressures (boundary layer effects become more important as the initial pressure is reduced). The acetylene-oxygen system has a property which makes it particularly convenient for qualitative studies: a strong chemiluminescence—brighter than the equilibrium thermal radiation from any part of the flow in the diluted system—accompanies the fire and thus marks this part of the flow. The CO/O_2 system with a small admixture of hydrogen has the advantage of historical familiarity and offers a chemical probe in the form of radiation in the blue continuum proportional to the concentration product $[CO][O]$, a feature extensively exploited by White (1961) as described in Chapter 3.

Two tube cross sections are used: round and rectangular. A round tube is easier to fabricate and thus historically the first used for observations. The lowest mode in the round tube—the so-called single spin—offers the strongest transverse wave observed in any system and has the unique feature of being the only system in which there are no collisions of transverse waves with each other or with the wall inasmuch as the single transverse wave present just rotates about the tube axis. Higher modes with transverse waves going in both directions around the tube can also be generated.

The rectangular tube offers the advantage of plane observation windows with no disturbance of the flow and a more nearly two-dimensional flow, particularly in the planar mode in which one family of transverse waves is suppressed by making one dimension of the cross section much smaller than the natural cell spacing. Even in the rectangular case, where both families of transverse waves are present, the waves do not interfere with each other very much and these advantages are still present although to a lesser extent. This has been the configuration of choice in some more recent work.

A number of observational tools are used. The book by Soloukhin (1966) has a good review. Much qualitative information can be gained from open-shutter photographs of a system like acetylene-oxygen, where the chemiluminescence of the fire records a good visual track. The simple soot film placed on the wall of the tube records a large amount of information with good resolution and has become a nearly indispensable tool. Acoustic tracks from sand grains or indentations on the foil show the local orientation of the front. Voitsekhovskii, Mitranov, and Topchian, hereafter abbreviated as VMT (1963) made very effective use of the rotating drum camera with velocity synchronization and a schlieren optical system; by proper choice of the slit angle and camera rotation speed, selected features of the structure are brought to rest on the film and thus singled out for study. For example, the triple-point structure is studied by using a wide slit oriented at the track angle to the tube axis and moving the film parallel to the slit at the velocity of the image of the triple point. More recently Strehlow and Crooker (1974), using similar techniques but with the advantage of a continuous laser source, were able to resolve much detail of the structure. Laser-illuminated snapshots are also much used. For example, Edwards, Hooper, Job, and Parry (1970) have made detailed interferograms of the structure, and the review by Lee, Soloukhin, and Oppenheim (1969) shows a sequence of snapshots through a thinly smoked foil to demonstrate directly that the track is indeed written by the triple point.

Transducers are also much used. Ionization gauges detect the arrival of the front and thus its velocity. Batteries of pressure transducers

Sec. 7A   OVERVIEW

mounted in the tube walls give pressure histories. The cross section of these can be made much smaller than the cell width. Platinum resistance gauges measure the temperature in similar fashion.

7A5. Calculations

The most common calculation used in interpreting the experiments is shock matching at the triple point. The flow is assumed steady in a frame attached to the triple point, so that the standard shock-polar analysis applies, Fig. 7.7. For background, see standard works such as Courant and Friedrichs (1948), or the paper of Oppenheim, Smolen, Kwak, and Urtiew (1970). The incoming flow velocity u_o and the inclination angle ϕ_1 of the incident portion of the front to it are experimentally measured. Ignoring for the moment the effects of chemical reaction, we have the shock polar CD which relates, for given incoming Mach number u_o/c_o (and polytropic gas index γ), the pressure ratio across the shock to the deflection angle θ of the particle path through it. The angle ϕ between the incoming flow and the shock (not shown in the figure) varies monotonically with arc length along the polar, from 0 at $p = p_o$ to 90° at the maximum pressure. Polar CD is first applied to the incident shock. In the example, the state behind it is point E, where $\phi = \phi_1$, the measured inclination. The next step is to repeat the process across the reflected shock. The polar for this is EF, with E the initial point (with θ now the total deflection through both incident and reflected shocks). The state behind the reflected shock must be on EF. But its pressure

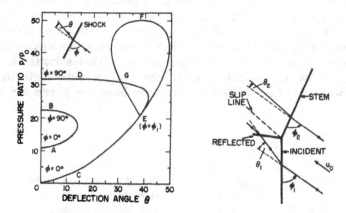

Fig. 7.7. Flow in the neighborhood of the triple point and shock polars for $2H_2O + O_2$, $p_o = 4.0$ kPa and $\phi_1 = 55°$. After Edwards, Parry, and Jones (1966).

STRUCTURE OF THE FRONT   Chap. 7

and deflection θ must be the same as that behind the stem shock, to which CD also applies. Hence their intersection at G gives this state (and the stem inclination ϕ_2). The entire flow around the triple point is thus determined.

If the stem shock is strong enough, the reaction zone behind it may be sufficiently short that treating the reaction there as instantaneous may be a better approximation. Under this assumption, the polar for the stem becomes that for a detonation, AB in the figure, and we look for its intersection with EF giving the state for the reflected shock. For the particular example shown, there is no solution under this assumption, but in other cases, generally with smaller ϕ_1, there is.

The other standard calculation is the collision of two transverse waves, as diagrammed in Fig. 7.8. In principle, knowledge of the entrance and exit angles α and α' determines all of the states around the triple points before and after collision. We will describe the collision process and summarize the results, considering only the symmetric case. For details, see Strehlow and Biller (1969), Strehlow, Adamczyk, and Stiles (1972), Oppenheim, Smolen, Kwak, and Urtiew (1970), and Urtiew (1976).

Our discussion refers to the upper labeled portion of the figure. The triple-point configuration is inverted by the collision. Before collision the upper segment M (above the track) of the leading shock is the Mach stem, the lower segment I is the incident shock, the slip line lies above the track, and the reflected shock R lies below it. After collision the upper segment I' of the lead shock, originally the Mach stem, is the incident shock, and the lower segment M' is the new Mach stem. The slip line now lies below the track and the reflected shock lies above it. Note that the original incident shock I is consumed in the collision and that the new Mach stem M' grows from it. The key property which makes the calculation possible is that the upper segment of the leading shock is unchanged by the collision, since the flow is supersonic in the neighborhood of the triple point. Thus we have p = p' and $\theta = \theta'$.

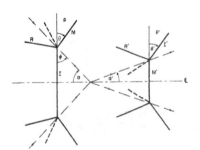

Fig. 7.8. Collision of two transverse waves.

Sec. 7A   OVERVIEW

To see that the specification of the entrance and exit angles α and α' determine all the states, recall from our earlier discussion that the complete state around a triple point is determined by the incoming flow speed u_0 and its inclination angle ϕ to the incident shock (in the steady frame). Express these in terms of the track angles α and α' and the axial velocity D

$$u_0 = D/\cos\alpha$$

$$\phi = 90° - \alpha.$$

The triple-point solution, which we here denote by a tilde, then allows us to express any other state quantity in terms of these. In particular we have before reflection

$$p = \tilde{p}\,(\alpha, D)$$

$$\theta = \tilde{\theta}\,(\alpha, D).$$

Similarly we have after collision

$$p' = \tilde{p}'\,(\alpha', D')$$

$$\theta' = \tilde{\theta}'\,(\alpha', D'),$$

the functions differing in detail because of the inversion through collision. From the constancy of the outside shock we have $p = p'$, and $\theta = \theta'$, so these become two simultaneous equations giving D and D' in terms of α and α'.

The sensitivity of this system of equations is best studied by posing the problem of an incoming triple point of given axial velocity D and entrance angle α. It turns out that the results for most of the properties of interest depend mostly on the value of α, and hardly at all on D. To a good approximation it is found that:

1. The entrance and exit angles α and α' are related to the invariant angle θ by

$$\alpha' = \alpha - \theta.$$

2. The transverse-wave strength (pressure ratio across the reflected shock minus one) is a nearly linear function of α, a typical result being that this strength varies from 0.15 to 1.4 as α changes from 25° to 50°.

3. The transverse-wave strength is changed only slightly by the collision.

In practice, the measured entrance angle is used to determine these properties and the measured exit angle is used as a check.

7B. MACROSCOPIC PROPERTIES

7B1. Structures

Detonations have a natural cell size one to two orders of magnitude greater than the reaction-zone length (of the hypothetical steady one-dimensional CJ detonation). If the detonation is unconfined, or if the tube diameter is very much larger than the cell size, random patterns like that of Fig. 7.1 are observed. With the initial pressure and degree of dilution chosen to make the cell size not too much smaller than the tube cross section, patterns of varying degrees of regularity are observed. Some examples in rectangular tubes are shown in Fig. 7.9.

Much of the work on this subject has been done by Strehlow and his coworkers, and is summarized in his reviews, Strehlow (1968b, 1969). No comprehensive explanation of the observations in terms of system properties has been given, although the work of Takai, Yoneda, and Hikita (1974) may offer a starting point. These authors discuss the change in system properties when H_2/O_2 is diluted with nitrogen, which

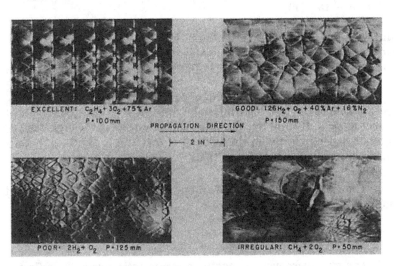

Fig. 7.9. Soot films showing different degrees of regularity. From Strehlow (1968b).

interferes with the chain reaction and destroys the regular structure; their results suggest that irregular patterns may be associated with systems near the detonation limit. In general, regular structure apears to be favored by low initial pressure and dilution with inert gases. Some of the observations, as summarized by Strehlow (1969) are

1. With O_2 as oxidizer, and with appropriate diluent and initial pressure, very regular patterns can be obtained from H_2, C_2H_2, C_2H_4, and CO, but not from NH_3, CH_4, C_2H_6 and C_3H_8.

2. Methane is the poorest performer, and shows an interesting variety of very irregular patterns under different conditions.

3. Dilution with argon is generally conducive to a more regular structure; in C_2H_6/O_2, helium is less effective, and in H_2/O_2, N_2 and CO_2 are ineffective or make things worse.

4. In H_2/O_2, fuel-lean mixtures are more regular than fuel-rich ones.

5. The tracks written by NH_3/O_2 and CO/O_2 are very weak.

The most extensively studied systems are those with regular patterns. Some examples are shown in Fig. 7.10. Fig. 7.10a shows soot films from the two sides and the end of the tube. There is sometimes an almost embarrassing amount of detail, reminiscent of elementary-particle tracks in photographic emulsions. It is interesting to note, Fig. 7.10c, the large number of cells which can still form a regular pattern.

All of these patterns are from the wider wall of the tube. The vertical lines mark the collision of the slapping wave(s) (those travelling across the narrower dimensions) with the wall. The interactions between waves of these two orthogonal families are evident in the jogs in the slapping wave imprints. The two orthogonal families are not in general in phase with each other.

Note that Figs. 7.10b and d are for the same mixture and initial pressure but the latter is in a narrower tube, which greatly increases the cell size and changes the structure. This change is discussed in Sec. 7C.

Fig. 7.11 is a diagram of a detonation with a single transverse wave in each family, showing the general shape of the front, and, in the cross section, the triple point and the fire.

In round tubes, the confluence line of the transverse wave lies along a tube radius, and rotates about the tube axis. Its intersection with the tube wall thus traces out a helical track as the front progresses, giving rise to the name "spinning detonation." Single spin, with just one

STRUCTURE OF THE FRONT   Chap. 7

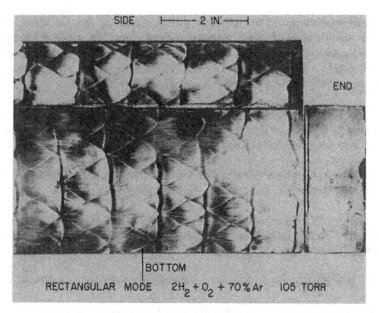

(a)

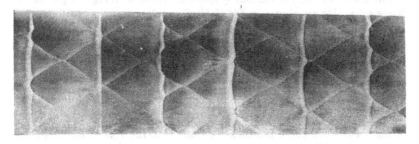

(b)

Fig. 7.10. Soot tracks of detonations in rectangular tubes. Each soot film covers the full width of the wider wall [part (a) includes the full width of the narrow wall and the entire end plate as well). All are in a 65 × 38 mm (3 1/4 × 1 1/2 in.) tube except (d), which is 65 mm × 6.3 mm (3 1/4 × 1/4 in.).
(a) $2H_2 + O_2 + 7Ar$, p_o = 14 kPa, Strehlow (1968b).
(b) $2H_2 + O_2 + 3Ar$, p_o = 7.7 kPa, Strehlow and Crooker (1974).
(c) $.15H_2 + .125O_2 + .7Ar + .025CH_4$, p_o = 20 kPa, Strehlow (private communication).
(d) $2H_2 + O_2 + 3Ar$, p_o = 7.7 kPa, Strehlow and Crooker (1974).

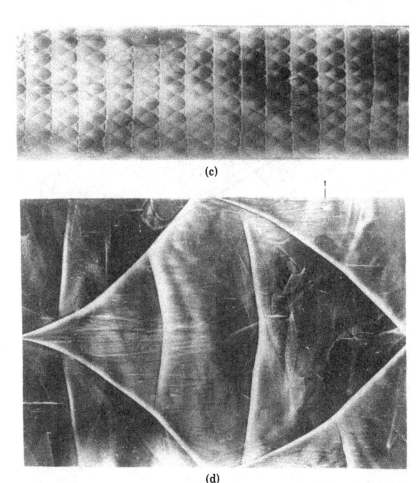

(c)

(d)

transverse wave, occurs near the detonation limit for the tube in question. An example is shown in Fig. 7.12. The track is seen to be a band having its own internal structure (to be discussed in Sec. 7C1) instead of a line. The fluctuating pattern of Fig. 7.12b is more common than the regular one of Fig. 7.12a, although both are obtained from the same system. Detonations with more transverse waves (approximately half rotate in each direction) give diamond-shaped patterns of moderate regularity more like those in rectangular tubes. In an annulus formed by inserting an axial rod two orthogonal systems of waves analogous to

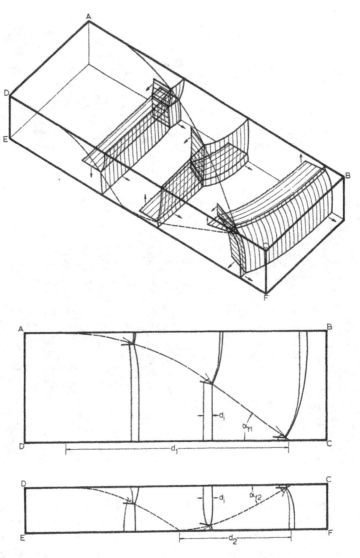

Fig. 7.11. Diagram of a detonation with two orthogonal transverse waves. From Steel and Oppenheim (1966).

those in a rectangular tube are obtained: one family moves along the circumference and the other along a radius. The wave system is sketched in Fig. 7.13.

In a class by themselves are a few instances of "galloping" detonations with the principal oscillation longitudinal, as in the calculations of Sec. 6C. There are two very different types, with periods (relative to the reaction time) differing by factors of about a hundred.

The short-period type corresponds to the calculations of Secs. 6B3 and 6C1. It has been observed only in a rather special situation: at the tip of the bow shock of a blunt body moving through a detonable gas mixture at velocities in the neighborhood of the CJ detonation velocity. Many pictures of these blunt-body flows have been taken. Some of the clearest are those of Lehr (1972), one of which is shown in Fig. 7.14. Different patterns are obtained at different projectile speeds. We do not enter here into the detailed explanation of the complete flow. All of the prominent periodic features are generated by an approximately one-dimensional oscillatory detonation located at the front tip of the shock. The reaction-zone thickness of this detonation is only a small fraction of the separation distance of the shock from the body. The regular corrugated pattern in the wake is essentially the trail of density- and contact-discontinuity fluctuations left by the oscillating detonation.

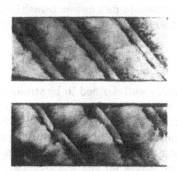

Fig. 7.12. Single spin: detonation in a round tube with one transverse wave in $0.067C_2H_2 + 0.100O_2 + 0.833Ar$, $p_o = 5.9$ kPa. From Schott (1965b).

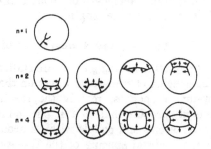

Fig. 7.13. End views of n = 1, 2 and 4 detonations in a round tube with axial rod. From VMT (1963).

STRUCTURE OF THE FRONT   Chap. 7

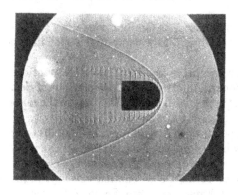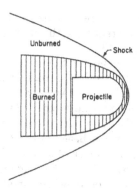

Fig. 7.14. Laser schlieren photograph of a blunt projectile moving into dilute hydrogen-oxygen at a velocity close to CJ detonation velocity. From Lehr (1972).

The longer-period type is seen in conventional detonation tubes, but is still comparatively rare. It has a very different mechanism. The most detailed study is that of Saint-Cloud, Guerraud, Brochet, and Manson (1972) on $C_3H_8 + 5O_2 + 10N_2$, $p_o = 40$ kPa. The detonation, in a square tube, is basically a 1 × 1 detonation which "fails" to a fire lagging behind the shock. The fire then accelerates and turns into a detonation again via the familiar mechanism of the deflagration-to-detonation transition. The spatial wavelength of about 2 m is some 10,000 times the (steady CJ) induction-zone length.

7B2. Spacing and Acoustic Coupling

In systems having a structure sufficiently well-defined to be studied, the spacing of transverse waves on the front is found to depend on the system properties, mainly the reaction-zone length, to which it is roughly proportional. In a detonation confined to a tube, the transverse waves excite and couple with its acoustic modes of vibration. The inherent natural spacing of the transverse waves on the front remains strongly in evidence but the tube acts as a constraint forcing a more or less regular structure.

In this section we consider two questions: (1) how is the inherent spacing related to the system properties, and (2) what is the nature of the coupling between the front and the tube? For the first question we limit ourselves here to the experimental observations. The a priori theories which at least suggest a spacing were described in Sec. 6B; and some semi-theoretical descriptions incorporating some of the observed proper-

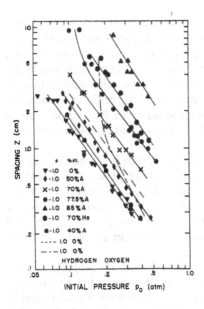

Fig. 7.15. Transverse-wave spacing in stoichiometric H_2/O_2/diluent systems vs. initial pressure. From Strehlow and Engel (1969).

ties, and comparison of predicted and observed spacings, are presented in Sec. 7D.

The second question is a problem in acoustics, with the front regarded as a driver exciting acoustic vibrations in the detonation products and of course influenced by them and by the constraint of the tube confinement. It is investigated by examining the acoustic modes and their interaction with the front, and comparing the predicted frequencies and wavelengths with experiment.

7B2.1. Spacing

Strehlow and Engel (1969) collected the available data on transverse wave spacing Z, in mixtures of H_2, CH_4, C_2H_4, and C_2H_2 with O_2, mostly in rectangular tubes, and correlated them with the system properties in various ways.

As shown in Fig. 7.15 for H_2/O_2, log Z is found to be inversely proportional to log p_0. The upward curvature of two of the curves at low pressure is ascribed to wall effects near the propagation limit.

Now the induction time τ is inversely proportional to the O_2 concentration

$$\tau = A \, [O_2]^{-1} \, e^{-T^\dagger/T}, \quad T^\dagger = 8210 \text{ K}, \quad A = 6.89 \times 10^{-11} \text{ s-mol/cm}^3$$

and thus approximately inversely proportional to the initial pressure at fixed system composition. The relation is of course not exactly linear because the increased degree of dissociation at lower p_o will reduce T somewhat. The obvious scaling is to divide by the induction-zone length I. The result is shown in Fig. 7.16, where I is the calculated induction-zone length for the hypothetical laminar one-dimensional CJ detonation. The ratio tends to approach a constant value for each mixture as the pressure is increased, ranging from $Z/I = 60$ at no dilution to $Z/I = 120$ at 85% dilution.

Correlation with the recombination zone length R is not quite as good. Since the recombination zone can have a very long tail (see Fig. 6.43) an arbitrary choice was made: The recombination time was defined as the time required for complete recombination at the initial recombination rate. The ratio Z/R was found to be a linear function of p_o, with Z/R changing with p_o (though much less rapidly than Z itself), and with values of Z/R ranging from 10 to 400.

The other systems show more complicated behavior. At large dilution Z/I still approaches approximately constant values at high pressure, with values on the order of 100 for C_2H_4/O_2, 60 to 90 for C_2H_2/O_2, and 20 or less for CH_4/O_2. With less dilution, Z/I becomes a strongly increasing function of pressure. The reader interested in the details is referred to the original paper.

Other things being equal, the cell size appears to be proportional to the sound speed. Two of the sets of data of Fig. 7.15 differ only in the choice of diluent: the systems $2H_2 + O_2 + 7Ar$ and $2H_2 + O_2 + 7He$. The main difference here is in the sound speed of the products, and the ratio

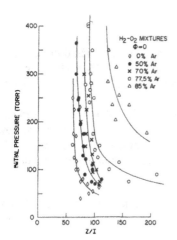

Fig. 7.16. The pressure dependence of the ratio of transverse wave spacing Z to calculated CJ induction zone length I for the H_2/O_2/diluent mixtures of Fig. 7.15. From Strehlow and Engel (1969).

Z/c, with c the CJ sound speed, turns out to be the same for the two systems over the entire range of p_o. Note that this ratio is just the (transverse) transit time of an acoustic wave over a cell width. We remark that Edwards (1969) has suggested that the different thermal conductivities of these two diluents could also have an effect by changing the rate of energy loss to the walls.

A similar study of $CO/H_2/O_2/Ar$ systems has been made by Libouton and van Tiggelen (1976). They varied the induction time by adding the chain-breaking inhibitor CF_3Br as well as by varying the initial pressure. The values of Z/l for these systems are similar to those described above.

In a perfectly regular structure, such as that of Fig. 7.10, the pattern repeats exactly down the tube. Fig. 7.17 shows the dependence of the mode number n (number of transverse waves across the tube) as a function of initial pressure for a large number of firings in such a system. The bars show the pressure range over which the particular mode number is observed. Over a few narrow ranges of pressure, one of two modes may be obtained in a given firing; at pressures near 9.3 kPa (70 mm) the value of n may be either 6 or 7. Note also that at higher n the even mode numbers (with the half cells at the two walls in phase with each other) are preferred.

7B2.2. Acoustic Coupling

The problem is that of acoustic vibrations in the detonation products driven by disturbances on the front. Throughout, the approximations are those of linear small disturbance, and the front, to the extent that it is treated, is regarded as a jump discontinuity with instantaneous reaction.

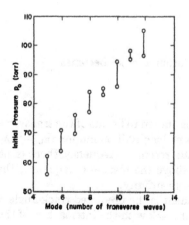

Fig. 7.17. Mode numbers vs. initial pressure in $2H_2 + O_2 + 3Ar$. From Strehlow, Adamczyk, and Stiles (1972).

The solutions for acoustic waves are the standard ones, and may be expressed in slightly different forms; our presentation is closest to that of Fay (1952). We assume a supported CJ detonation, neglecting the Taylor rarefaction wave, so that the state is just the uniform CJ state of the detonation products, moving at the CJ particle velocity. We use an overbar for this state and plain symbols for perturbations on it. We work in the frame in which the products are at rest; in this frame the front has velocity $\bar{c}$. We consider the simplest case first: a flat channel of height ℓ with motion in only one direction, as in Fig. 7.5, with z distance along and y distance across the channel, and with y = 0 at the lower wall. We use n for the mode number, λ_n and λ_L for the transverse and longitudinal wavelengths (with corresponding wave numbers $k_n = 2\pi/\lambda_n$ and $k_L = 2\pi/\lambda_L$), and w and u for the transverse and longitudinal particle velocities.

The general solution satisfying the wall boundary conditions w = 0 at y = 0 and y = ℓ is

$$p = -\bar{\rho}\phi_t$$

$$u = \phi_z$$

$$w = \phi_y$$

$$\phi = A \cos(\omega t + k_L z) \cos(k_n y)$$

$$k_n = n\pi/\ell \quad \text{or} \quad \lambda_n = 2\ell/n, \quad \text{n integral,}$$

where A has dimensions of (length)^2/(time). The effective wave number k is

$$k^2 = k_n^2 + k_L^2$$

so that the wavelength-frequency relation $\omega = k\bar{c}$ becomes

$$\omega = (k_n^2 + k_L^2)^{1/2}\bar{c}.$$

Note that if the angular frequency ω is chosen to be one of the transverse resonant frequencies $\omega = k_n\bar{c}$, there is no longitudinal motion ($k_L = 0$, $\lambda_L = \infty$). For given n, with ω below the resonant frequency there is no steady solution. With ω increasing above the resonant frequency, the longitudinal wavelength decreases from infinity.

The main comparison with experiment is the predicted track angle α (Fig. 7.5). We identify the triple-point track with the intersection of the crest of the acoustic solution (or of one of its traveling-wave components in the rectangular case) with the front. Evaluating ϕ at the front $z = \bar{c}t$ we have

$$\phi(z = \bar{c}t) = \cos(\omega_f t)\cos(k_n y)$$

$$\omega_f = \omega + k_L \bar{c},$$

so that ω_f is the angular frequency at the front. (Thus in a frame attached to the front we would have pure transverse vibration with angular frequency ω_f.) Denoting by c^* the lateral wave velocity on the front we have

$$c^* = \omega_f/k_n = \bar{c}\,[(1 + k_L^2/k_n^2)^{1/2} + k_L/k_n]$$

so that $c^*/\bar{c}$ depends only on the wave-number ratio k_L/k_n. The track angle α (laboratory frame) is then just

$$\tan\alpha = c^*/D.$$

To determine the wave-number ratio and thus complete the solution we need to apply a longitudinal boundary condition. In principle we do this by specifying the impedance p/u (a constant) at the front. We define a dimensionless impedance Z

$$Z = (1/\bar{\rho}\bar{c})(p/u)_{z=\bar{c}t},$$

and find by substituting for p and u

$$Z = -\omega/k_L\bar{c} = 1 - \omega_f/k_L\bar{c}.$$

From the original wavelength-frequency relation we can then replace ω by the wave numbers and get the desired expression for the wave-number ratio in terms of Z

$$Z = -(1 + k_n^2/k_L^2)^{1/2}.$$

We may also use this result to express ω_f, and thus c^*, in terms of Z

$$c^* = \omega_f/k_n = \bar{c}\,[(Z-1)/(Z+1)]^{1/2}.$$

In practice we do not know enough about the front to specify Z there and instead resort to estimating k_L from experimental photographs. It

STRUCTURE OF THE FRONT   Chap. 7

turns out that k_L is much smaller than k_n and Z is of the order of -20, so that c^* is only a few percent larger than $\bar{c}$.

In a round tube of radius r_o and radial coordinate r we have

$$\phi = \cos(wt + k_L z + n\theta) J_n(k_{nm}r)$$

$$k_{nm} = \zeta_{nm}/r_o,$$

in which the argument of the first cosine function has the added term $n\theta$ and the second cosine function is replaced by the nth-order Bessel function. Here ζ_{nm} is the mth zero of the first derivative of J_n, that is, it satisfies

$$J_n'(\zeta_{nm}) = 0,$$

and ϕ has n nodes in θ and m nodes in r. All the remaining equations for the rectangular case apply if we replace k_n by k_{nm}, except for the fact that the relation between ω_f/k_{nm} and c^* has an added factor ζ_{nm}/n as given in

$$c^* = (\zeta_{nm}/n)\, \omega_f/k_{nm} = \bar{c}\,(\zeta_{nm}/n)\,[(Z-1)/(Z+1)]^{1/2}.$$

The first few ζ_{n1} are

n	1	2	3	4	5
ζ_{n1}/n	1.84	1.53	1.40	1.34	1.27

As $n \to \infty$, $\zeta_{n1}/n \to 1$, and the equation for c^* approaches the rectangular result. We will limit our discussion to $m = 1$, which appears to match the experiments. Note that at the periphery, the pressure peak travels faster than the speed of sound, almost twice as fast in the fundamental mode. How nature constructs the driver (front with transverse wave) to accomplish this will be seen in the next section.

In round tubes the pitch/diameter ratio is sometimes used instead of the track angle. The pitch is defined as the longitudinal spacing of a helical track on the tube wall written by a particular disturbance on the front (regarding the helix as a screw thread, "pitch" has its usual meaning). The pitch/diameter ratio is related to the track angle by

$$(\text{pitch/diameter ratio}) = \pi/\tan\alpha.$$

Let us now visualize these solutions and how they would look to a smear camera viewing them through a slit on the tube parallel to the tube axis. To simplify matters let us suppose a medium which lights up (only) at the peak pressure. Consider first the rectangular tube, Fig. 7.18. The smear camera looks up into the bottom of the tube (the bottom and top will also be referred to as the near and far sides). Ignore for the present the diagonal tracks and circles. For the fundamental mode we have n = 1 with no longitudinal component, that is, $k_L = 0$; the bottom and top walls light up alternately, the period being the time between successive lightings at the same wall. Successive transverse pressure profiles over a half cycle are sketched at the right. The camera record is a sequence of horizontal lines, with those from the far side dimmer (shown dashed) from absorption. With a longitudinal component present, the first cosine function has a phase angle $k_L z$ depending on z. The lighting is no longer simultaneous at each wall but travels along it with phase velocity $-\omega k_L$, so that the lines on the film are inclined as shown; the longitudinal wavelength can be inferred from their tilt.

The n=2 pattern can be constructed by stacking two half-scale n=1 patterns. For $k_L = 0$ the lighting sequence becomes bottom, middle, top, middle, bottom, etc.; both the (vertical) spacing of the lines on the film and the period are half those for n=1. Again the line type indicates the effect of absorption: solid, chain, and dashed lines for near side, middle,

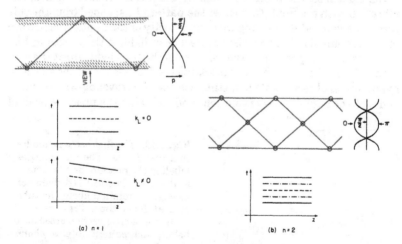

Fig. 7.18 Acoustic modes in a rectangular tube, with schematic smear-camera records. Pressure profiles at the indicated phase angles are shown at the right.

STRUCTURE OF THE FRONT
Chap. 7

and far side. The effect of $k_L \neq 0$ (not shown) is simply to tilt the lines on the film as before.

In the round tube, Fig. 7.19, the transverse pressure profile for $n=1$ is similar to that in the rectangular tube, that is, $J_1(\zeta_{11}r/r_0)$ is like $J_1(\zeta_{11})\sin[(\pi/2)r/r_0]$, slightly damped. The peak-pressure point is at the wall and rotates with angular velocity ω, so that for $k_L = 0$ the system may be visualized as a rotating cylinder with a strip of light (parallel to the axis) attached to its outer surface. The camera record is thus exactly like that for the rectangular tube for $n=1$, $k_L=0$. For $k_L \neq 0$, the surface of constant phase is

$$\theta = -k_L z,$$

a helix of pitch $\lambda_L = 2\pi/k_L$, so that the strip of light is now a helix around the periphery instead of a line parallel to the axis. The camera record is again just like that for the rectangular case for $n=1$, $k_L \neq 0$. For $n > 1$ we have additional peripheral nodes from the $n\theta$ term in the argument of the first cosine term of ϕ. Thus for $n=2$, $k_L=0$ we would have two strips of light on opposite sides of the cylinder, and for $k_L \neq 0$ two helices. A feature not present in the rectangular solution is the arbitrary direction of rotation; the complete solution may be a linear combination of solutions rotating in both directions.

Consider next the pattern formed by the intersection of the acoustic vibration with the front, that is, let the pattern be scanned from the side through a vertical slit moving with the velocity of the front. This pattern in the rectangular tube consists of flashes of light at the circles in Fig. 7.18 marking the position of the front when the tube lights up at the center or edges. The diagonal lines mark the track as seen from the side by a hypothetical observer able to distinguish the two traveling-wave components of the standing-wave solution, and following the pressure peak of

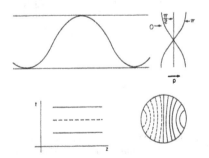

Fig. 7.19. The first acoustic modes in a round tube. The phase angles labeling the pressure profiles represent either time at fixed diameter or angle of rotation about the tube axis at fixed time. Pressure contours over a cross section are shown below; the pattern rotates about the tube axis.

the one having (currently) positive pressure. This could, of course, also be the path of a single narrow pulse (such as the triple point) moving back and forth between the walls. In the round tube we have immediately a pattern quite similar to experiment: one or more points of light (spin "heads") rotating around the periphery.

Discussions of the acoustic problem in the context of detonations (and comparisons with experiment) have been given by Manson (1946, 1947), Fay (1952), and Chu (1956). Manson considered only transverse vibrations. Fay added discussion of longitudinal vibrations and of the longitudinal impedance boundary condition at the front. Both of these works involve relatively straightforward application of standard acoustic theory. Chu analyzed the problem in more detail, but to get results by his approach, had to limit consideration to the slightly artificial situation of a piston-supported overdriven detonation. A brief account of his work follows.

The system is perturbed by a wave, generated by the piston, which overtakes the front. To generate the wave the piston's face is configured into the form of the solution and rotated with angular velocity greater than or equal to the tube's fundamental natural frequency (recall that this is the minimum frequency which will propagate a longitudinal wave). The stability and shape of the detonation front and the nature of the wave reflected back from it after overtake is studied.

The front is found to be unconditionally stable, a result which could have been obtained immediately from the shock stability theory of Sec. 6A4. Aside from the front stability, the principal results are the frequency and phase of the reflected wave relative to the incident one and the (time-dependent) shape of the front. The results are in good agreement with the main features of the smear-camera record of a spinning detonation in a round tube: the shape of the trace from the front and the phase of the horizontal stripes from the transverse vibrations in the burned gas are reproduced.

Aside from prominent qualitative features, such as the horizontal stripe in the round-tube smear-camera record, the principal comparison is between the calculated and experimental average track angle α. Recall that α is given by

$$\tan \alpha = c^*/D,$$

with the transverse velocity c^* related to the unperturbed sound speed (in the products) $\bar{c}$ by

STRUCTURE OF THE FRONT   Chap. 7

$$c^* = \bar{c}\,[(1 + k_L^2/k_n^2)^{1/2} + k_L/k_n], \quad \text{rectangular tube}$$

$$c^* = \bar{c}\,(\zeta_{nm}/n)\,[(1 + k_L^2/k_{nm}^2)^{1/2} + k_L/k_{nm}], \quad \text{round tube}$$

$$k_{nm} = \zeta_{nm}/r_o,$$

where k_L is the longitudinal wave number. In the rectangular case, n is the mode number and k_n is the transverse-wave number. In the round case, ζ_{nm} is the mth root of the first derivative of the Bessel function J_n, and r_o is the tube radius. Typically k_L is much less than k_n or k_{nm}, so that for the rectangular case c^* is in practice close to $\bar{c}$ for all n.

Refinements, such as the Taylor wave, frozen vs. equilibrium sound speed, and deviations of the unperturbed state, are ordinarily neglected. This is reasonable since, as we remarked earlier, it probably doesn't make sense to push this type of comparison too far. The sound speed $\bar{c}$ is thus taken to be the (equilibrium) sound speed at the calculated CJ state. To a good aproximation, this is just

$$\bar{c}/D = \gamma/(\gamma + 1) = \bar{p}/\bar{\rho}_o.$$

Some values are

γ	c/D	rectangular	round, n=1
1.1	0.524	27.5°	44.0°
1.2	0.545	28.6°	45.1°
1.4	0.583	30.3°	47.0°
1.67	0.625	32.0°	49.0°
2.	0.667	33.7°	50.8°
3.	0.75	36.9°	54.1°

The values of α here are for pure transverse vibration. Note that α is rather insensitive to γ in the physically reasonably range; apparently what is often done is to estimate γ rather than calculate the CJ state. As an example we may take the calculation of Table 5.3: for $2H_2 + O_2 + 9Ar$, $p_o = 10.1$ kPa, the calculated values (equilibrium γ and sound speed throughout) are $\gamma = 1.215$, $\bar{c}/D = 0.565$, $\alpha = 29.5°$; the approximation $\bar{c}/D = \gamma/(\gamma+1)$ gives $\bar{c}/D = 0.548$, $\alpha = 28.7°$.

A short summary of comparisons with experiment may be found in VMT (1969). For the round tube, m is usually taken to be 1, other values not improving the agreement. Fay (1952) found the calculated angles for $n \leq 5$ in a number of systems, assuming transverse vibration only, to deviate from experiment in both directions but on the whole to be 3 to 6% low, which he ascribed to the omission of longitudinal vibration. Duff (1961) reached a similar conclusion for H_2/O_2: As n increases from 1 to 9, the observed track angle decreases from 46° to 34°. The calculated angle decreases from about 45° to 32° over the same range. VMT state that the measured angles for very large n range from 31° to 35°, compared to the typical calculated value for $n = \infty$ of 29°, in apparent contrast with the slight trend of deviation with n found by Duff. For detonation in an annulus, produced by inserting a concentric rod in the tube, the predicted track angle is greater; inserting a rod of half the outer diameter, for example, increases the calculated track angle from about 45° to 53°. Here again there is reasonable agreement with experiment.

In rectangular tubes, VMT give as a typical observed value $\alpha = 30°$, very close to the calculated value of 29°. Edwards, Parry, and Jones (1966) find good agreement with the calculated (transverse only) value of 30° for $n > 8$, but with the observed value rising to as high as 40° for lower n. We remark that the patterns of Fig. 7.10b and c have $\alpha \simeq 30°$ and $\alpha \simeq 29°$, respectively.

7B3. The Transverse Wave

We depart slightly here from our subject, the mechanism of propagation of steady detonations, to consider, so to speak, the life cycle and habitat of transverse waves. Whenever we use the term "transverse wave" here, the added phrase "in the current environment" should be understood, since its behavior depends strongly on that of the shock along which it is running.

Fig. 7.20 shows the birth of transverse waves in a shock-initiated detonation as studied by Strehlow and Cohen (1962) and Strehlow, Liaugminas, Watson, and Eyman (1967). An incident shock I, too weak to initiate reaction, comes in from the right and reflects from the wall as the reflected shock R. Reaction behind the reflected shock appears first in the fluid which has been hot the longest, that is, at the wall. The activation energy is high enough to produce rapid acceleration of the reaction by the heat released. The rapid energy release generates a compression wave which runs over the remaining material before it has time to react appreciably. Typically this overtaking wave O quickly turns into a shock closely followed by fire and begins to build up toward a CJ detonation in the heated and compressed material behind the reflected shock. Before this build-up is complete, it overtakes the reflected shock at point C, and merges with it to form an overdriven detonation which slowly

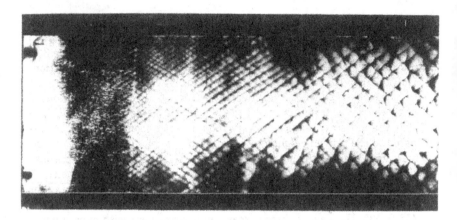

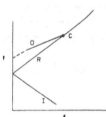

Fig. 7.20. Reflected shock initiation of $2H_2 + O_2 + 7Ar$, $p_o = 6.7$ kPa; incident shock Mach number is 2.16. From Strehlow, Liaugminas, Watson, and Eyman (1967).

decays to the CJ state. The process has been calculated by Gilbert and Strehlow (1966). The calculations of Fickett, Jacobson, and Wood (1970) in their investigation of oscillatory detonation show that the details of such processes can be quite complicated.

On the soot film, the transverse wave tracks typically first appear in the neighborhood of the overtake point C, all at about the same time and with a remarkably uniform transverse spacing. This regularity may be due to a particular initial perturbation: a rather large-amplitude transverse acoustic wave generated by the 0.1 mm gap between the end plate and tube. Initially the transverse waves are quite weak, as indicated by the lack of refraction where they collide. It takes some time for them to grow to their full strength and establish the regular structure. Note that the cells as recorded on the soot film are shortened because the detonation is moving into material accelerated to the left by the incoming shock I.

Another situation in which new transverse waves appear is the application of a strong perturbation to an existing front with established transverse waves. Gordeev (1976a) points out that the perturbation upsets the three-shock balance around each existing triple point causing it

Sec. 7B   MACROSCOPIC PROPERTIES

to split into two or more new ones. (The same type of splitting also takes place in nonreactive materials.) Typical perturbations are rarefactions from openings in the walls, or a sudden transition to the same gas at much higher initial density (such as that produced by cooling a section of the detonation tube in liquid nitrogen). With less drastic perturbations, such as transition from one mixture to a slightly different one with higher mode number, new transverse waves can take a long time to appear, as shown by the quantitative studies of Strehlow, Adamczyk, and Stiles (1972).

In a detonation front whose area is increasing, new transverse waves must apparently be born continually to maintain the spacing if the detonation is not to fail. A striking example is the comparison of propagating and failing cylindrically expanding detonations in Lee, Soloukhin, and Oppenheim (1969). The propagating detonation creates new transverse waves as required to maintain a constant spacing; the failing detonation creates no new transverse waves. Essentially the same process can be seen in divergent channels; see for example Strehlow and Salm (1976).

With the exception of those in single spin, transverse waves in an established detonation are not steady structures, but transient decaying phenomena requiring continual regeneration by collision to maintain their existence. The detailed structure studies described in the next section show that the detonation in each cell is in the process of failing, only to be revived by the next collision. The failure studies described below lead to the same conclusion.

Can the transverse wave survive overdrive? The stability results of Sec. 6A indicate that the instability persists to very large overdrive. But this is for a simple system with fixed exothermicity. As Gordeev (1976b) points out, real systems become endothermic at some degree of overdrive, since the degree of dissociation of the products increases with temperature. In practice, overdrive is produced in three ways, all transient: by overinitiating a detonation and letting it decay to CJ, by letting a CJ detonation run into a convergent channel, and by implosion. Extensive studies of the first type have been made by Gordeev (1976b). The typical result was that transverse waves appeared at some point in the decay; he took this to be the point at which the system becomes exothermic. The convergent channel experiments of Strehlow, Adamczyk, and Stiles (1972) show the transverse waves disappearing as the wave becomes more overdriven. For a convergent-divergent channel such as that used to produced the laminar detonation of Fig. 7.4, any weak transverse waves managing to survive the passage through the convergent section tend to die out in the partial failure at the entrance to the divergent section (see below). In implosion, the other method of producing increasing overdrive, the transverse waves are probably reinforced by the instability of the (nonreactive) shock (Whitham, 1974, p. 309). The cylindrical implosion experiments of Fujiwara, Mizoguchi, and Sugimura (1971), Gavrilenko, Topchian, and Yasakov (1967), and Knystautas and Lee (1971), show that, although the transverse waves become weaker and more closely spaced (and a few may disappear), they persist essentially to collapse.

Finally, we consider detonation failure. The beautiful pictures presented in Soloukhin (1965) of failure at an abrupt increase in area show that the transverse waves near the edges, moving out behind the expanding wave, lose their collision partners and rapidly decay. Those near the center continue to undergo collision for a while and hold out longer against the rarefactions coming in from the sides. Only if there are a certain minimum number (roughly ten or more) of transverse waves across the channel, are there enough collisions in the middle so that a central hot region survives to reinitiate the detonation. Strehlow and Salm (1976), in their detailed studies of failure at the entrance to a divergent channel, reach a somewhat similar conclusion, with the minimum number of transverse waves a function of the entrance angle.

Finally, we mention one result from the quantitive studies of transverse-wave decay rate by Strehlow, Adamczyk, and Stiles (1972): When a detonation in $H_2/O_2/He$ runs into a following section filled with an inert gas, the cellular pattern persists for a long time, decaying only slowly; the transverse-wave strength (pressure ratio across the reflected shock minus one) decays at the rate of about 7 percent per cell length.

7B4. The Sonic Surface

As mentioned in Chapter 3, early experiments demonstrated that the flow behind the front passes through a sonic transition to become distinctly supersonic, with Mach numbers of 1.1 or larger. Vasiliev, Gavrilenko, and Topchian (1972) used a novel experimental technique to locate the sonic surface. They mounted a thin plate perpendicular to the front with its edge lying along a tube diameter and took a sequence of schlieren snapshots of the bow wave generated by passage of the detonation wave over the plate. Such a bow wave in steady flow is detached from the plate when the incoming flow is supersonic. From their photographs they were able to find the point at which the wave began to detach itself from the plate; they took this to be the position of the sonic surface. Over several different mixtures of gases and a range of initial pressures, the ratio of the distance x of the sonic surface from the

Sec. 7C   DETAILS OF STRUCTURE

front to the cell size Z varied from about 1 to 3. The ratio x/Z was found to be a smooth, nearly linear function of d/Z, the ratio of tube diameter to cell size, with d/Z ranging from 4 to 30.

7C. DETAILS OF STRUCTURE

We first define some terms. The two structures epitomizing those observed are the simplified *weak* and *strong* structures shown in Fig. 7.21. In the weak structure the relatively weak reflected shock of the front triple point A constitutes the entire transverse wave. In the strong structure the small front piece of the transverse wave is this same reflected shock, but most of it, segment BC, is a detonation. Following Strehlow and Biller (1969) we define the transverse-wave strength S as that of the reflected shock at the front triple point A

$$S = (p_3/p_1 - 1),$$

evaluated at A. We will speak of strong structures of varying strength. The stronger the structure, the larger is S, the stronger is BC (ranging from underdriven to strongly overdriven), and the greater is the transverse velocity, with correspondingly larger track angle. The soot track left by the strong structure usually shows three lines written by triple points A, B, and C, with A the strongest. The detonation segment BC can have its own transverse waves which move back and forth, giving a diamond-shaped pattern to the band between B and C. Gordeev (1976b) argues that this will be the case so long as BC is not overdriven into the endothermic region, as discused in Sec. 7B3. The track left by the weak structure is the single line from point A.

Detonations themselves are classified as *marginal* and *ordinary*. A marginal detonation is just what the name suggests: one in a system near the detonation limit, so that it would fail if the pressure were a little lower or the tube a little smaller. Detonations which are not marginal are ordinary. Of course this is a continuous rather than a discrete grading, but we may set down some numbers (Strehlow and Crooker [1974], Edwards, Hooper, Job, and Parry [1970]) as a guide:

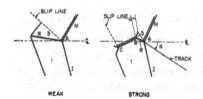

Fig. 7.21. Weak and strong transverse-wave structures.

	Ordinary	Marginal
mean propagation velocity $\bar{D}/D_j$	~1	0.8 − 0.9
variation through the cell of D/D_j	1.2 − 0.85	1.4 − 0.7
transverse-wave strength S	0.5	1.5
transverse-wave spacing Z/Z_{normal}	1	2

By Z_{normal} we mean the "normal" spacing defined by the straight lines of Fig. 7.15. The second and fourth lines of this table of course do not apply to single spin, which has no cells and only one transverse wave. We mention one apparent anomaly: Strehlow and Crooker (1974) state, without giving details, that the limiting case in their 83 × 38 mm (3 1/4 × 1 1/2 in.) tube has the properties of an ordinary detonation.

The systems studied in detail are all marginal: single spin in round tubes, 1 × 1 detonations in rectangular tubes, and planar detonations in narrow rectangular tubes. Ordinary detonations have too many cells for detailed study of the structure, but some quantitative information about them can be obtained by analysis of soot films.

Where the two structures of Fig. 7.21 occur is not entirely settled. The situation is roughly as follows. Structures of the strong type are generally associated with propagating detonations. The more marginal the detonation, the stronger the structure. Nature faces the greatest challenge in single spin, where the acoustic solution requires a tip (peripheral) transverse-wave speed of 1.84 times the sound speed. The challenge is met by the strongest structure of all, with the detonation segment BC highly overdriven.

Structures of the weak type are seen in separately generated Mach reflection like that of Fig. 7.4, and in detonations in the process of failing. In a propagating marginal detonation, the structure appears to be the weak form immediately after the transverse-wave collision, but soon changes over to the strong one. Other instances of transitions from one form to the other are given below. There is some indirect evidence that ordinary detonations have the strong structure, but, if so, they are probably the weakest strong structures of all, writing only a single line for their track on the soot film.

7C1. Marginal Detonation in a Round Tube (Single Spin)

The structure of single spin has been studied in some detail, most carefully and comprehensively by VMT (1963) and Schott (1965b). They worked in different systems but found essentially the same structure. The unique simplification of a strong, nearly steady transverse wave without collisions makes this perhaps the easiest system to study.

Figure 7.22 is a model of the front, Fig. 7.23 is a drawing extending some distance behind, and Fig. 7.24 and Table 7.1 give the details of the structure near the front. The model of Fig. 7.22 rotates as it proceeds down the tube. The next two figures are in the rotating frame attached to the triple point at the wall, in which the flow is steady, and show the structure as seen by an observer looking into the tube. The segment BC of the transverse wave (Fig. 7.23) is an overdriven detonation propagating into the unreacted material which has passed through the incident shock. The Mach stem at A is tilted so that the normal velocity of the material into it corresponds to an overdriven detonation with $D = 1.4\ D_j$. Consequently its reaction zone is very short. Proceeding around the tube (upwards in the diagram) on the leading shock, we find the fire lagging farther and farther behind until it joins the end of the transverse wave at C. Behind the front we have then two interleaved helical strips of material, strip 2 having passed through the fire and strip 5 through the transverse wave. There is a considerable gradient across

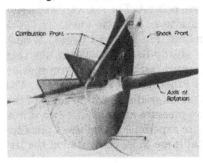

Fig. 7.22. Model of the front in single spin. From Schott (1965a).

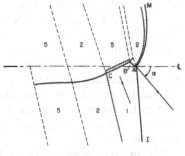

Fig. 7.23. Overview of transverse-wave structure. Axial distance is horizontal and circumferential distance vertical; about half the tube circumference is shown. In $12.5\%(C_2H_2 + 1.5\ O_2) + 87.5\%Ar$, $p_o = 2.4$ kPa. From Schott (1965b).

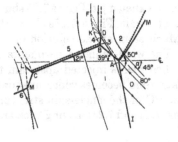

Fig. 7.24. Detailed structure of single spin near the tube wall; about one-third of the tube circumference is shown. In $2CO + O_2$, $p_o = 10.1$ kPa. From VMT (1963).

STRUCTURE OF THE FRONT   Chap. 7

Table 7.1 STATES IN THE SINGLE SPIN STRUCTURE OF FIG. 7.24

Region	p/p_0	ρ/ρ_0	T(K)	$q(m/s)^a$	M=q/c
0	.1	.1	293	2404	8.35
1	19.3	4.92	1140	1950	2.97
2	54.5	4.88	3550	491	0.446
3	54.5	10.48	1500	1720	2.31
4	170	23.8	2020	1290	1.5
5	170	14.5	3730	663	0.616
6	~15				
7	~30				

[a] q is the flow speed (in the steady frame attached to the triple point).

strip 2. The hottest material of all is that on the left edge of strip 2 which has passed through the stem just above point A. The rearward extension of BC is a shock which rapidly decays to an acoustic wave.

The more detailed structure given here, Fig. 7.24 and Table 7.1, both from VMT (1963), was obtained by steady-flow calculations like that described in Sec. 7A4, using the observed track angle and propagation velocity, and is closely confirmed by experimental measurements. The picture is oversimplified in that it ignores gradients within some of the regions (in region 1, for example, the pressure at point C is two-thirds that at B). Except for the approximate experimental values of pressure in regions 6 and 7, the states in the table are those calculated at the triple points A and B.

We turn our attention first to the main segment BC of the transverse wave. This is an overdriven detonation (shock closely followed by fire) moving into region 1 with a detonation velocity of 1950 m/s compared to the CJ velocity in this material of 1720 m/s. It is bounded on each end by a triple shock structure, for each of which it is the Mach stem. (The details at the left end, LCM, are not clearly seen in the photographs and are partly guessed). It may be helpful to compare this structure to the possibly more familiar one occurring in an exhaust jet. Figure 7.25 shows the jet produced at the end of a divergent nozzle. Figure 7.25a shows the case of no reaction, with a central Mach stem at the collision of the shocks from the edges of the nozzle. The wind-tunnel experiments of Gross and Chinitz (1960) show that when fuel is injected upstream to produce a standing detonation, the Mach stem becomes wider, as shown in Fig. 7.25b. The detonation segment BC of the single-spin structure has transverse waves (not shown in Figs. 7.23 and 7.24) moving back and

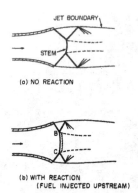

Fig. 7.25. Jet formed at the end of a nozzle. The standing detonation BC in the reactive case is to be compared with the corresponding segment BC of the transverse wave in Fig. 7.24. From Gross and Chinitz (1960).

forth on it. These produce the familiar diamond pattern within the band which is the track of BC, as seen in Fig. 7.12.

Returning to Fig. 7.24, point A is the triple point on the front, with stem AM, incident shock AI, reflected shock AB, and slip line AD. The reflected shock from triple point B meets slip line AD at D; the resulting interaction produces the centered rarefaction DFK, which deflects the flow as shown. (The flow in region 4 is supersonic and DK is the lead characteristic of the rarefaction.) The extent and location of reaction in region 4 are not clear: the calculated temperature of 2020 K is quite high but the particles are quickly cooled by the rarefaction.

In the triple-point matching calculations for points A and B, better results were obtained by treating the stems as detonations with instantaneous reaction; it is these results that are shown in Table 7.1. No calculations were done for point C.

As mentioned earlier, the structure discussed so far is that near the outer wall. As we move toward the center, the transverse wave gets weaker and there are some additional complications. The confluence line shown as the triple point above is apparently itself a three-shock configuration as shown by the end-plate soot film of Fig. 7.26 (this complication is not included in the model of Fig. 7.22). We omit from our discussion some additional details of the structure and some minor difficulties of interpretation.

The track fluctuations sometimes seen have received considerable study, see Steel and Oppenheim (1966), Denisov and Troshin (1971), Manzhalei and Mitrofanov (1973), Gordeev (1974), and Topchian and Uljanitskij (1976). The most common mechanism appears to be a transition between weak and strong structures, often triggered by a weak counter-rotating wave. An example of such a transition is shown in Fig. 7.27. This soot film appears to have a weak structure at the upper left,

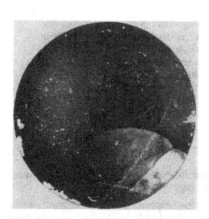 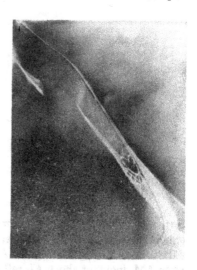

Fig. 7.26. Single-spin end-plate soot imprint showing that the confluence line is itself a three-shock configuration. From Schott (1965b).

Fig. 7.27. Soot film showing weak-strong-weak structure transition in single spin in $2H_2 + O_2$, $p_o = 6.1$ kPa. From Steel and Oppenheim (1966).

with a sharp trace and a track angle of 30°. A transition to the strong structure, with a banded track and a track angle of 50°, takes place rather abruptly. The strong structure then decays back into the weak one: points B and C (Fig. 7.24) recede rapidly from the front, with B overtaking and coalescing with C, thus removing the detonation segment and leaving just the shock AB of the weak structure. VMT (1963) have observed essentially the same strong-to-weak transitions in failing planar detonations, as described in the next section.

We remark that Jones (1975) has considered the effect of the boundary layer on single spin, and concluded that it is minor.

7C2. Marginal Detonation in Rectangular Tubes

Compared to single spin, this geometry has the disadvantage that the structure changes as the wave moves through the cell, and also differs from one cell to the next. But the variation across the tube along the line of sight is less, and can be virtually eliminated by making the tube narrow enough.

We consider mainly the planar (narrow tube) case. The boundary-layer effect across the narrow dimension is important here. A recent study is that of Strehlow and Salm (1976). The boundary layer grows

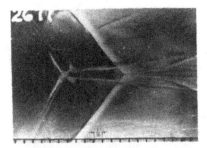

Fig. 7.28. Magnified view of a collision from a soot film like that of Fig. 7.10d. A 2 × 0 detonation in 50%($2H_2 + O_2$) + 50%Ar, p_o = 6.7 kPa. From Crooker (1969).

Fig. 7.29. Overview of planar detonation structure. (The 2 × 0 detonation of Figs. 7.10d and 7.28.) From Strehlow and Crooker (1974)).

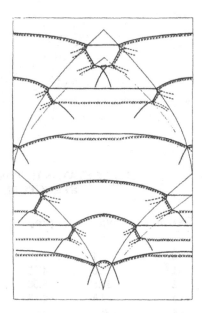

rapidly behind the front and soon fills the tube. Using an approximate theory, these authors calculated the pressure profile behind the front, adjusting the friction coefficient so as to match the experimental profile. They also estimated the rate of boundary-layer growth and summarized the characteristic lengths and relevant properties as follows: for 50%($2H_2 + O_2$) + 50%Ar at p_o = 12 kPa in a 6.3 mm (1/4 in.) wide channel, the calculated one-dimensional CJ reaction-zone length is 2 mm, the transverse-wave spacing Z is 60 mm, and the boundary layer fills the tube 85 mm behind the front. Thus boundary-layer effects across the narrow dimension are clearly important, giving rise to the marginal character. These effects are of course accepted in exchange for the favorable geometry, the best obtainable for detailed studies of the structure.

7C2.1. Structure

The structure is shown in Fig. 7.28 - 7.30 and Table 7.2. Fig. 7.28 is a magnified view of the collision portion of a soot film like that of Fig. 7.10d. Figure 7.29 is a drawing of the structure through a complete cell. Figure 7.30 and Table 7.2 give the detailed structure.

As seen in Figs. 7.10d, slapping waves are present in planar detonation, but are weak. Apparently they are generated at the collisions of the principal waves and die out rapidly between these collisions. As shown

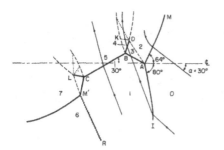

Fig. 7.30. Detailed structure of planar detonation at the average track angle of 30°. In 4 × 0 to 8 × 0 detonation in 97%(2CO + O_2) + 3%H_2 at p_o = 33 kPa. From VMT (1963).

Table 7.2 STATES IN THE PLANAR DETONATION STRUCTURE OF FIG. 7.30[a]

Region	p/p_o	ρ/ρ_o	T(K)	q(m/s)[b]	M=q/c
0	1	1	293	1960	6.80
1	17.5	4.85	1060	1480	2.35
2	39.5	6.06	1920	345	0.41
3	39.5	8.83	1310	1270	1.82
4	109	18.6	1705	812	1.18
5	109	16.9	1900	430	0.517
1	16	4.68	1000	1170	1.91
2	29.5	5.67	1530	300	0.396
5	65.0	12.7	1500	430	0.57

[a]The upper part of the table is for the overall average track angle of 30°; the lower, for the typical track angle of 42° just before collision.

[b]q is the flow speed (in the steady frame attached to the triple point).

in Fig. 7.29, the structure immediately after collision appears to be of the weak type; details cannot be resolved because of the short reaction time, but the track is single, with the double track of the strong structure often appearing rather abruptly about one-fourth of the way to the next collision. The diamond pattern within the band from strong transverse waves on the detonation segment BC, typical in single spin, is absent here. The induction zone on the axis becomes resolvable at about the center of the cell and has become quite long at the end.

The detailed structure given in Fig. 7.30 and Table 7.2 was obtained in the same way as that for single spin. The values given are again from triple-point calculations at points A and B from the measured track angle and wave (triple point A) velocity, confirmed by pressure

Sec. 7C DETAILS OF STRUCTURE

measurements and photographs, except that here all waves are treated as nonreactive. The fire behind the stem AM and transverse wave segment BC is not shown because its position is uncertain. Again the structure at point C is partly guessed, and no calculations are done for it. The most striking difference from the single spin structure is that the detonation BC is now unsupported instead of overdriven, with the position of the fire uncertain, but appreciably displaced from the shock. The solution for the detonation BC is essentially that of Chapter 5, Sec. G for a detonation with lateral divergence, with the slip lines from B and C serving as the channel boundaries, and a transition from subsonic to supersonic flow on some surface behind BC.

As the wave proceeds through the cell the structure becomes weaker. The track angle increases (because the front is slowing down), the Mach number M_4 drops below one, and the rarefaction front DK overtakes and erodes shock BC. Before this happens, the rarefaction transmitted into region 5 could weaken BC if the sonic transition in region 5 has not occurred before point K. The observed result as the wave proceeds is that the segment AB lengthens, and BC weakens. The main fire RM' of course lags farther behind the front as the front gets weaker and the induction time increases.

The states just before collision at a typical track angle of 42° are also listed in Table 7.2. We remark that immediately after collision the pressure behind BC jumps to about $200p_o$. Details of the collision process are discussed by Crooker (1969) and Subbotin (1975b). Subbotin suggests that the collision produces both front- and rear-facing jets which ignite pockets of unburned gas to produce the double-bow pattern seen in recent photographs. An example is indicated by the chain lines in the drawing of Fig. 7.31.

VMT (1963) extended their study to the case of failing transverse waves, removing the collision partners by running the detonation into a divergent section, Fig. 7.32a. The decay process described above continues and the structure turns into a structure of the weak type which they call type II, Fig. 7.32b. Points B and C recede from the front, but B recedes faster and overtakes and merges with C, forming the weak structure, the process being essentially the same as that described in the alternating-structure type of single spin discussed above. This same weak structure is seen in an occasional cell in propagating detonations. Subbotin (1975a) discusses this subject in more detail.

The structure of 1 × 1 detonation in square or rectangular tubes is similar to that of planar detonations. These have been extensively studied by Edwards, Hooper, Job, and Parry (1970), and Lundstrom and Oppenheim (1969). The structure drawn by Edwards, et al., is similar to

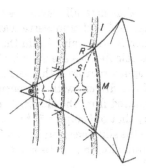 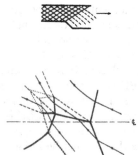

Fig. 7.31. Sketch of the double-bow pattern produced by transverse-wave collision in 30%(2H$_2$ + O$_2$) + 70%Ar at 13.3 kPa in a 27 × 27 mm tube. From Takai, Yoneda, and Hikita (1974).

Fig. 7.32. Structure of a failing planar detonation. (a) Failing transverse waves (dashed lines) in a divergent channel. (b) The weak (VMT type II) structure of the failing waves. From VMT (1963).

Fig. 7.30 with the segment AB longer relative to BC and details at C omitted. An interesting special case presented by Lee, Soloukhin and Oppenheim (1969) is the attainment of single spin in a rectangular tube, with a single transverse wave with strongly banded trace rotating around the periphery.

A periodic weak-to-strong transition observed in a 1 × 1 rectangular detonation has been analyzed in detail by Urtiew (1969). The soot film is shown in Fig. 7.33. The abrupt transition at point A is triggered by the collision of the slapping wave with the wall. A faint trace from the rear triple point C (Fig. 7.30) can be detected before the collision. After the collision the track angle increases abruptly, and transverse waves appear almost immediately on the detonation segment BC, Fig. 7.30, of the strong structure.

We close this section with brief descriptions of two interesting details of structure which have not been explained or integrated into the general picture.

• *Second Compression Wave.* A "second compression wave" parallel to and just behind or in the tail of the main fire is mentioned by Strehlow and Crooker (1974) and studied in some detail by Edwards, Hooper, and Meddins (1972), in a planar 4 × 0 detonation in 40%(2H$_2$ + O$_2$) + 60%Ar at p_o = 10.7 kPa in a 76 × 6.3 mm (3 × 1/4 in.) tube. One or more strong longitudinal density maxima extending uniformly across the wave ride along on the tail of the entire fire (behind both the incident shock and

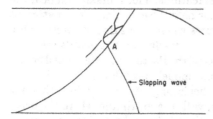

Fig. 7.33. Abrupt weak-to-strong transitions triggered by slapping-wave collisions in a 1 × 1 rectangular detonation in $H_2 + O_2$, $p_o = 47$ kPa, 19 × 44.5 mm tube, wide wall shown. From Urtiew (1969).

the Mach stem). This "wave" remains attached to the fire and retains a constant amplitude as the wave front decays. A similar phenomenon was observed in a one-dimensional wave in $2CO + O_2$ at 3.3 kPa in a diverging channel. The likely conclusion is that it should be predictable by a one-dimensional model of flow behind a decaying wave. A definite explanation of this phenomen has not been found; the authors discuss qualitatively some of the possibilities, including a suggestion by Soloukhin of a longitudinal oscillation like that of Sec. 6C, but confined to the region of the fire.

• *Blast Waves from Hot Spots.* Lu, Dabora, and Nicholls (1969) worked at relatively high pressure, 101 kPa, in a relatively small tube, 13 × 91 mm (0.5 × 0.3 in.); a typical case of interest here is a 1 × 1 detonation in $98.3\%(2CO + O_2) + 1.7\%H_2$. They saw a succession of weak blast waves moving back into the products, apparently produced by repeated explosions of hot spots near the front. The strength of these blast waves was comparable to that of the transverse waves. The authors did not study the phenomenon in great detail or draw any further conclusions. The spacing of the blast waves appears to be comparable to the tube dimensions; it seems likely that the explosions are an integral part of the process and repeat at the fundamental period. A similar one-shot phenomenon is the "Explosion-in-the-Explosion" which is a characteristic feature of one type of deflagration-to-detonation transition; see, for example Lee, Soloukhin, and Oppenheim (1969).

7C2.2. Variation through the Cell

The variations of some of the (calculated) pressures of Fig. 7.30 through the cell are shown in Fig. 7.34. The dashed line in the first part of the cell is a reminder that the structure is not resolved there. Variations of front velocity along the centerline for several systems are shown in Fig. 7.35. Strehlow and Crooker's data show the scatter obtained in a number of firings. Values for the planar system studied by VMT (1963) described above are similar to those of Steel and Oppenheim (1966).

The axial velocity history observed by Takai, Yoneda, and Hikita (1974), Fig. 7.36, shows two unusual features: a local maximum near the beginning of the cell, and a sharp rise near the end. They do not attempt to explain why they see these features when others do not; presumably their measurements have higher spatial resolution, but details are lacking. The local maximum tends to confirm Barthel's model of cell structure described in Sec. 7D. This model postulates an initiation mechanism after collision at the beginning of the cell like that for homogeneous initiation (Sec. 7B3), with the maximum where the overtaking wave reaches the front. The rise at the end remains a mystery.

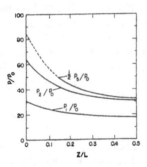

Fig. 7.34. Pressures for the detonation of Fig. 7.30 vs. normalized distance through the cell. From VMT (1963).

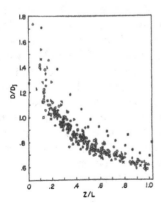

Fig. 7.35. Axial front velocity vs. normalized distance through the cell. Open symbols and X's, detonation of Figs. 7.28, 7.29, Strehlow and Crooker (1974); •, 1×1 in $50\%(2H_2 + O_2) + 50\%N_2$, $p_o = 7.9$ kPa, 25×38 mm ($1 \times 1\ 1/2$ in.) tube, Lundstrom and Oppenheim (1969); ■, 1×1 in $20\%\,2(H_2 + O_2) + 71\%N_2$, $p_o = 13.3$ kPa, 19.1×44 mm ($3/4 \times 1\ 3/4$ in.) tube, Steel and Oppenheim (1966). From Strehlow and Crooker (1974).

Sec. 7C   DETAILS OF STRUCTURE

Strehlow and Crooker also calculated induction-zone lengths on the assumption that each particle remains in the state produced by its passage through the shock. The calculated length, Fig. 7.37, increases by a factor of 3000 through the cell.

A more realistic estimate of induction time can be made by using the mathematical form of standard nonreactive blast wave theory to model the flow. A blast wave is the flow produced by an instantaneous release of energy on a point, line, or sheet source, producing a self-similar flow with sphere, cylinder, or slab symmetry. In the strong-shock limit for a polytropic gas the motion of the front is given by (with the energy release at $x = t = 0$)

$$D = ax/t$$

for

$$x = (Ke/\rho_0)^{a/2} t^a$$

$$a = 2/(\alpha + 3)$$

$\alpha = 0, 1, 2$ for slab, cylinder, sphere.

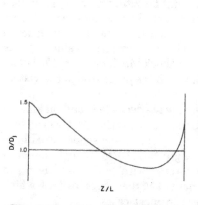

Fig. 7.36. Axial velocity through the cell for the detonation of Fig. 7.31. From Takai, Yoneda, and Hikita (1974).

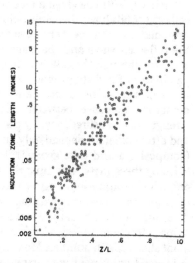

Fig. 7.37. Calculated induction zone length vs. normalized distance through the cell ($L \cong 5"$) for the detonation of Figs. 7.28 and 7.29. From Crooker (1969).

Here x is distance from the source, k is a dimensionless constant, and e is the energy per unit area for the slab, per unit length for the cylinder, and total energy for the sphere. There are also simple equations for the complete flow field, although their exact solution in convenient independent variables requires the iterative solution of a nonlinear equation.

Lundstrom and Oppenheim (1969) found that their measurements of front position in a 1 × 1 detonation in 50%($2H_2 + O_2$) + 50%N_2 at p_o = 7.9 kPa in a 25 × 38 mm (1 × 1 1/2 in.) tube were well reproduced by the blast-wave equation if they took α = 1/2, that is, by a hypothetical blast wave midway between the slab and cylinder cases. The blast-wave origin is outside the cell at Z/L = −0.28. With this more realistic model it is clear that the detonation is failing, with the induction time becoming infinite about 0.7 of the way through the cell.

The blast-wave model, though useful, is not quantitatively correct. Edwards, Hooper, Job, and Parry (1970) have shown that the blast-wave pressure profiles differ appreciably from experiment. Strehlow (1970) has shown that to first approximation the point of origin and energy of the approximating blast wave depend only on the transverse-wave strengths, and that blast-wave energies evaluated from observed transverse-wave strengths are of the order of 100 times the energy available from detonating all the material in the cell.

Urtiew (1976) has given a nice summary of the invariants through the cell. Fig. 7.38 shows the triple point moving through a cell. To a good approximation both the transverse-wave strength and the angle ϕ between the reflected wave and the track (the same as angle ϕ_1 of Fig. 7.7) are constant through the cell. The triple-point structure may be pictured as an object of fixed shape moving along a curved track, rotating as it goes to keep the angle ϕ constant. Its strength of course undergoes a monotone decrease between collisions. The extreme values of the strength (as measured, say, by the axial shock Mach number D/c_0 before and after collision) depend only on ϕ, which is a parameter of the system (composition and tube geometry).

Using these facts, assuming the blast-wave decay law, and making an empirical assumption, Urtiew was able to express the blast-wave point of origin and the decay exponent as functions of the single variable ϕ. The variation of normalized wave strength $M(Z/L)/\bar{M}$ through the cell is a one-parameter function with parameter ϕ, with $\bar{M}$ the time-averaged axial shock Mach number. Measurement of $\bar{M}$ then gives $M(Z/L)$, which is in good agreement with experiment in most cases.

7C3. Ordinary Detonation

Most of the information on ordinary (non-marginal) detonations comes from soot films. As explained in Sec. 7A5, the entrance angle (between the colliding tracks) determines the transverse-wave strength to a good approximation. One can also find the change in transverse-wave strength and in front velocity through the collision, and check the conclusions by measuring the exit angle.

Strehlow and Biller (1969) measured the track angles for a large number of collisions in the systems $2H_2 + O_2$, $30\%(2H_2 + O_2) + 70\%Ar$, $15\%(2C_2H_2 + 5O_2) + 85\%\,Ar$, and $25\%(C_2H_4 + 3O_2) + 75\%Ar$, at 6.7 to 47 kPa initial pressure in their 83 × 38 mm (3 1/4 × 1 1/2 in.) tube. The data base consisted largely of soot films accumulated in previous investigations. The mode numbers are not given, but, according to Edwards, Hooper, Job, and Parry (1970), they range from 6 to 21.

A typical set of results is shown in Fig. 7.39. There is a large variation of transverse-wave strength, but the mean value is essentially independent of pressure. Following individual transverse waves through many

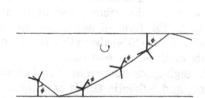

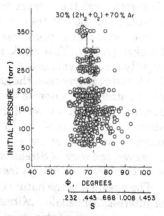

Fig. 7.38. The triple point rotates counterclockwise as it proceeds so as to keep the angle ϕ between the reflected shock and the track approximately constant. From Urtiew (1976).

Fig. 7.39. Entrance angle Φ and corresponding transverse wave strengths S (pressure ratio minus one) for a number of collisions in $30\%(2H_2 + O_2) + 70\%Ar$ over a range of initial pressure. From Strehlow and Biller (1969).

collisions showed that the strength of a particular wave oscillates strongly about some mean value, and that the amplitude of this oscillation varies markedly from wave to wave. Further details on this point may be found in Strehlow, Adamczyk, and Stiles (1972). The average value of the entrance angle and corresponding transverse-wave strength is nearly the same for all systems. The average entrance angle (twice the track angle for a symmetric collision) appears to be very close to 70°; the corresponding transverse wave strengths S (pressure ratio across the shock minus one) all lie within the range 0.44 to 0.48. The mean strength value 0.46 ± 0.05 for the $H_2/O_2/Ar$ system is given in a later paper, Strehlow (1970). Through a typical collision in this same mixture, the transverse-wave strength decreases by 5%, while the axial front velocity increases 48%. As stated earlier, a typical variation in axial D through the cell is 1.2 D_j to 0.85 D_j.

The structure has not been directly observed. VMT (1969) believe that it is essentially the same as that seen in marginal (rectangular and planar) detonations. They point out that there is enough reaction in the transverse waves so that they can be recorded by their own light. Also they state that the average track angle in ordinary detonations is 33°, greater than the 30° they found in their studies of planar detonations (these detonations were less marginal than those of Strehlow and Crooker [1974]). This track angle of 33° may be compared with the slightly smaller values of 30° and 29° which one can measure on Fig. 7.10b and c. In any case, the argument is that the greater track angle implies a stronger transverse wave and thus suggests the strong structure. Strehlow (1969) has favored the weak structure, on the grounds that the observed tracks are single lines, and that the strong structure should leave some trace of its existence in the form of additional marks. However, the work of Strehlow and Crooker (1974) shows that the strong structure does not always leave its mark on the soot film.

The question of the nature of the high-mode-number limit is of course of interest. Manzhalei, Mitrofanov, and Subbotin (1974) have made a systematic study of this question, increasing the initial pressure to 1000 kPa. They found that the structure persists essentially unchanged. This is not surprising since, as Strehlow and Biller (1969) point out, all of the characteristic lengths—induction-zone length, shock thickness, and slipstream thickness—are inversely proportional to the initial pressure.

7D. COMPARISON OF THEORY AND EXPERIMENT

A few comparisons between theory and experiment have been or can be made; some are qualitative because the calculations, coming first, were not done for the systems observed. The topics are (1) the onset of

Sec. 7D   COMPARISON OF THEORY AND EXPERIMENT

instability with increasing heat of reaction, (2) fast gallop (one-dimensional oscillation) in the projectile bow wave, and (3) cell size and structure in the free-running detonation. Some of the "theories" of cell size are not a priori, but take as their starting point the main features of the observed structure.

7D1. Onset of Instability

Soloukhin and Brochet (1972), studied the onset of instability with increasing heat of reaction and shock strength in $x(1/2\ NH_3 + 1/2\ O_2) + (1 - x)Ar$, $x = 0.1$ to 0.3, in a 30×15 mm tube at $p_o = 1.01$ to 7.1 kPa. The tube was operated as a shock tube with a hydrogen driver section, and the shock strength was varied over the range $M = 4.4$ to 6 by changing p_o. (For $x = 0.3$, M_J is about 2.3; these detonations are strongly overdriven.) The one-dimensional flow was found to be stabilized by increasing the shock strength or by decreasing the heat of reaction (by decreasing x). Three points on the stability boundary in mole fraction of diluent x and shock temperature T were found:

x	T(K)
0.1	<1700
0.2	2200-2300
0.3	2300-2400.

This system has an induction zone; the first appearance of instability is in the neighborhood of the fire, with the shock and most of the induction zone unaffected. The crude Shchelkin theory, Sec. 6A2, predicts this stability boundary reasonably well. We may also compare Erpenbeck's result, Fig. 6.12, from an exact treatment of the stability problem but for first-order Arrhenius instead of branching-chain kinetics. Erpenbeck places the stability boundary at $q/RT_o = 5$ to 10 for $E^\dagger = 0$ depending on the degree of overdrive f, and at $q/RT_o \cong 0.3$, almost independent of f, for $E^\dagger = 50\ RT_o$. The NH_3/O_2 system studied here has $E^\dagger = 70\ RT_o$, and $q/RT_o \cong 1.9$ and 2.4 on the stability boundary at $\times = 0.2$ and $\times = 0.3$. Thus its critical heat of reaction is at least ten times that for the Erpenbeck system and its dependence on shock strength (degree of overdrive) is greater.

7D2. Fast Gallop

The one-dimensional oscillation which we call fast gallop, Sec. 6B3, is seen in only one experimental situation—at the tip of a projectile bow wave, Fig. 7.14. There are two calculations which can be compared with experiment. We will use the symbols

343

STRUCTURE OF THE FRONT   Chap. 7

τ = induction time

l = induction-zone length

$x_{1/2}$, $t_{1/2}$ = half-reaction distance, time.

Alpert and Toong (1972) have collected the data for fast-gallop periods in projectile bow waves for several different H_2/O_2 systems, and compared them to calculated induction times. With some scatter, they find two periods, 1.7τ and 4.6τ. The two sets of calculations are those of Alpert and Toong (1972) with a square-wave model, and those of Fickett, Jacobson, and Schott (1972) with a more realistic model. Alpert and Toong's square-wave model was the usual one—a constant-state induction zone with the correct activation energy, followed by instantaneous reaction. They used an approximate solution of the flow problem. Since the real steady reaction zone for this dilute system has a very long recombination tail, the square-wave model may be a rather poor approximation. Fickett et al. (see Fig. 6.43) used a one-reaction model with the correct induction-zone activation energy and about the right decrease in exothermicity with increasing drive, and did the flow problem essentially exactly by the method of characteristics. They used two rate models: rate 1, giving a steady profile with the desired long-tailed recombination zone but with the induction zone about twice too long, and rate 2, giving more nearly a square wave. But they were unaware of Toong's work and calculated for a somewhat more dilute system than any of those observed.

Both the Alpert and Toong model and the Fickett et al. rate 1 give both short and long periods, while the Fickett et al. rate 2 gives only the short period. Also the Alpert and Toong model and the Fickett et al. rate 1 give rather low-amplitude oscillations; the Fickett et al. rate 2 gives larger amplitudes, especially for the longer period.

The calculated periods are compared with experiment in Table 7.3. Surprisingly, the cruder the model, the better the results, with Toong and Alpert in agreement with the observations. Note however, that when the ratio of induction time to recombination time is changing, as between Fickett et al. rates 1 and 2, the period is more closely related to the half-reaction time than to the induction time. Nevertheless, the periods predicted by the more realistic rate 1 are clearly too long.

We remark that the quoted figure for the observed cell length in the "Discussion" section of Fickett et al. is in error. In comparing their (one-dimensional) spatial period with the cell length in a detonation with two-dimensional structure, they quoted the incorrect value of 9 mm for the observed cell length in $2H_2 + 9Ar$ at $p_o = 13.3$ kPa (0.1 atm). The

Table 7.3 FAST-GALLOP PERIODS

Source	Model	Time		Space	
		Short	Long	Short	Long
Observed[a]		1.7τ	4.6τ	2.6I	6.9I
Alpert and Toong	Square-Wave	1.6τ	3.8-5.4τ	2.4I	5.7-8.1I
Fickett et al.	Rate 1[b]	$3.1 t_{1/2}$ or 22τ	none	$9.6 x_{1/2}$ or 11I	None
Fickett et al.	Rate 2[b]	$1.7 t_{1/2}$ or 2.1τ	$7 t_{1/2}$ or 8.8τ	$5.5 x_{1/2}$ or 9.8I	$28 x_{1/2}$ or 50I

[a] Alpert and Toong (1972).
[b] Rate 1 is fairly realistic; rate 2 is more nearly a square wave; see Fig. 6.43.

correct value is 60 mm, or 189I. The calculated one-dimensional longitudinal wavelengths of Table 7.3 are smaller than this, ranging from 2.4I to 118I.

7D3. Cell Size

There are several theories which predict transverse wave spacing. For convenience we name them as follows:

1. Exact stability theory and extensions (Erpenbeck and others)

2. Ray shock contact (Strehlow)

3. Ray hot spot (Barthel)

4. Acoustic (Strehlow)

5. Initiation (Barthel).

The word "ray" is short for "geometrical acoustics." The first three theories, described in Chapter 6, are a priori treatments. The last two, described here, take the main features of the observed structure as their starting point. We first describe these more empirical treatments and then compare all of the predictions with experiment.

7D3.1. Acoustic Theory

Strehlow (1970) presents an "acoustic theory for spacing," to be distinguished from the a priori geometrical acoustic theory of Sec. 6B2 but with some points of connection. A structure of the observed form is assumed, with the spacing to be determined. The condition postulated to determine the spacing is that an additional "test" transverse acoustic wave on the front propagating through the existing structure, Fig. 7.40, neither grow nor decay. The processes which affect the acoustic wave's amplitude are (1) amplification by chemical reaction, (2) passage

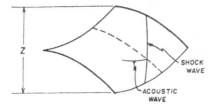

Fig. 7.40. The "test" transverse acoustic wave (track marked by dashed line) of the acoustic theory. From Strehlow (1970).

through the steep pressure gradient behind the front (recall the blast-wave approximation), and (3) collision with the (regular) transverse wave at the cell boundary.

The test wave is taken to be a step function oriented normal to the front, with strength S_a and with associated rays trapped in the reaction zone as described in Sec. 6B2. The first effect, chemical-reaction amplification, is given approximately by the result of that section as

$$S_a = S_a^0 \, e^{kt}$$

with S_a^0 the acoustic-wave strength (pressure ratio across the wave minus one) just after it enters the cell, t the elapsed time from that point, and k a function of the state behind the front at the critical plane. For simplicity, the CJ state is used throughout for its evaluation. For the total amplification across the cell, t is the transit time of the test wave across the cell. The second effect, passage through the blast-wave gradient, is an amplification small enough to be neglected. The third effect, collision with the transverse wave, is simplified to a one-dimensional problem, Fig. 7.41, by neglecting the transverse wave's slip line and taking its reflected shock (strength S) normal to the front. Passage of the test wave through the reflected shock reduces its strength.

Denoting post-collision states by primes, the solution of the matching problem is

$$\Delta p = \Delta p'; \; \Delta p \equiv p_{1a} - p_1, \; \Delta p' \equiv p_{2a} - p_2$$

$$S_a'/S_a = (1 + S)^{-1},$$

where the subscripts 1, 2, 1a, and 2a refer to the figure and S is the strength of the regular transverse wave. The pressure jump across the acoustic wave is unchanged, but its strength is reduced since its initial pressure is increased (from p_1 to p_2).

We now have in hand the two effects to be balanced. The increase in the strength of the acoustic wave by chemical amplification as it

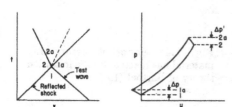

Fig. 7.41. One-dimensional collision of the test wave with the reflected shock of the regular transverse wave in the acoustic theory. From Strehlow (1970).

proceeds across the cell is equated to the decrease in its strength by collision at the cell boundary. Doing this and using the mean cell properties to estimate the transit time gives for the spacing Z

$$Z = 4k^{-1}M_j c_j (\tan \alpha) \ln(1 + S)$$

with α the mean track angle, and $M_j = D_j/c_o$.

7D3.2. Initiation Theory

Barthel (1972) presents a spacing theory which we will call the "initiation theory." It focuses attention on the new detonation at the beginning of a cell, having in mind the one-dimensional initiation process diagrammed in Fig. 7.20. Fig. 7.42 is a hypothetical snapshot before any material has reacted. The shaded region is heated material processed by the shock; it lies inside the tracks by virtue of its forward motion. Processes near the end of the previous cell and in the adjacent cells to the sides are neglected. The material at the rear apex reacts first, sending forward a reactive cylindrical compression wave, which later overtakes the front. The next consideration is the effect of the overtake on the front. Observations of isolated triple points generated by wedge reflection in both reactive and nonreactive mixtures (reported in this paper and in Liaugminas, Barthel, and Strehlow [1973]) shed some light on this process. Calculation of the overtake shows that if it occurs in the first half of the cell, while the front is still the Mach stem, the transverse wave is strengthened, but if it occurs in the second half of the cell, where the front has become the incident shock, the transverse wave is weakened.

The postulated condition to determine the spacing is that the overtake occur in the first half of the cell, so that it strengthens the transverse wave. To get numbers, the initiation is treated as one-dimensional behind a constant velocity shock. The overtaking wave originates at the apex after the calculated induction time for the initial shock (Mach stem) strength, and then moves forward with acoustic velocity. From an estimated or measured initial shock strength, one can

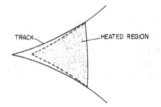

Fig. 7.42. Snapshot before any material has reacted, for the initiation theory of Barthel (1972).

Sec. 7D   COMPARISON OF THEORY AND EXPERIMENT

calculate the distance to overtake; the postulated spacing condition requires the cell length to be at least twice this. Alternatively, one can use the measured cell length to set a lower limit on the initial shock strength. The author suggests that, at least in the $H_2/O_2/Ar$ system, an upper limit is set by the strength at which the net exothermocity changes sign due to dissociation; Fickett, Jacobson, and Schott (1972) set this at $D = 1.4 D_j$.

The work of Takai, Yoneda, and Hikita (1974) provides some experimental confirmation of this assumed mechanism of initiation, as pointed out by Strehlow in an appended comment. Their measured axial velocity, Fig. 7.36, starts at $D = 1.5 D_j$, drops rapidly, and then rises to a local maximum about one-fifth of the way through the cell. They attribute the first rapid fall to the initial endothermicity, and the local maximum to the start of exothermic reaction, presumably via an initiation mechanism like that described above.

7D3.3. Comparison with Experiment

Table 7.4 gives a summary of comparisons with experiment. All the comparisons that have been made are for the system $H_2/O_2/Ar$. Detailed comparisons with experiment have only been made for theories 2, 3, and 4; see Strehlow (1969) and Strehlow, Maurer, and Rajan (1969) for 2, Barthel (1972) for 3, and Strehlow (1970) for 4. The ray shock contact theory (the original geometrical acoustic theory) gives spacings which are much too small. The other two are reasonably good, except that the

Table 7.4 CALCULATED VS. EXPERIMENTAL[a]
TRANSVERSE-WAVE SPACING (CELL HEIGHT) Z

Theory	Z_{calc}/Z_{exp}
1. Exact Stability	1.2 (see text)
2. Ray shock contact	1/30 to 1/120
3. Ray hot spot	1.7 to 2.5
4. Acoustic	1/20 for 0% Ar 1 to 3 for 50 to 85% Ar
5. Initiation	>1/6

[a]In $H_2/O_2/Ar$.

acoustic theory is poor at low dilution. All three give about the right dependence on pressure. The initiation theory 5 gives only a minimum value, and has been compared with experiment at only one point.

The exact stability theory typically gives only a maximum value, with the range of unstable wavelengths extending from this value to zero. Also, it has been applied only to an idealized system (polytropic gas with first-order Arrhenius rate). The system chosen by Erpenbeck for the nonlinear extension of the theory, Sec. 6B1, under the constraints imposed there gives values like $6x_{1/2}$. To try to make a correspondence with a system with branching-chain kinetics, we may look at Fig. 6.43, and take the induction-zone length I something like 1/20 of the half-reaction-zone length $x_{1/2}$. This rough correspondence would give a Z_{calc} of 120I from the value of $6x_{1/2}$, close to the experimental value of 100I for the dilute system.

7E. LIQUIDS AND SOLIDS

Liquids and solids are much harder to study than gases. Solids (other than single crystals) are not transparent, so we can't look inside them. Liquids are transparent, and for this reason are more studied. The favorite system is nitromethane, usually diluted with acetone to slow down the reaction and make the structure large enough to be readily observed. There are several differences from gases which might be expected to affect the structure. The main one is that there is no such thing as a rigid confining wall. The wall compressibility is an important effect, particularly with lower-impedance materials such as glass, Plexiglas, and air.

The most recent comprehensive review of this subject is in the book by Dremin, Savrov, Trofimov, and Shvedov (1970). An interesting result is the existence of several liquids—nitroglycerin, dinitroglycerin, tetranitromethane, and 87/13 nitromethane/benzene—in which the detonation is apparently one-dimensional in accord with the simple theory. The evidence comes from the analysis of light reflected from the front, which is thereby found to be smooth on the scale of the wavelength of visible light. In tetranitromethane a triangular reaction zone 2 mm long (reaction time ~0.5 μs) is resolved by the electromagnetic foil method. This method, with a time resolution of about 0.1 μs, fails to reveal a reaction zone in the other explosives. We will say no more here about one-dimensional detonations but concentrate instead on materials with structure like that seen in gases.

Although the structure is superficially like that in gases, there are some important differences and several puzzling features. A consistent picture has not yet emerged. We review the evidence and discuss the main questions, adding some ideas of our own. Recent short review discussions may be found in Urtiew (1975) and Persson and Persson (1976).

At the present state of knowledge it seems to us important to distinguish between light or medium confinement on the one hand, and heavy confinement on the other. Particularly with heavy confinement, it may also be important to distinguish between the structure in the interior and at the edge. With light or medium confinement, large scale disturbances, usually called failure waves, continually form at the edge and propagate into the interior. Their structure is fairly well understood, and is different from the triple-shock transverse-wave structure seen in gases. However, they do seem to interact with each other in somewhat the same way, forming fairly regular patterns resembling those on gas soot films. With heavy confinement, there are no obvious failure waves forming at the edge, and the general structure is at least superficially like that in gases. However, the structure of the transverse waves is not known. Some evidence suggests that it is like that of the large failure waves instead of the gas triple-shock configuration. It also appears to us that the edge and the interior may constitute two somewhat independent systems, with more regular and larger transverse-wave spacing at the edge.

The differences between gas- and condensed-phase properties which might be expected to affect the structure are discussed in Sec. 7E1. The light-confinement case is considered in Sec. 7E2, with emphasis on the failure-wave structure. The heavy-confinement case is discussed in Sec. 7E3, subdivided into the interior and the edge. In Sec. 7E4, we summarize the picture as well as it is known and review the evidence bearing on the remaining questions.

7E1. Differences from Gases

Detonations in liquids and solids differ from those in gases in several important ways which might be expected to affect the structure. The compressibility of the walls lengthens the reaction zone at the edge and produces a configuration which is a poor reflector of transverse waves from the interior. A detonation running in shocked but unreacted material is dimmer than a normal detonation, not brighter, as in gases. Such a detonation also runs faster than normal, since detonation velocity is a strong function of initial density, not almost independent of it, as in gases.

The edge is a poor reflector for transverse waves. The finite wall impedance allows some energy to be transmitted into the wall, and the lateral expansion induces a flow configuration at the edge which is a poor reflector in its own right. High-impedance materials are of course better reflectors than low-impedance ones, with the case of matching impedance probably somewhat special, since then to first approximation there is no reflection at all. Relative to a typical liquid explosive like nitromethane, brass and steel are of much higher impedance, glass is slightly higher, plastics are about the same, and air (with the explosive confined in a tube of thin plastic film) is much lower. As might be expected, there is a sharp difference in behavior between the high-impedance confining materials like brass and steel, and the others. Material-strength properties of course play some part; it may well be that the fracture properties of glass are important.

The flow configuration at the edge of the charge, as proposed by Bdzil (1976), is diagrammed in Fig. 7.43. Compared to gases, for which the shock is nearly flat, the shock here is quite curved and meets the wall more obliquely. The less compressible the unreacted explosive, the greater the angle between the shock and the deflected wall. In the unconfined case the shock is more curved, the reaction zone is longer, and the sonic locus meets the wall at the shock instead of behind it. Aside from transmission into the wall, one expects this edge configuration to be a poor reflector for transverse waves reaching it from the interior. Much of the energy from such a wave will be transmitted to the rear along the wall. The reflection that does take place tends to smear out the reflected wave, for the flow at the edge is more nearly sonic so that transverse disturbances propagate slowly, and reflection takes a long time.

The second effect is the relative brightness of a detonation wave traveling in shocked but unreacted material. In the condensed phase it is dimmer than normal as demonstrated by the classic one-dimensional shock-initiation experiments (Campbell, Davis, and Travis [1961], Johansson and Persson [1970], Sec. 2.1). This is mainly a density effect; the theoretical explanation (Fickett, Wood, and Salsburg [1957]) is that the energy partitioning depends on the density. With increasing density, more energy goes into intermolecular repulsion and less into translation, so that the temperature drops. This effect appears in the failure wave described in the next section, where (re-)initiation similar to that in the one-dimensional experiment plays an important part. Whether this is important to the observation elsewhere is not known. We sketch in Fig. 7.44 the strong triple-point structure for gases and mark on it the relative brightness which the three reaction zones would have if the same structure applies here. The stem is still the brightest part, but is now backed up by a transverse detonation wave which is dimmer, not brighter, than normal.

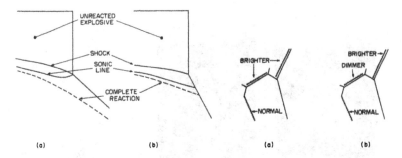

Fig. 7.43. Edge configuration in (a) unconfined, and (b) confined condensed-phase explosive. From Bdzil (1976).

Fig. 7.44. Relative brightness of the three reaction zones in a triple-shock configuration in (a) gases, and (b) liquids.

The third effect is the well-known increase of detonation velocity with density in the condensed phase, which will obviously be of some importance. In the strong triple-point structure, for example, the transverse detonation wave would have a higher-than-normal velocity into the shocked-but-unreacted material in the reaction zone behind the initial shock. The overtaking component of the failure wave described in the next section moves at a higher than normal velocity for the same reason.

7E2. Light Confinement

The only waves whose structure has been studied in any detail are the large-scale failure waves from the edge in low impedance—glass or Plexiglas—confinement.

Figure 7.45 shows photographs and diagrams of the process. The discussion will refer mostly to the cross-section drawing, Fig. 7.45d, in which the reaction-zone thickness of the normal detonation is neglected. At time 1 a small perturbation extinguishes the reaction at a point on the wall. A rarefaction wave, whose head follows the trajectory AA, spreads out from this point. To the left of AA we have a nonreactive shock, oblique because of its lower (normal) velocity. At time 2 the particle at P' (originally at P) explodes and initiates an overtaking detonation wave. Running in shocked unreacted material, this wave first overtakes the front at time 2 at point B on the wall. As it proceeds, it eats up the oblique nonreactive shock from the left along the intersection BB. Just left of the intersection we have an overdriven or Mach-stem region. Region AB (Fig. 7.45c and d) is dimmer than normal because its light is that of the overtaking detonation running in shocked, unreacted

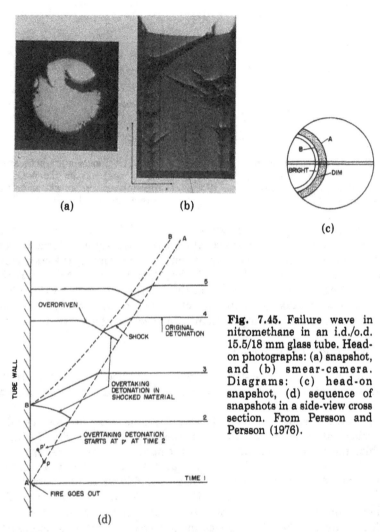

Fig. 7.45. Failure wave in nitromethane in an i.d./o.d. 15.5/18 mm glass tube. Head-on photographs: (a) snapshot, and (b) smear-camera. Diagrams: (c) head-on snapshot, (d) sequence of snapshots in a side-view cross section. From Persson and Persson (1976).

material. The overdriven region is of course brighter than normal. These regions are clearly seen in the photographs. In the head-on photograph, Fig. 7.45a, the pattern is asymmetrical, with the initiation point of the overtaking detonation to the left of that of the rarefaction. This may be due to interaction with a neighboring failure wave. A more detailed study of failure waves may be found in Dremin and Persson (1977).

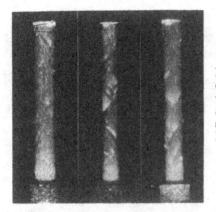

Fig. 7.46. Open-shutter photograph of edge waves in nitromethane in glass tubes of i.d./o.d. of (a) 19.5/22, and (b) and (c) 15.5/17.9 mm, essentially the same as Fig. 7.45. From Persson and Bjarnholt (1970).

Failure waves can start at many points on the wall, and apparently couple with each other to form a structure with some degree of regularity, which appears to contain some persistent transverse waves. Figure 7.46 shows an open-shutter view, from the side, of nitromethane in glass. Figure 7.46a is a slightly larger diameter tube and does not show the large failures seen in Fig. 7.46b and c, which have the same diameter tube as that of Fig. 7.45. Watson (1970) has photographed failure waves in 97/3 (vol.%) nitromethane/acetone which are concentric with the tube axis and thus appear to have been initiated on a circumferential ring.

Similar experiments in solids with very low-impedance (gas) confinement show a very regular structure, whose cell size depends in the expected way on reaction time. An example is shown in Fig. 7.47. A square cross-section (50 mm) stick of explosive is covered on one side by a thin (2 mm) layer of argon confined by a thin plastic film, and the whole assembly is placed in a box of propane (which does not light up from the shocks it receives). The luminosity of the shocks in the argon is a sensitive function of the pressure in the transverse waves; an open-shutter photograph reveals the structure. The cell size is reduced by applying confinement to the remaining sides of the charge, or by substituting pressed TNT, with its shorter reaction zone, for cast.

The regularity of the structure is surprising. Although there is no direct evidence that these are failure waves like those in liquids, the very low-impedance (gas) confinement suggests that they are. If so, the extreme regularity seems surprising. Additional information on the interaction between the expanding explosive products and the confining gas layer has been given by Held (1972).

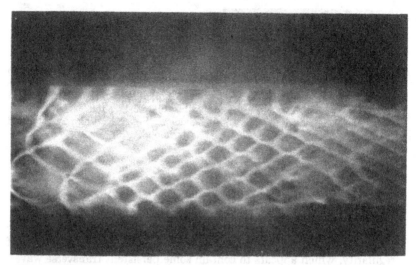

Fig. 7.47. Edge waves in cast TNT of 50 mm square cross section, recorded by the argon-flash technique (open-shutter photograph). From Howe, Frey, and Melani (1976).

7E3. Heavy Confinement

It appears to us that there are some grounds for regarding the structure at the edge and in the interior as partially decoupled, and we subdivide this section accordingly.

7E3.1. Interior

The interior structures observed to date are all irregular. The irregularity more nearly resembles that in gases due to lack of confinement (Fig. 7.1, spherical detonation in acetylene/oxygen) than the more severe inherent irregularity (Fig. 7.9, methane/oxygen in a rectangular tube).

Figure 7.48 is a head-on snapshot of the luminosity of the front from an unsuccessful attempt to achieve a regular structure by using a dense (tungsten) tube of square cross section, and choosing the dilution so that the cell size (10-15 cells across the tube) is not too far below the largest possible for steady propagation. The edge cell size obtained from wall imprints (see below) is shown for comparison.

By photographing an "impedance mirror," Mallory (1967) showed that there are also appreciable pressure variations across the front. The im-

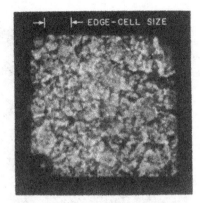

Fig. 7.48. Head-on snapshot of the front in 80/20 (vol. %) nitromethane/acetone. Tube 12.7 mm square interior, 300 mm long; walls: 1/2 mm of tungsten backed by 6 mm of brass; booster: 50 mm Composition B, point initiated. Measured detonation velocity: 5700 m/s, confirming that the detonation is at most slightly overdriven. Image-intensifier camera, 12 ns exposure. The circular blob in the lower left corner is from the triggering probe. By W. C. Davis.

pedance mirror is a Plexiglas sheet on the end of the tube coated on the side toward the explosive with a thin layer of evaporated aluminum. Pressure variations in the explosive wrinkle the mirror, as shown in the snapshots of Fig. 7.49.

A head-on smear-camera photograph of the front, Fig. 7.50, reveals a wealth of detail, with the broader transverse waves having the trailing bright band in common with the failure waves. Such photographs must be interpreted with some care; as explained in Appendix 7A, the recording system essentially selected only those transverse waves moving in a particular direction. This shot, like that of Fig. 7.48, was an unsuccessful attempt to produce a regular structure.

As mentioned earlier, the structure is apparently not present in all explosives, Dremin, Savrov, Trofimov, and Shvedov (1970), Chapter 3, Sec. 6, report that for several liquid explosives, including nitroglycerin and tetranitromethane, analysis of light reflected from the front reveals a smooth surface (so that any irregularities must be smaller than the wavelength of visible light). They also report than in some liquids, including pure nitromethane, but not nitromethane/acetone, the structure can be removed by overdrive. Mallory (1976) has shown that the structure appears after some delay in a one-dimensional shock initiation; presumably the process is much like that in gases, Fig. 7.20.

Observations at different dilutions, such as those of Fig. 7.49, show that the cell size is at least roughly proportional to reaction-zone length. By watching the surface of his impedance mirrors over a time interval after they are struck by the shock, Mallory (1976) found that the structure decays rapidly behind the front and that the decay time is roughly proportional to the reaction time. This observation is similar to White's (Sec. 3A) in gases.

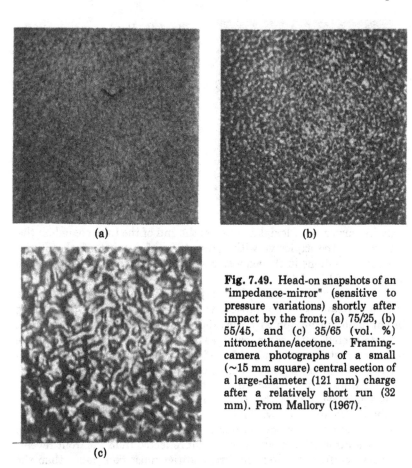

Fig. 7.49. Head-on snapshots of an "impedance-mirror" (sensitive to pressure variations) shortly after impact by the front; (a) 75/25, (b) 55/45, and (c) 35/65 (vol. %) nitromethane/acetone. Framing-camera photographs of a small (~15 mm square) central section of a large-diameter (121 mm) charge after a relatively short run (32 mm). From Mallory (1967).

7E3.2. Edge

Tracks on the tube wall can be recorded by means similar to the soot-film technique in gases. Urtiew, Kusubov, and Duff (1970) found matte-finish stainless steel to be a good medium. The tracks are recorded both by indentations from the excess pressure of the transverse waves and by differential deposition of soot from them. The records are, unfortunately, not nearly so sharp and contrasty as those on gas soot films.

One such record, together with a gas soot film for comparison, is shown in Fig. 7.51. Slapping waves appear to be absent. The track angles are around 45°, and the waves appear to pass through each other with little interaction. Oppenheim et al. (1970) have successfully calculated the collision by assuming a less common type of triple-point

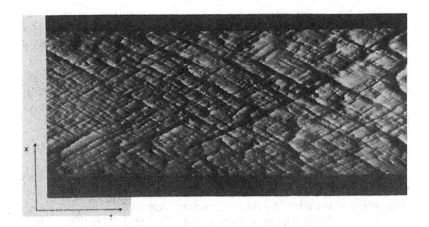

Fig. 7.50. Head-on smear-camera photograph of the front in 73/27 (vol. %) nitromethane/acetone in 20 × 80 mm rectangular brass tube. From Persson and Persson (1976).

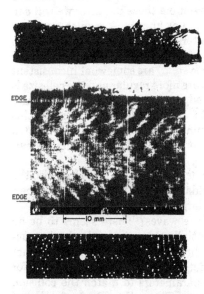

Fig. 7.51. Wall traces from nitromethane/acetone (vol. %): (a) 75/25 in a 20 mm square tube, (b) 80/20 in a 15 mm square tube (enlarged view). (c) Soot film from gaseous hydrogen/oxygen (in a 25.4 × 38.1 mm tube at p_o = 11.7 kPa) for comparison. From Urtiew, Kusubov, and Duff (1970), and Oppenheim et al. (1970).

structure. Whether or not the transverse waves reflect off the adjoining walls in the same way that they do off each other is not too clear from the records. Although no perfectly regular patterns have been found, the patterns are fairly regular, apparently much more so than those of the interior. The spacing appears to be significantly larger than that of the interior. The edge transverse-wave spacing for 80/20 (vol. %) nitromethane/acetone is, according to Urtiew, Kusubov, and Duff (1970), about 2.5 mm. We have marked off this spacing on the edge of the head-on view, Fig. 7.48, for comparison.

There are some similarities to gases. The behavior of the transverse waves in an expanding channel is similar (Urtiew and Kusubov [1970]). Also, triple-shock configurations generated at corners by the interior propagate about as expected (Urtiew, Kusubov, and Duff [1970]).

7E4. Discussion

A review of the evidence reveals some puzzling features. The only waves whose structure is known are the relatively large failure waves originating at the edge in light confinement. Apparently they can couple together into fairly regular patterns (Figs. 7.46 and 7.47). The experiment with solids, Fig. 7.47, gives a very regular pattern. With very light (gas) confinement, one would expect these to be failure waves like those studied in detail in liquids, but there is no direct evidence that they are.

Heavy confinement removes the large failure waves. The remaining waves, which at least superficially resemble those in gases, we will call "transverse waves." Their structure is not known. A regular structure (pattern of transverse waves) in the interior has not so far been achieved, even though conditions presumably conducive to it have been tried. The observations at the edge (in the same system) are somewhat inconsistent with those in the interior: wall tracks are more regular and the cell size is larger, suggesting to us that the edge waves might be a separate phenomenon more or less decoupled from the interior. Supporting this view is the apparent absence of slapping wave imprints on the wall. On the other hand, a different structure at the edge is not prominently evident in the head-on photographs (Figs. 7.48 and 7.50).

A few observed track angles are collected in Table 7.5. Here also the edge waves in liquids (but not in solids!) seem to fall in a class of their own, with track angles appreciably larger than the others.

What is known about the transverse waves? There seems to be no reason why they could not have essentially the same triple-shock structure as that seen in gases. Oppenheim et al. (1970) have in fact calculated a slightly different variety of triple-shock structure for nitromethane/acetone, choosing the parameters to match the collision and track angles observed in wall tracks. The wall tracks from tubes of reasonable cross section show no failure waves. (These can be obtained under heavy confinement by making the cross section small enough, and leave a characteristic raised imprint due to their low pressure.) Some, at least, of the interior waves do look like failure waves in that they consist

Table 7.5 TRACK ANGLES

1. Interior transverse waves in liquids	33°
2. Edge transverse waves in liquids	40-48°
3. Failure waves in liquids	25-35°
4. Failure (?) waves in solids	28°

Sources:
1. Fig. 7A-2
2. Fig. 7.51
3. Persson and Persson (1976), from photographs like those of Fig. 7.45
4. Fig. 7.47

of a dark band followed by a narrow bright stripe; possibly both kinds of waves populate the interior.

Some of the apparent contradictions might be resolved by regarding the interior and edge as two separate systems. The interior structure would then be irregular because of the poor reflection properties of the edge. The longer reaction zone at the edge would be regarded as a separate annular channel with its own more or less planar regular structures like those of Fig. 7.5. This would explain the larger cell size and more regular structure. The larger track angles at the edge would follow from the smaller mode number there.

APPENDIX 7A: INTERPRETATION OF SMEAR-CAMERA PHOTOGRAPHS

A head-on smear-camera photograph of the front in a liquid explosive, such as that of Fig. 7.50, seems to show waves moving in only two directions, both having the same orientation with respect to, say, a horizontal axis in the field of view. In reality, with such an irregular structure, there are transverse waves moving in all directions, but the peculiarities of the recording mechanism in effect select only a particular direction to appear on the film. Thus, care in interpreting such pictures is required, and it is well to understand how this comes about.

Figure 7A-1 shows the film moving vertically upward with velocity S past the stationary slit image (of width W) and the image of a transverse wave (idealized as a dark band of width w) oriented at an angle to the slit and moving normal to itself with velocity V. It is seen that a transverse wave with orientation $\cos \theta = V/S$ will cast a stationary image on the film and thus be recorded with the least blurring. By "stationary image" we mean that the image of the band has zero component of velocity normal to itself on the film. It does, of course, slide along its own length, so any variations in structure along its length will be blurred.

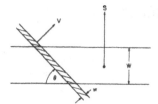

Fig. 7A-1. Diagram of a smear-camera film recording a dark wave. The film moves upward with velocity S past the stationary slit image of width W. The dark-wave image, width w, is oriented at angle θ and moves normal to itself with velocity V.

Fig. 7A-2. Smear-camera photographs of the light from the front in 84/16 (vol. %) nitromethane/acetone, at two different writing speeds (a) $\theta = 50°$, (b) $\theta = 0°$ (see text for definition of θ). From T. P. Cotter, Jr., private communication, (1955).

There is still some blurring in the direction normal to the band because both its own width and that of the slit are finite; a little analysis shows this blurring to be proportional to $(W/w)(\cos\theta - V/S)$. If the moving bands were brighter than the background, the range of orientations recorded could be increased by increasing the exposure; with dark bands this is precluded by the upper limit on useful contrast. The wave velocity V is readily found by measuring θ on the film and knowing the camera writing speed S, i.e.,

$$V = S\cos\theta.$$

For nitromethane/acetone, V turns out to be about 4 mm/μs, roughly the speed of sound in the unreacted material. Fig. 7A-2 shows two photographs, one with the writing speed chosen to give $\theta = 50°$, and one $\theta = 0°$ (V = S). The second records waves with orientations of $\theta = 0°$ to about 30° because the cosine function varies slowly in this range.

Bibliography

ADAMS, G. K. and COWPERTHWAITE, M.
    1965. Explicit solutions for unsteady shock propagation in chemically reacting media. In *Fourth Symposium (International) on Detonation*, pp. 502-511. Washington, D.C.: Office of Naval Research — Department of the Navy (ACR-126). [102]

ADAMSON, T. C.
    1960. On the structure of plane detonation waves. *Phys. Fluids* 3:706-714. [192]

AGUILAR-BARTOLOME, F.
    1972. *Los explosivos y sus aplicaciones*. Madrid: Servicio de Publicaciones de la J.E.N. [11]

ALPERT, R. L. and TOONG, T. Y.
    1972. Periodicity in exothermic hypersonic flows about blunt projectiles. *Astronaut. Acta* 17:539-560. [244, 275, 277, 344, 345]

ARKHIPOV, V. N.
    1962. Rarefaction waves in a relaxing gas mixture. *Zh. Prik. Mekh. Tekh. Fiz.* 1962 (4):40-45. [114]

ASLANOV, S. K., BUDZIROVSKII, V. N., and SHCHELKIN, K. I.
    1968. An investigation of gasdynamic stability of a detonation wave of arbitrary profile. *Dokl. Akad. Nauk SSSR* (Phys. Sect.) 182:53-55. [246]

BARTHEL, H. O.
    1972. Reaction zone-shock front coupling in detonations *Phys. Fluids* 15:43-50. [348, 349]
    1974. Predicted spacings in hydrogen-oxygen-argon detonations. *Phys. Fluids* 17:1547-1553. [273, 275]

BARTHEL, H. O., and STREHLOW, R. A.
    1966. Wave propagation in one-dimensional reactive flows. *Phys. Fluids* 9:1896-1907. [9, 267, 271, 272, 274]

BDZIL, J. B.
    1976. Perturbation methods applied to problems in detonation physics. In *Sixth Symposium (International) on Detonation*, pp. 352-369. Washington, D.C.: Office of Naval Research — Department of the Navy (ACR-221). [199, 203, 352, 353]

BDZIL, J. B., and DAVIS, W. C.
    1975. *Time-dependent detonations*. Los Alamos Scientific Laboratory Report LA-5926-MS. [69]

BIBLIOGRAPHY

BEATTIE, J. A.
1955. Thermodynamics of real gases and mixtures of real gases. In *Thermodynamics and Physics of Matter,* ed. F. D. Rossini, Sec. C. Princeton: Princeton University Press. [32]

BECKER, R.
1922. Impact waves and detonations. *Z. Phys.* 8:321-362. (English translation: NACA TM 505 and TM 506, March 1929.) [3]

BERETS, D. J., GREENE, E. F., and KISTIAKOWSKY, G. B.
1950. Gaseous detonations. I. Stationary waves in hydrogen-oxygen mixtures. *J. Am. Chem. Soc.* 72:1080-1086. [5]

BERTHELOT, M., and VIELLE, P.
1881. On the velocity of propagation of explosive processes in gases. *C. R. Hebd. Sceances Acad. Sci.* 93:18-21. [2]
1882. On explosive waves. *C. R. Hebd. Sceances Acad. Sci. 94:149-152, 94:882. [2]*

BETHE, H., and TELLER, E.
1942. Deviations from equilibrium in shock waves. Ballistics Research Laboratory Report BRL-X-117. (Available from TDIE, Oak Ridge, as Rept. No. NP-4898.) [112]

BIRD, P. F., DUFF, R. E., and SCHOTT, G. L.
1964. *HUG, a FORTRAN-FAP code for computing normal shock and detonation wave parameters in gases.* Los Alamos Scientific Laboratory Report LA-2980. [153]

BORISOV, A. A.
1974. On the origin of exothermic centers in gaseous mixtures. *Acta Astronaut.* 1:909-920. [110]

BOWEN, J. R.
1967. Approximations to the structure of a laminar detonation wave by the method of matched asymptotic expansions. *Phys. Fluids* 10:290-296. [192]

BROCHET, C., MANSON, N., ROUZE, M. and STRUCK, W.
1963. Influence of the initial pressure on the velocity of stable detonations in stoichiometric mixtures of propane-oxygen and acetylene-oxygen. *C. R. Hebd. Sceances Acad. Sci.* 257: 2412-2414. [5, 59]

BROWNE, W. G., WHITE, D. R., and SMOOKLER, G. R.
1969. A study of the chemical kinetics of shock heated $H_2/CO/O_2$ mixtures. In *Twelfth Symposium (International) on Combustion,* pp. 557-567. Pittsburgh: The Combustion Institute. [151]

CAMERON, I. G.
1963. The application of the von Neumann method to the steady one-dimensional detonation wave. J. Soc. Ind. Appl. Math. 11:398-412. [193]

CAMPBELL, A. S.
1951. Thermodynamic properties of reactive gas mixtures. *J. Franklin Inst.* 251:437-452. [76]

BIBLIOGRAPHY

CAMPBELL, A. W., DAVIS, W. C., and TRAVIS, J. R.
   1961. Shock initiation of detonation in liquid explosives. *Phys. Fluids* 4:498-510. [352]

CAMPBELL, C., and WOODHEAD, D. W.
   1927. The ignition of gases by an explosion wave. Part I. Carbon monoxide and hydrogen mixtures. *J. Chem. Soc.* 1927:1572-1578. [4]

CHAPMAN, D. L.
   1899. On the rate of explosion in gases. *Philos. Mag.* 47:90-104. [3]

CHEN, P. J., and GURTIN, M. E.
   1971. Growth and decay of one-dimensional shock waves in fluids with internal state variables. *Phys. Fluids* 14:1091-1094. [101]

CHERNYI, G. G.
   1975. On problems involving gas flow associated with gasdynamics. *Acta Astronaut.* 2:839-965. [242]

CHERNYI, G. G., and MEDVEDEV, S. A.
   1970. Development of oscillations associated with attenuation of detonation waves. *Astronaut. Acta* 15:371-375. [246, 251, 277, 285, 286]

CHIU, K. W., and LEE, J. H.
   1976. A simplified version of the Barthel model for transverse-wave spacings in gaseous detonation. *Combust. Flame* 26:353-361. [275]

CHOU, P. C. and HOPKINS, A. K., eds.
   1972. *Dynamic response of materials to intense impulsive loadings*. Ohio: Wright-Patterson Air Force Materials Laboratory. [91, 92, 101]

CHU, B. T.
   1956. Vibration of the gas column behind a strong detonation wave. In *Gas Dynamics Symposium on Aerothermochemistry*, pp. 95-111. Evanston: Northwestern University Press. [321]

CLARKE, J. F., and McCHESNEY, M.
   1976. *Dynamics of relaxing gases*. London and Boston: Butterworths. [76]

COURANT, R., and FRIEDRICHS, K. O.
   1948. *Supersonic flow and shock waves*. New York: Interscience. [89, 91, 242, 202, 295, 303]

COWAN, R. D.
   1958. Properties of the Hugoniot function. *J. Fluid Mech.* 3:531-545. [89]

COWAN, R. D., and FICKETT, W.
   1956. Calculation of the detonation properties of solid explosives with the Kistiakowsky-Wilson equation of state. *J. Chem. Phys.* 24:932-939. [32]

COWPERTHWAITE, M.
   1968. Properties of some Hugoniot curves associated with shock instability. *J. Franklin Inst.* 285:275-284. [254]
   1969a. Explicit solutions for the build-up of an accelerating reactive shock to a steady-state detonation wave. In *Twelfth Symposium (International) on Combustion*, pp. 753-759. Pittsburgh: The Combustion Institute. [102]
   1969b. Relationships between incomplete equations of state. *J. Franklin Inst.* 287:379-387. [123]

BIBLIOGRAPHY

1971. Two dimensional steady-state detonation waves. In *Thirteenth Symposium (International) on Combustion*, pp. 1111-1117. Pittsburgh: The Combustion Institute. [227]

COWPERTHWAITE, M., and ADAMS, G. K.
1967. Explicit solutions for steady- and unsteady-state propagation of reactive shocks at constant velocity. In *Eleventh Symposium (International) on Combustion*, pp. 703-711. Pittsburgh: The Combustion Institute. [102]

COWPERTHWAITE, M., and ZWISLER, W. H.
1973. TIGER computer program documentation. *Stanford Research Institute Publication No. Z106*. [32]
1976. The JCZ equations of state for detonation products and their incorporation into the TIGER code. In *Sixth Symposium (International) on Detonation*, pp. 162-170. Washington, D.C.: Office of Naval Research — Department of the Navy (ACR-221). [33]

CRAIG, B. G.
1965. Measurements of the detonation front structure in condensed-phase explosives. In *Tenth Symposium (International) on Combustion*, pp. 863-867. Pittsburgh: The Combustion Institute. [73]

CROOKER, A. J.
1969. *Phenomenological investigation of low mode marginal planar detonations*. University of Illinois (Urbana) Technical Report AAE 69-2 (Ph.D. thesis). [298, 333, 335, 339]

DABORA, E. K., NICHOLLS, J. A., and MORRISON, R. B.
1965. The influence of a compressible boundary on the propagation of gaseous detonations. In *Tenth Symposium (International) on Combustion*, pp. 817-830. Pittsburgh: The Combustion Institute. [212]

DAVIS, H. T.
1960. *Introduction to nonlinear differential equations*. Washington, D.C.: United States Atomic Energy Commission. [167, 179, 180]

DAVIS, W. C.
1976. Magnetic probe measurements of particle velocity profiles. In *Sixth Symposium (International) on Detonation*, pp. 637-641. Washington, D.C.: Office of Naval Research — Department of the Navy (ACR-221). [69]

DAVIS, W. C., CRAIG, B. G., and RAMSAY, J. B.
1965. Failure of the Chapman-Jouguet theory for liquid and solid explosives. *Phys. Fluids* 8:2169-2182. [6, 71, 73]

DAVIS, W. C., and VENABLE, D.
1970. Pressure measurements for Composition B-3. In *Fifth Symposium (International) on Detonation*, pp. 13-21. Washington, D.C.: Office of Naval Research - Department of the Navy (ACR-184). [69, 73]

DEAL, W. E.
1957. Measurement of Chapman-Jouguet pressure for explosives. *J. Chem. Phys.* 27:796-800. [24, 26, 68]
1958. Measurement of the reflected Shock Hugoniot and isentrope for explosive reaction products. *Phys. Fluids* 1:523-527. [26]

BIBLIOGRAPHY

DENISOV, Yu. N., and TROSHIN, Ya. K.
  1959. Pulsating and spinning detonation of gaseous mixtures in tubes. *Dokl. Akad. Nauk SSSR (Phys-Chem. Sec.)* 125:110-113. [10]
  1971. The fine structure of spinning detonation. *Combust. Flame* 16:141-145. [331]
DOBRATZ, BRIGITTA M., ed.
  1972. *Properties of chemical explosives and explosive simulants.* Lawrence Livermore Laboratory Report UCRL-51319, revised July 31, 1974. [29]
DOERING, W.
  1943. On detonation processes in gases. *Ann. Phys.* 43:421-436. [4, 13]
DREMIN, A. N.
  1968. Critical phenomena in the detonation of liquid explosives. In *Twelfth Symposium (International) on Combustion*, pp. 691-698. Pittsburgh: The Combustion Institute. [293]
DREMIN, A. N., and PERSSON, P. A.
  1978. The nature of dark waves in liquid homogeneous explosives with unstable detonation front. *Sixth International Colloquium on Gas Dynamics*, in press. [354]
DREMIN, A. N., SAVROV, S. D., TROFIMOV, V. S., and Shvedov, K. K.
  1970. *Detionatsionnyye Volny v Kondensirovannykh Sredakh (Detonation Waves in Condensed Media).* Moscow:Izd-vo Nauka (English translation: Wright Patterson Air Force Base, Dayton, Ohio, FTD-HT-23-1889-71). [11, 350, 357]
DUFF, R. E.
  1958. Calculation of reaction profiles behind steady-state shock waves. I. Application to detonation waves. *J. Chem. Phys.* 28:1193-1197. [153]
  1961. Investigation of spinning detonation and detonation stability. *Phys. Fluids* 4:1427-1433. [10, 323]
  1965. Summary of papers on condensed phase detonation. In *Fourth Symposium (International) on Detonation* pp. 198-201. Washington, D.C.: Office of Naval Research — Department of the Navy (ACR-126). [73]
DUFF, R. E., and FINGER, M.
  1965. Stability of a spherical gas detonation. *Phys. Fluids* 8:764. [293]
DUFF, R. E., and HOUSTON, E.
  1955. Measurement of the Chapman-Jouguet pressure and reaction zone in a detonating high explosive. *J. Chem. Phys.* 23:1268-1273. [26]
DUFF, R. E., and KNIGHT, H. T.
  1958. Further comments on the letter of Fay and Opel. *J. Chem. Phys.* 29:956-957. [67]
DUFF, R. E., KNIGHT, H. T., and RINK, J. P.
  1958. Precision flash x-ray determination of density ratio in gaseous detonations. *Phys. Fluids* 1:393-398. [5, 60, 64, 67]
DUVALL, G. E.
  1962. Shock wave stability in solids. In *Ondes de Detonation*, pp. 337-52. Paris: Editions du Centre National de La Recherche Scientifique. [254]

BIBLIOGRAPHY

EDWARDS, D. H.
1969. A survey of recent work on the structure of detonation waves. In *Twelfth Symposium (International) on Combustion*, pp. 819-828. Pittsburgh: The Combustion Institute. [315]

EDWARDS, D. H., HOOPER, G., JOB, E. M., and PARRY, D. J.
1970. The behavior of the frontal and transverse shocks in gaseous detonation waves. *Astronaut. Acta* 15:323-333. [302, 327, 335, 340]

EDWARDS, D. H., HOOPER, G., and MEDDINS, R. J.
1972. Instabilities in the reaction zone of detonation waves. *Astronaut. Acta* 17:475-485. [336, 341]

EDWARDS, D. H., JONES, T. G., and PRICE, B.
1963. Observations on oblique shock waves in gaseous detonations. *J. Fluid Mech.* 17:21-34. [5, 60, 61, 67]

EDWARDS, D. H., PARRY, D. J., and JONES, A. T.
1966. On the coupling between spinning detonation and oscillation behind the wave. *Br. J. Appl. Phys.* 17:1507-1510. [303, 323]

ENGELKE, R.
1974. Ray trace acoustics in unsteady inhomogeneous flow. *J. Acoust. Soc. Am.* 56:1291-1292. [267]

ERPENBECK, J. J.
1961. Two-reaction steady detonation. *Phys. Fluids* 4:481-492. [176, 182]
1962a. Stability of steady-state equilibrium detonations. *Phys. Fluids* 5:604-614. [9, 233, 237, 245, 246]
1962b. Stability of step shocks. *Phys. Fluids* 5:1181-1187. [253]
1963. Structure and stability of the square-wave detonation. In *Ninth Symposium (International) on Combustion*, pp. 442-453. New York: Academic Press. [238, 245, 250]
1964a. Steady detonations in idealized two-reaction systems. *Phys. Fluids* 7:1424-1432. [176, 178, 184, 189]
1964b. Stability of idealized one-reaction detonations. *Phys. Fluids* 7:684-696. [10, 54, 233, 234, 246, 249]
1965. Stability of idealized one-reaction detonations: zero activation energy. *Phys. Fluids* 8:1192-1193. [233, 246, 248]
1966. Detonation stability for disturbances of small transverse wavelength. *Phys. Fluids* 9:1293-1306. [233, 236, 246, 248, 249, 267]
1967. Nonlinear theory of unstable one-dimensional detonations. *Phys. Fluids* 10:274-288. [234, 259, 264]
1968. Steady-state analysis of quasi-one-dimensional reactive flow. *Phys. Fluids* 11:1352-1370. [224]
1969a. Theory of detonation stability. In *Twelfth Symposium (International) on Combustion*, pp. 711-721. Pittsburgh: The Combustion Institute. [233, 238, 245, 250]
1969b. Steady quasi-one-dimensional detonations in idealized systems. *Phys. Fluids* 12:967-982. [212, 224]
1970. Nonlinear theory of two-dimensional detonation. *Phys. Fluids* 13:2007-2026. [259, 266]

BIBLIOGRAPHY

EVANS, M. W., and ABLOW, C. M.
    1961. Theories of detonation. *Chem. Rev.* 61:129-178. [13, 137]

EWALD, R., SCHMOLINSKE, E., SCHULTZ-GRUNOW, F., and STRUCK, W. G.
    1972. Mach stem induced detonation and its relation to the structure of detonation waves. *Astronaut. Acta* 17:467-73. [294]

EYRING, H., POWELL, R. E., DUFFY, G. H., and PARLIN, R. B.
    1949. The Stability of detonation. *Chem. Rev.* 45:69-181. [210]

FAY, J. A.
    1952. Mechanical theory of spinning detonation. *J. Chem. Phys.* 20:942-950. [316, 321, 323]
    1959. Two-dimensional detonations: velocity deficit. *Phys. Fluids* 2:283-289. [67, 200]

FAY, J. A., and OPEL, G.
    1958. Two-dimensional effects in gaseous detonation waves. *J. Chem. Phys.* 29:955-956. [5, 60, 67]

FICKETT, W.
    1962. *Detonation properties of condensed explosives calculated with an equation of state based on intermolecular potentials.* Los Alamos Scientific Laboratory Report LA-2712. [33, 72, 148]
    1963. Calculation of the detonation properties of condensed explosives. *Phys. Fluids* 6:997-1006. [33]
    1970. Mechanism of propagation of steady detonation. In *Explosive Chemical Reactions Seminar*, pp. 223-54. U.S. Army Research Office (Durham, NC) report ARO-D 70-4. [294]

FICKETT, W., JACOBSON, J. D., and SCHOTT, G. L.
    1972. Calculated pulsating one-dimensional detonations with induction-zone kinetics. *AIAA J.* 10:514-516. [277, 283, 344, 349]

FICKETT, W., JACOBSON, J. D., and WOOD, W. W.
    1970. *The method of characteristics for one-dimensional flow with chemical reaction.* Los Alamos Scientific Laboratory Report LA-4269. [112, 114, 278, 282, 283, 284, 285, 286, 324]

FICKETT, W. and RIVARD, W. C.
    1974. *Test problems for hydrocodes.* Los Alamos Scientific Laboratory Report LA-5479. [47]

FICKETT, W., and SCHERR, L. M.
    1975. *Numerical calculation of the cylinder test.* Los Alamos Scientific Laboratory Report LA-5906. [203]

FICKETT, W., and WOOD, W. W.
    1958. A detonation-product equation of state obtained from hydrodynamic data. *Phys. Fluids* 1:528-534. [26, 27, 28, 29]
    1966. Flow calculations for pulsating one-dimensional detonations. *Phys. Fluids* 9:903-916. (Details of this work and additional calculations are given in Fickett, Jacobson, and Wood [1970].) [10, 277, 278, 279, 280, 282, 284, 286]

BIBLIOGRAPHY

FICKETT, W., WOOD, W. W., and SALSBURG, Z. W.
  1957. Investigations of the detonation properties of condensed explosives with equations of state based on intermolecular potentials. I. RDX with fixed product composition. *J. Chem. Phys.* 27:1324-1329. [352]

FINGER, M., LEE, E., HELM, F. H., HAYES, B., HORNIG, H., McGUIRE, R., KAHARA, M., and GUIDRY, M.
  1976. The effect of elemental composition on the detonation behavior of explosives. In *Sixth Symposium (International) on Detonation*, pp. 710-722. Washington, D.C.: Office of Naval Research — Department of the Navy (ACR-221). [33]

FOWLES, G. R.
  1976. Conditional stability of shock waves — a criterion for detonation. *Phys. Fluids* 19:227-238. [254]

FRANKLIN, J. N.
  1968. *Matrix theory.* Englewood Cliffs: Prentice Hall. [120]

FRIEDRICHS, K. O.
  1946. *On the mathematical theory of deflagrations and detonations.* Naval Ordance Laboratory Report NAVORD-79-46. [192]

FUJIWARA, T., MIZOGUCHI, K., and SUGIMURA, T.
  1971. Smoked film record for oxyhydrogen converging detonations. *J. Phys. Soc. Japan* 31:621-622. [326]

FUJIWARA, T., and TSUGE S.
  1972. Quasi-one-dimensional analysis of gaseous free detonation. *J. Phys. Soc. Japan* 33:237-241. [228]

GARR, L. J., and MARRONE, P. V.
  1963. *Inviscid, non-equilibrium flow behind bow and normal shock waves.* Cornell Aeronautical Laboratory Report QM-1626-A-12 (I and II). [153]

GARRIS, C. A., TOONG, T. Y., and PATUREAU, J. P.
  1975. Chemi-acoustic instability structure in irreversible reacting systems. *Acta Astronaut.* 2:981-997. [110]

GAVRILENKO, T. P., TOPCHIAN, M. E., and Yasakov, V. A.
  1967. Transverse waves in a converging cylindrical detonation wave. *Fiz. Goreniya Vzryva* 3:501-504. [326]

GETZINGER, R. W., BOWEN, J. R., OPPENHEIM, A. K. and BOUDART, M.
  1965. Steady detonations in gaseous ozone. In *Tenth Symposium (International) on Combustion*, pp. 779-784. Pittsburgh: The Combustion Institute. [59]

GILBERT, R. B., and STREHLOW, R. A.
  1966. Theory of detonation. *AIAA J.* 4:1777-1783. [324]

GORANSON, R. W.
  1946. *Method for determining equations of state and reaction zones in the detonation of high explosives and its application to pentolite, composition B, baratol, and TNT.* Los Alamos Scientific Laboratory report LA-487. [26]

BIBLIOGRAPHY

GORDEEV, V. E.
   1974. Spin without a transverse detonation front. *Dokl. Akad. Nauk SSSR* (Phys. Chem. Sect.) 215:363-365. [331]
   1976a. The cause of multiplication of discontinuities in a detonation. *Dokl. Akad. Nauk SSSR* (Phys. Sect.) 226:288-291. [324]
   1976b. Limiting velocity of supercompressed detonation and stability of shocks in detonation spin. *Dokl. Akad. Nauk SSSR* (Phys. Chem. Sect.) 226:619-622. [325, 327]

GREENE, E. F., and TOENNIES, J. P.
   1964. *Chemical reactions in shock waves*. London: Edward Arnold. [56]

GROSS, R. A., and CHINITZ, W.
   1960. A study of supersonic combustion. *J. Aerosp. Sci.* 27:517-534. [330, 331]

GRUSCHKA, H. D., and WECKEN, F.
   1971. *Gasdynamic theory of detonation*. New York: Gordon and Breach. [11, 76]

HELD, M.
   1972. Method of measuring the fine structure of detonation fronts in solid explosives. *Astronaut. Acta* 17:599-606. [355]

HERZFELD, K. F., and LITOVITZ, T. A.
   1959. *Absorption and dispersion of ultrasonic waves*. New York: Academic Press. [110]

HILL, R., and PACK, D. C.
   1947. An investigation, by the method of characteristics, of the lateral expansion of gases behind a detonating slab of explosive. *Proc. R. Soc. London Ser. A* 191:524-541. [212]

HIRSCHFELDER, J. O., and CURTIS, C. F.
   1958. Theory of detonations, I. Irreversible unimolecular reaction. *J. Chem. Phys.* 28:1130-1147. An error in this paper is corrected in Wood (1961). See also Hirschfelder, et al. *Phys. Fluids* 4:260-263 (1961). [192]

HIRSCHFELDER, J. O., CURTISS, C. F., and BIRD, R. B.
   1964. *Molecular Theory of Gases and Liquids* New York: Wiley. [32]

HOSKIN, N. E., ALLEN, J. W. S., BAILEY, W. A., LETHABY, J. W., and SKIDMORE, I. C.
   1965. The motion of plates and cylinders driven by detonation waves at tangential incidence. In *Fourth Symposium (International) on Detonation*, pp. 14-26. Washington, D.C.: Office of Naval Research — Department of the Navy (ACR-126). [203, 212]

HOWE, P., FREY, R., and MELANI, G.
   1976. Observations concerning transverse waves in solid explosives. *Comb. Sci. and Technol.* 14:63-74. [277, 286, 356]

HURWITZ, H., and KAMLET, M. J.
   1969. The chemistry of detonations. V. A simplified method for calculation of pressures of C-H-N-O explosives on K-W isentropes. *Isr. J. Technol.* 7:431-440. [35]

IL'KAEVA, L. A., and POPOV, N. A.
   1965. Hydrodynamic solutions for one-dimensional perturbations of an unstable detonation wave. *Fiz. Goreniya Vzryva* 1, No. 3:20-26. [245, 251, 277, 286, 288]

BIBLIOGRAPHY

JACOBS, S. J.
   1956. *Energy of detonation.* United States Naval Ordnance Laboratory Report NAVORD-4366. [36]
   1960. Recent advances in condensed media detonations. *J. Am. Rocket Soc.* 30:151-158. [25]
   1969. On the equation of state for detonation products at high density. In *Twelfth Symposium (International) on Combustion,* pp. 501-510. Pittsburgh: The Combustion Institute. [33]

JOHANSSON, C. H., and PERSSON, P. A.
   1970. *Detonics of high explosives.* London: Academic Press. [11, 352]

JONES, H.
   1947. A theory of the dependence of the rate of detonation of solid explosives on the diameter of the charge. *Proc. R. Soc. London Ser. A* 189:415-427. [211]
   1949. The properties of gases at high pressures which can be deduced from explosion experiments. In *Third Symposium on Combustion, Flame, and Explosion Phenomena,* pp. 590-594. Baltimore: Williams and Wilkins. [6, 28, 69, 70]
   1975. The dynamics of spinning detonation waves. *Proc. R. Soc. London Ser. A* 348:299-316. [110, 332]

JOST, W.
   1946. *Explosion and combustion processes in gases.* New York: McGraw-Hill. [4]

JOUGUET, E.
   1905. On the propagation of chemical reactions in gases. *J. de Mathematiques Pures et Appliquees* 1:347-425, continued in 2:5-85 (1906). [3]
   1917. *Mecanique des explosifs.* Paris: Octave Doin et Fils. [3, 101]

KAMLETT, M. J., and ABLARD, J. E.
   1968. Chemistry of detonation II. Buffered equilibria. *J. Chem. Phys.* 48: 36-42. [34]

KAMLETT, M. J., and DICKINSON, C.
   1968. Chemistry of detonations III. Evaluation of the simplified calculational method for Chapman-Jouguet detonation pressures on the basis of available experimental information. *J. Chem. Phys.* 48: 43-50. [34]

KAMLET, M. J., and HURWITZ, H.
   1968. Chemistry of detonations. IV. Evaluation of a simple predictional method for detonaton velocities of C-H-N-O explosives. *J. Chem. Phys.* 48:3685-3692. [34]

KAMLET, M. J., and JACOBS, S. J.
   1968. Chemistry of detonations. I. A simple method for calculating detonation properties of C-H-N-O Explosives. *J. Chem. Phys.* 48:23-25. [34]

KIRKWOOD, J. G., and WOOD, W. W.
   1954. Structure of a steady-state plane detonation wave with finite reaction rate. *J. Chem. Phys.* 22:1915-1919.[6, 76]

KNYSTAUTAS, R., and LEE, J. H.
   1971. Experiments on the stability of converging cylindrical detonations. *Combust. Flame* 16:61-73. [326]

BIBLIOGRAPHY

KOUMOUTSOS, N. G., and KOVITZ, A. A.
 1963. Detonation wave structure near the hot boundary. *Phys. Fluids* 6:1007-1015. [192]

KURY, J. W., HORNIG, H. C., LEE, E. L., McDONNEL, J. L., ORNELLAS, D. L., FINGER, M., STRANGE, F. M., and WILKINS, M. L.
 1965. Metal acceleration by chemical explosives. In *Fourth Symposium (International) on Detonation*, pp. 3-13. Washington, D.C.: Office of Naval Research — Department of the Navy (ACR-126). [28]

KURZROCK, J. W.
 1966. *Exact numerical solutions of the time-dependent compressible Navier-Stokes equations*. Cornell Aeronautical Laboratory Report AG-2026-W-1. [200]

LAMBOURN, B. D., and HOSKIN, N. E.
 1970. The computation of general problems in one-dimensional flow by the method of characteristics. In *Fifth Symposium (International) on Detonation*, p. 501 (abstract only). Washington, D.C.: Office of Naval Research — Department of the Navy (ACR-184). [24]

LEBEDEV, Yu. A., MIROSCHNICHENKO, E. A., and KNOBEL', Yu. K.
 1970. *Termokhimia nitrosoedinenii* (Thermochemistry of nitro-compounds). Moscow: Izd. "Nauka." [73]

LEE, E., FINGER, M., and COLLINS, W.
 1973. *JWL equation of state coefficients for high explosives*. Lawrence Livermore Laboratory Report UCID-16189. [28, 29]

LEE, E. L., HORNIG, H. C., and KURY, J. W.
 1968. *Adiabatic expansion of high explosive detonation products*. Lawrence Radiation Laboratory Report UCRL-50422. [28, 203]

LEE, J. H., SOLOUKHIN, R. I., and OPPENHEIM, A. K.
 1969. Current views on gaseous detonation. *Astronaut. Acta* 14:565-584. [302, 325, 337]

LEHR, H. F.
 1972. Experiments on shock-induced combustion. *Astronaut. Acta* 17:589-597. [311, 312]

LEVINE, H. B., and SHARPLES, R. E.
 1962. *Operators manual for RUBY*. Lawrence Radiation Laboratory Report UCRL-6815. [32]

LEWIS, B., and FRIAUF, J. B.
 1930. Explosions in detonating gas mixtures. I. Calculation of rates of explosion in mixtures of hydrogen and oxygen and the influence of rare gases. *J. Am. Chem.* 52:3905-3920. [5]

LIAUGMINAS, R., BARTHEL, H. O., and STREHLOW, R. A.
 1973. Mach stem structure in exothermic systems. *Combust. Flame* 20:19-31. [348]

LIBOUTON, J. C., and van Tiggelen, P. J.
 1976. Influence of the composition of the gaseous mixture on the structure of detonation waves. *Acta Astronaut.* 3:759-769. [315]

BIBLIOGRAPHY

LIGHTHILL, M. J.
  1957. Dynamics of a dissociating gas. Part I. Equilibrium flow. *J. Fluid Mech.* 2:1-12. [102]

LU, P. L., DABORA, E. K., and NICHOLLS, J. A.
  1969. The structure of H_2-CO-O_2 detonations. *Comb. Sci. and Technol.* 1:65-74. [337]

LUNDSTROM, E. A., and OPPENHEIM, A. K.
  1969. On the influence of non-steadiness on the thickness of the detonation wave. *Proc. R. Soc. London Ser. A* 310:463-478. [335, 338, 340]

MADER, C. L.
  1961. *Detonation performance calculations using the Kistiakowsky-Wilson equation of state.* Los Alamos Scientific Laboratory Report LA-2613. [96]
  1965. *A study of the one-dimensional, time-dependent reaction zone of nitromethane and liquid TNT.* Los Alamos Scientific Laboratory Report LA-3297. [10, 227, 288]
  1967a. *FORTRAN BKW.* Los Alamos Scientific Laboratory LA-3704. [32]
  1967b. *The time-dependent reaction zones of ideal gases, nitromethane, and liquid TNT.* Los Alamos Scientific Laboratory Report LA-3764. [10, 278, 288, 289, 290]

MADER, C. L., and CRAIG, B. G.
  1975. *Nonsteady-state detonations in one-dimensional plane, diverging, and converging geometries.* Los Alamos Scientific Laboratory Report LA-5865. [69]

MADER, C. L., and GAGE, W. R.
  1967. *FORTRAN SIN — a one-dimensional hydrodynamics code for problems which include chemical reaction, elastic-plastic flow, spalling, and phase transitions.* Los Alamos Scientific Laboratory Report LA-3720. [24]

MALLARD, E., and Le CHATELIER, H.
  1881. On the propagation velocity of burning in gaseous explosive mixtures. *C. R. Hebd. Sceances Acad. Sci.* 93:145-148. [2]

MALLORY, H. D.
  1967. Turbulent effects in detonation flow: diluted nitromethane. *J. Appl. Phys.* 38:5302-5306. [356, 358]
  1976. Detonation reaction time in diluted nitromethane. *J. Appl. Phys.* 47:152-156. [357]

MANSON, N.
  1946. On the structure of so-called helical detonation waves in gaseous mixtures. *C. R. Hebd. Sceances Acad. Sci.* 222:46-51. [321]
  1947. *Propagation des detonations et des deflagrations dans les melangees gaseux* (Propagation of detonations and deflagrations in gaseous mixtures). Paris: L'Office National d'Etudes et des Recherches Aeronautiques et L'Institut Francais des Petroles. [321]

BIBLIOGRAPHY

1958. A new relation in the hydrodynamic theory of detonation waves. *C. R. Hebd. Sceances Acad. Sci.* 246:2860-2864. [6, 69, 71]

MANZHALEI, V. I., and MITROFANOV, V. V.
   1973. The stability of detonation shock waves with a spinning configuration. *Fiz. Goreniya Vzryva* 9:703-710. [331]

MANZHALEI, V. I., MITROFANOV, V. V., and SUBBOTIN, V. V.
   1974. Measurement of inhomogenieties of a detonation front in gas mixtures at elevated pressure. *Fiz. Goreniya Vzryva* 10:102-110. [342]

MORDUCHOW, M., and PAULLAY, A. J.
   1971. Stability of normal shock waves with viscosity and heat conduction. *Phys. Fluids* 14:323-331. [253]

NUNZIATO, J. W., KENNEDY, J. E., and HARDESTY, D. R.
   1976. Modes of shock wave growth in the initiation of explosives. In *Sixth Symposium (International) on Detonation*, pp. 47-59. Washington, D.C.: Office of Naval Research — Department of the Navy (ACR-221). [102]

OPPENHEIM, A. K., and ROSCISZEWSKI, J.
   1963. Determination of the detonation wave structure. In *Ninth Symposium (International) on Combustion*, pp. 424-434. New York: Academic Press. [192, 199, 200]

OPPENHEIM, A. K., SMOLEN, J. J., KWAK, D., and URTIEW, P. A.
   1970. On the dynamics of shock intersections. In *Fifth Symposium (International) on Detonation*, pp. 119-136. Washington, D.C.: Office of Naval Research — Department of the Navy (ACR-184). [303, 304, 358, 359, 360]

OPPENHEIM, A. K., URTIEW, P. A., and LADERMAN, A. J.
   1964. Vector polar method for the evaluation of wave interaction processes. *Archiwum Budowy Maszyn* 11: 441-495. [242]

PACK, D. C., and WARNER, F. J.
   1965. Whitham's shock-wave approximation applied to the initiation of detonation in solid explosives. In *Tenth Symposium (International) on Combustion*, pp. 845-853. Pittsburgh: The Combustion Institute. [102]

PEEK, H. M., and Thrap, R. G.
   1957. Gaseous detonations in mixtures of cyanogen and oxygen. *J. Chem. Phys.* 26:740-745. [5, 59]

PERSSON, P. A., and BJARNHOLT, G.
   1970. A photographic technique for mapping failure waves and other instability phenomena in liquid explosives detonation. In *Fifth Symposium (International) on Detonation*, pp. 115-118. Washington, D.C.: Office of Naval Research — Department of the Navy (ACR-184). [355]

PERSSON, P. A., and PERSSON, G.
   1976. High resolution photography of transverse wave effects in the detonation of condensed explosives. In *Sixth Symposium (International) on Detonation*, pp. 414-425. Washington, D.C.: Office of Naval Research — Department of the Navy (ACR-221). [351, 354, 359, 361]

BIBLIOGRAPHY

PETRONE, F. J.
1968. Validity of the classical detonation wave structure for condensed explosives. *Phys. Fluids* 11: 1473-1478. [73]

PUKHNACHEV, V. V.
1963. The stability of Chapman-Jouguet detonations. *Dokl. Akad. Nauk. SSSR.* (Phys. Sect.) 149:798-801. [54, 237, 250]

RICHARDS, G. T.
1965. A numerical calculation of two-dimensional steady-state detonations by the method of characteristics. Lawrence Radiation Laboratory Report UCRL-14228. [203, 212]

RIVARD, W. C., VENABLE, D., FICKETT, W., and DAVIS, W. C.
1970. Flash x-ray observation of marked mass points in explosive products. In *Fifth Symposium (International) on Detonation*, pp. 3-11, Washington, D.C.: Office of Naval Research — Department of the Navy (ACR-184). [69]

SAINT-CLOUD, J. P., GUERRAUD, Cl., BROCHET, C., and MANSON, N.
1972. Some properties of very unstable detonations in gaseous mixtures. *Astronaut. Acta* 17:487-498. [312]

SCHOTT, G. L.
1960. Kinetic studies of hydroxyl radicals in shock waves. III. The OH concentration maximum in the hydrogen-oxygen reaction. *J. Chem. Phys.* 32:710-716. [62]

1965a. Structure, chemistry, and instability of detonation in homogeneous, low-density fluids — gases. In *Fourth Symposium (International) on Detonation*, pp. 67-77. Washington, D.C.: Office of Naval Research — Department of the Navy (ACR-126). [56, 329]

1965b. Observation of the structure of spinning detonation. *Phys. Fluids* 8:850-865. [10, 311, 328, 329, 332]

SHCHELKIN, K. I.
1959. Two cases of unstable combustion. *Zh. Eksp. Teor. Fiz.* 36:600-606. [9, 245, 250]

1965. Unidimensional instability of detonation. *Dokl. Akad. Nauk SSSR* (Phys. Chem. Sect.) 160:1144-1146. [245, 250]

SHCHELKIN, K. I. and TROSHIN, Ya. K.
1963. *Gazodinamika Goreniya* (Gasdynamics of combustion). Moscow: Izdatel'stvo Akademii Nauk SSSR. (English translation: NASA-TT-F-23, 1964). [242]

SHIELDS, Paul C.
1964. *Linear algebra*. Reading: Addison-Wesley. [120]

SICHEL, M.
1966. A hydrodynamic theory for the propagation of gaseous detonations through charges of finite width. *AIAA J* 4:264-272. [212]

SKIDMORE, I. C.
1967. The physics of detonation. *Science Prog, Oxf.* 55:239-257. [25, 28]

BIBLIOGRAPHY

SOLOUKHIN, R. I.
- 1965. Multiheaded structure of gaseous detonation. *Combust. Flame* 9:51-58. [326]
- 1966. *Shock waves and detonation in gases.* Baltimore: Mono Book Co. [147, 302]

SOLOUKHIN, R. I., and BROCHET, C.
- 1972. The development of instabilities in a shocked exothermic gas flow. *Combust. Flame* 18:59-64. [343]

SPALDING, D. B.
- 1963. Contribution to the theory of the structure of gaseous detonation waves. In *Ninth Symposium (International) on Combustion,* pp. 417-423. New York: Academic Press. [192]

STANYUKOVICH, K. P.
- 1955. *Neustanovivshiesya dvizheniya sploshnoi sredy (Unsteady motion of continuous media).* Moscow: Gostekhizdat. (English translation: London: Pergamon Press, 1960.) [6, 69]

STEEL, G. B., and OPPENHEIM, A. K.
- 1966. *Experimental study of the wave structure of marginal detonation in a rectangular tube.* University of California Berkeley Report AS66-4. [310, 331, 332, 338]

STRATTON, J. A.
- 1941. *Electromagnetic theory,* pp. 333-340. New York: McGraw Hill. [96]

STRAUSS, W. A., and SCOTT, J. N.
- 1972. Experimental investigation of the detonation properties of hydrogen-oxygen and hydrogen-nitric oxide mixtures at initial pressures up to 40 atmospheres. *Combust. Flame* 19:141-143. [59]

STREHLOW, R. A.
- 1968a. *Fundamentals of combustion.* Scranton: International Textbook. [11, 31, 76, 147, 151]
- 1968b. Gas phase detonations: recent developments. *Combust. Flame* 12:81-101. [56, 299, 306, 308]
- 1969. The nature of transverse waves in detonations. *Astronaut. Acta* 14:539-548. [306, 307, 342, 349]
- 1970. Multi-dimensional detonation wave structure. *Astronaut. Acta* 15:345-357. [340, 342, 346, 347, 348]

STREHLOW, R. A., ADAMCZYK, A. A., and STILES, R. J.
- 1972. Transient studies on detonation waves. *Astronaut. Acta* 17:509-527. [304, 315, 325, 326, 342]

STREHLOW, R. A., and Biller, J. R.
- 1969. On the strength of transverse waves in gaseous detonations. *Combust. Flame* 13:577-582. [304, 327, 341, 342]

STREHLOW, R. A., and Cohen, A.
- 1962. Initiation of detonation. *Phys. Fluids* 5:97-101. [323]

STREHLOW, R. A., and Crooker, A. J.
- 1974. The structure of marginal detonation waves. *Acta Astronaut.* 1:303-315. [302, 308, 327, 328, 333, 336, 338, 342]

BIBLIOGRAPHY

STREHLOW, R. A., and ENGEL, C. D.
1969. Transverse waves in detonation II: structure and spacing in H_2-O_2, C_2H_2-O_2, C_2H_4-O_2, and CH_4-O_2 systems. *AIAA J.* 7:492-496. [313, 314]

STREHLOW, R. A., and FERNANDES, F. D.
1965. Transverse waves in detonations. *Combust. Flame* 9:109-119. [267, 269, 270]

STREHLOW, R. A., LIAUGMINAS, R., WATSON, R. H., and EYMAN, J. R.
1967. Transverse wave structure in detonations. In *Eleventh Symposium (International) on Combustion*, pp. 683-691. Pittsburgh: The Combustion Institute. [323, 324]

STREHLOW, R. A., MAURER, R. E., and Rajan, S.
1969. Transverse waves in detonations I: spacing in the hydrogen-oxygen system. *AIAA J.* 7:323-328. [274, 349]

STREHLOW, R. A., and SALM, R. J.
1976. The failure of marginal detonations in expanding channels. *Acta Astronaut.* 3:983-994. [325, 326, 332]

SUBBOTIN, V. A.
1975a. Two kinds of transverse structures in multi-front detonations. *Fiz. Goreniya Vzryva* 11:96-102. [335]
1975b. Collisions of transverse waves in gas detonations. *Fiz. Goreniya Vzryva* 11:486-491. [335]

SWAN, G. W., and FOWLES, G. R.
1975. Shock wave stability. *Phys. Fluids* 18:28-35. [253, 254]

TAKAI, R., YONEDA, K., and HIKITA, T.
1974. Study of detonation wave structure. In *Fifteenth Symposium (International) on Combustion*, pp. 69-78. Pittsburgh: The Combustion Institute. [306, 336, 338, 339, 349]

TAYLOR, J.
1952. *Detonation in condensed explosives.* London: Clarendon Press. [13, 31]

TAYLOR, G. I., and TANKIN, R. S.
1958. Gas dynamical aspects of detonations. In *Fundamentals of Gas Dynamics*, ed. H. W. Emmons, pp. 622-685. High Speed Aeronautics and Jet Propulsion, vol. 3. Princeton: Princeton University Press. [32]

THOMPSON, P. A.
1972. *Compressible fluid dynamics.* New York: McGraw Hill. [91]

TOPCHIAN, M. E., and ULJANITSKIJ, V. Yu.
1976. On the nature of one of the modes of the spinning detonation istability. *Acta Astronaut.* 3:771-779. [331]

TRUESDELL, C.
1969. *Rational thermodynamics.* New York: McGraw-Hill. [92]

TSUGE, S., FURUKAWA H., MATSUKAWA, M., and NAKAGAWA, T.
1970. On the dual property and the limit of hydrogen-oxygen free detonation waves. *Astronaut. Acta* 15:377-386. [212, 228]

TSUGE, S.
1971. The effect of boundaries on the velocity deficit and the limit of gaseous detonations. *Comb. Sci. Technol.* 3:195-205. [228, 229]

BIBLIOGRAPHY

URTIEW, P. A.
- 1969. Reflections of wave intersections in marginal detonations. *Astronaut. Acta* 15:335-343. [336, 337]
- 1975. From cellular structure to failure waves in liquid detonation. *Combust. Flame* 25:241-245. [351]
- 1976. Idealized two-dimensional detonation waves in gaseous mixtures. *Acta Astronaut.* 3:187-200. [304, 340, 341]

URTIEW, P. A., and KUSUBOV, A. S.
- 1970. Wall traces of detonation in nitromethane-acetone mixtures. In *Fifth Symposium (International) on Detonation*, pp. 105-114. Washington, D.C.: Office of Naval Research — Department of the Navy (ACR-184). [360]

URTIEW, P. A., KUSUBOV, A. S., and DUFF, R. E.
- 1970. Cellular structure of detonation in nitromethane. *Combust. Flame* 14:117-122. [358, 359, 360]

VAN ZEGGEREN, F., and STOREY, S. H.
- 1970. *The computation of chemical equilibria*. London: Cambridge University Press. [31]

VASILIEV, A. A., GAVRILENKO, T. P., and TOPCHIAN, M. E.
- 1972. On the Chapman-Jouguet surface in multi-headed gaseous detonations. *Astronaut. Acta* 17:499-502. [326]
- 1973a. Chapman-Jouguet condition for real detonation waves. *Fiz. Goreniya Vzryva* 9:309-315. [60]
- 1973b. Pressure in the detonation front in gases. *Fiz. Goreniya Vzryva* 9:710-716. [60]

VINCENTI, W. G., and KRUGER, C. H.
- 1965. *Introduction to physical gas dynamics*. New York: John Wiley and Sons. [76, 147]

VMT — see Voitsekhovskii, Mitrofanov, and Topchian.

VOITSEKHOVSKII, B. V., MITROFANOV, V. V., and TOPCHIAN, M. E.
- 1963. *Struktura fronta detonatsii v gazakh* (Structure of a detonation front in gases). Novosibirsk: IZD-VO Sibirsk, Otdel. Akad. Nauk SSST. [English translation: Wright-Patterson Air Force Base Report FTD-MT-64-527 (AD-633, 821), 1966]. [10, 302, 311, 328, 329, 330, 332]
- 1969. Structure of the detonation front in gases. *Fiz. Goreniya Vzryva* 5:385-395. [322, 342, 334, 335, 336, 338]

VON NEUMANN, J.
- 1942. Theory of detonation waves. In *John von Neumann, collected works. Vol. 6.*, ed. A. J. Taub. New York: Macmillan. [4, 8, 13, 133, 174]

WAGMAN, P.
- 1968. *Selected values of chemical thermodynamic properties*. U.S. National Bureau of Standards Technical Note 270-3. [73]

WATSON, R. W.
- 1970. Dark waves in liquid explosives: their role in detonation failure. In *Fifth Symposium (International) on Detonation*, pp. 169-174, Washington, D.C.: Office of Naval Research — Department of the Navy (ACR-184). [355]

WECKEN, F.
 1959. *A method for the determination of the detonation pressure of condensed explosives.* Institute Franco - Allemand de Recherches de Saint Louis Report 4a/59. [71]
 1965. Non-ideal detonation with constant lateral expansion. In *Fourth Symposium (International) on Detonation,* pp. 107-116. Washington, D.C.: Office of Naval Research — Department of the Navy (ACR-126). [227]

WEGENER, P. P., ed.
 1969. *Non-equilibrium flows,* parts I and II. New York: Marcel Decker. [76]

WHITE, D. R.
 1961. Turbulent structure of gaseous detonation. *Phys. Fluids* 4:465-480. [5, 56, 62, 63, 67, 301]
 1967. Density induction times in very lean mixtures of D_2, H_2, C_2H_2, and C_2H_4, with O_2. In *Eleventh Symposium (International) on Combustion,* pp. 147-154. Pittsburgh: The Combustion Institute. [301]

WHITE, D. R., and CARY, K. H.
 1963. Structure of gaseous detonation. II. Generation of laminar detonation. *Phys. Fluids* 6:749-750. [297]

WHITHAM, G. B.
 1974. *Linear and nonlinear waves.* New York: John Wiley and Sons. [102, 233, 326]

WILKINS, M. L.
 1965. The use of one- and two-dimensional hydrodynamic machine calculations in high explosives research. In *Fourth Symposium (International) on Detonation,* pp. 519-526. Washington, D.C.: Office of Naval Research — Department of the Navy (ACR-126). [203]
 1969. *Calculations of elastic-plastic flow.* Lawrence Livermore Laboratory Report UCRL-7322, Revision I. [25]

WILKINS, M., FRENCH, J., and GIROUX, R.
 1962. *A computer program for calculating one-dimensional hydrodynamic flow; KO code.* Lawrence Radiation Laboratory Report UCRL-6919. [24]

WOOD, W. W.
 1961. Existence of detonation for small values of the rate parameter. *Phys. Fluids* 4:46-60. [192]
 1963. Existence of detonations for large values of the rate parameter. *Phys. Fluids* 6:1081-1090. [192, 197, 198, 199]

WOOD, W. W., and FICKETT, W.
 1963. Investigation of the Chapman-Jouguet Hypothesis by the "inverse method." *Phys. Fluids* 6:648-652. [6, 69, 71]

WOOD, W. W., and KIRKWOOD, J. G.
 1954. Diameter effect in condensed explosives. The relation between velocity and radius of curvature in the detonation wave. *J. Chem. Phys.* 22:1920-1924. [204, 210, 226]
 1957. Hydrodynamics of a reacting and relaxing fluid. *J. Appl. Phys.* 28:395-398. [76]

BIBLIOGRAPHY

WOOD, W. W., and PARKER, F. R.
    1958. Structure of a centered rarefaction wave in a relaxing gas. *Phys. Fluids* 1:230-241. [114]

WOOD, W. W., and SALSBURG, Z. W.
    1960. Analysis of steady-state supported one-dimensional detonations and shocks. *Phys. Fluids* 3:549-566. [94, 146, 176, 190, 224]

ZAIDEL', R. M.
    1961. The stability of detonation waves in gaseous mixtures. *Dokl. Akad. Nauk SSSR* (Phys. Chem. Sect.) 136:1142-1145. [245]

ZAIDEL', R. M., and ZELDOVICH, Ya. B.
    1963. One-dimensional instability and attenuation of detonation. *Zh. Prikl. Mekh. Tekh. Fiz.* 1963(6):59-65. (English Translation: Wright-Patterson Air Force Base, Dayton, Ohio, FTD-MT-64-66, p. 85.) [245, 250]

ZELDOVICH, Ya. B.
    1940. On the theory of the propagation of detonation in gaseous systems. *Zh. Eksp. Teor. Fiz.* 10:542-568. (English translation: NACA TM 1261, 1960.) [4, 13, 66]

ZELDOVICH, Ya. B., and KOMPANEETS, S. A.
    1955. *Teoriya detonatsii (Theory of detonations)*. Moscow: Gostekhizdat. (English translation: New York: Academic Press, 1960.) [13]

Index

Acceleration wave, 92
Acoustic modes in tube or channel, 315-323
Acoustic ray theory, waves in reaction zone, 267-275
Acoustic wave theory, wave trapping, 269-271
Activation energy, 45

Binary mixture of polytropic gases, 105-108
Boron explosive compounds, 96
Boundry layer of gas detonation, 66-68
Bow shock, detachment locates sonic surface, 60-61, 326-327

Chapman-Jouguet theory, definition and description, 3-4, *See also* CJ point; Simplest theory
Characterization of explosive, 38-40
Chemical equilibrium, condition for, 107, 139
Chemical reactions: as source term in equations, 8; thermicity coefficient, 8; in language of fluid mechanics, 79-88; independent set, 80-82, 86; arbitrary independent set, 85-86, 118-121; accessible regions of λ-space, with examples, 86-88; development of equations, 114-121; ozone, 117-121; hydrogen, 151-154
Central stream tube, 199-229
CJ point: definition, 19; conditions at, 19-20; frozen and equilibrium, 139-140
Complex theory: discussion, 6-9; table of cases treated, 136
Computer codes: hydrodynamics, 24-25; equations of state, 32
Condensed-phase explosives, detonation formulas, 54
Constant-volume detonation, 17, 45
Curved-front detonation, 199-229

Critical points: in phase plane, 7; definition, 167; node and saddle, 180-181

D-discussion: definition, 14; simplest theory, 21; exclusion of $D < D_J$, 142
Derivatives: subscript notation, 88; dot notation, 90
Detonation: power comparison, 1; in complicated flow configurations, 2
Detonation theory, definition of terms, 13-15
Detonation velocity: eigenvalue, 8, 133, 135, 168-173, 189-190; CJ, dependence on energy and density, constant-γ, 26, constant-β, 28, Jones's relation, 30, Kamlet 34-35; effect of initial state variation, derivation of equations, 69-71, experiments, 71-74
Divergent-flow detonation: theory, 199-229; applications and results, 226-229

Eigenvalue detonation: description, 133, 135; with two irreversible reactions, 168-173; numerical comparison with CJ, 189-190
Electric generating capacity of US, 1
Elementary reaction, 81
Entropy production equation, 94
Equations of motion: with chemical reaction, 76-102; discussion, 89-93; characteristic form, 91-93; steady, 98-100; for slab, cylinder, and sphere, 124-126
Equations of state: for applications, 24-42; constant-γ, 25-26, 52-54; first-order expansion about reference curve, 26-27; reference curve for, 26-27; constant-β, 27-28; constant-α, 28; general constant-γ, 28; JWL, 28-29; thermal properties, 29,

383

INDEX

121-124; calibration, 30; choice for practical problems, 30-31, 41-42; with explicit chemistry, 31-35; Kistiakowsky-Wilson, 32; computer codes for, BKW, RUBY, TIGER, 32; calibration, with explicit chemistry, 33-34; LJD, 33; JCZ, 33; Kamlet's fit to BKW, 34-35; accessible regions of the p-v plane, 40-41; warnings about use for applications, 41-42; minimum assumptions, 88-89; well-behaved, 89; simple material models, 102-108; consequences of assumptions, 137-139; equilibrium, 139-140
Eta η, 100
Experiments: plane steady detonation as a limit, 1-2
--gases, comparison with theory: discussion; 4, 5-6, 56-68; summary, 58-59; detonation velocity, 59; x-ray densitometry, 60; characteristic velocity, 60-61; pressure, 60-61; Mach number, 61-62; CO-O concentration product, 62-64; review of evidence, 65-68
--liquids and solids, comparison with theory: 6, 68-74; extrapolation to plane, steady detonation, 68-69
Extended theory: discussion, 6-9; table of cases treated, 136
Euler equations. See Equations of motion

Failure, role of transverse waves, 326
Failure waves in liquid and solid explosives, 351, 353-355
Fickett-Jacobs cycle, 35-38
Fine structure: effect on comparison with theory, 57; spark interferograms, 62-63
Final state: definition, 3, 13; simplest theory, 21
Flow equations. See Equations of motion

Galloping detonation: definition, 231; numerical calculations, 276-288; ahead of blunt projectile, 311-312; comparison of experiment and theory, 343-346
Gruneisen coefficient, 27, 28, 29

History, 2-11
Hugoniot curves: definition, 17, 99; polytropic gas, formula, 18; partial reaction, 43-44; frozen and equilibrium, 110-111, 139-140; state properties on, 138; envelope of crossing, 174; deflagration branch, 239-243
Hydrodynamic stability: history and review, 9-11; definition, 230; general theory, 233-238; summary of results, 246-252; analysis with nonlinear terms, 259-267
Hydrogen/oxygen: numerical example, 151-162; divergent detonation in, 228

Initiation, formation of transverse waves, 323-324
Instability: experimental, 10; two-dimensional calculations, 288-290; experimental onset with heat of reaction, 343
Integral curve: definition, 134; properties, 145; for two reversible reactions, 178-183

Jacobs's detonation engine, 36

Laboratory reference frame, 97, 100
Lambda λ, 17, 43, 77, 79-88, 83
Lighthill gas, 102, 108
Locus of tangents, 20, 138
Longitudinal instability: galloping detonation, 231; in square wave model, 275-277; numerical calculations, 276-288

Mach stem, in transverse waves, 294-296, 298-300, 303-306
Marginal detonation, 327-340
Master equation, 78, 91
Mole change: polytropic gases, 105-108; pathological detonation, 174-176

Navier-Stokes equations, 192-195
Nitromethane: initial state variation, 71-73; pressure measurement, 72-73; failure waves, 353-355, 358-360; transverse waves, 356-357

Ordinary detonation, 327-328, 341-342
Oscillatory detonation: approximate theories, 258-276; numerical calculations, 276-288
Ozone: reactions, 57-58, 117-120; effect of transport properties, 199-200; divergent detonation in, 228-229

Pathological detonation, von Neumann's, caused by mole decrement, 133, 174-176
Pathological locus, with arbitrary number of reactions, 191

384

INDEX

Pathological point, 172, 176, 179-183
Phase portrait, 164
Piston problem, 14, 21-24
Polytropic gas: detonation formulas, 52-53; mixture with single heat capacity, 103-105; mixture with different heat capacities, 105-108
Precursor wave, 96
Pressure-distance profiles: two irreversible reactions, 173; two reversible reactions, 187-188
Progress variable λ, 17, 43, 77, 79-88, 83

Quasi-one-dimensional detonation, 224-226
Quasistatic cycle for detonation, 35-38

Radial velocity derivative, approximations, 208-212
Rankine-Hugoniot relations, 16-18, 99
Rarefaction wave: in nonreacting gas, joined to reaction zone, 23-24; in reactive mixture, 114; joining to reaction zone, 143; joining to reaction zone when rate is reversible, 146-147
Rayleigh line: definition and derivation, 16-17, 99; state properties on, 138
Reaction rate: state dependence assumptions, 44, 88; Arrhenius, 45, 102, 107-108; hypothetical solid explosive, 47; transformation to language of fluid dynamics, 82-86; simple forms, 105; irreversible, 108; combined for several elementary reactions, 116-117; reversible, effects on properties, 139-140; magnitude of effect of reversibility, 147-151
Reaction space vectors, 97
Reaction zone: computed example for gas, 45-50; temperature maximum in, 46, 51; computed example, solid, 46-51

Separatrix, 180-181
Shock-change equation: discussion, 101-102; derivation, 131-132
Shock reference frame, 97, 100
Shock polar, for triple point matching, 303-304
Shock wave: in reactive mixture, 110-114; diffuse, 112, 114; hydrodynamic stability, 252-258; breakup into two waves, 255-258
Sigma σ, 78, 91, 95-96, 107
Simplest theory: description, 16-24; application, 24-42

Slapping waves, 300
Solar power level, 1
Sonic locus: definition, 20, 138; in phase plane, 166-167
Sonic parameter η, 100
Sound speed: condition that it be real, 89
--frozen and equilibrium: definition, 95-96; in mixture with a single C_p, 104; in a mixture with different values of C_p, 108; effect on propagation of disturbances, 109; derivation of relationship between, 126-130; magnitude of the effect of reversibility, 147-151
Sound waves: in reactive mixture, 110; amplification of, 110, 269-271
Specific heats, from incomplete equation of state, 123-124
Spinning detonation, 4, 328-332
Square-wave model: stability of, 231; hydrodynamic stability, 239-246; conditions for existence of perturbation solutions, 244; galloping detonation in, 275-277, 344
Stability, reaction without flow, 109-110
Stability analysis, 230
Steady reference frame, 97, 100
Steady solutions, 97-100
Strong point, 3, 19

Tangency condition, 19-20
Taylor wave, 23
Temperature: maximum in reaction zone, 46, 51; from incomplete equation of state, 121-123; detonation, dependence on initial density, 352
Tetranitromethane, reaction zone, 350
Thermicity σ: definition, 78, 91; alternate forms for, 95; polytropic gases, 96, 107; volume change contribution to, 96
TNT: initial state variation, 73-74; edge failure waves, 355-356
Transport effects, 191-199
Transverse waves: nonlinear stability analysis, 263-267; amplification, 269-271; two dimensional calculations, 288-290; overview, 291-306; definition of terms, 298-300, 327; experimental methods, 301-303; macroscopic properties, 306-327; regular and irregular structures, 306-309; acoustic modes in confining tube or channel, 315-323; life cycle and habitat, 323-326; details of structure, 327-342; strong and weak structures, 327-328; in

385

INDEX

rectangular tubes, 332-340; variation of properties through the cell, 338-340; comparison of theory and experiment, 342-350; in liquids and solids, 350-361; dark, not bright, in liquids, 352; discussion of, in liquids and solids, 360-361; interpretation of smear camera photographs of liquids, 361-363
--spacing: ray theory of, 271-275; hot spot theory of, 273-275; rough criteria for, 295; related to reaction zone, 313-315; comparison of theory and experiment, 346-350
--track entrance and exit angles: in gases, 305, 321-323, 341-342; in liquids and solids, 360-361
Triple point, in transverse waves, 295-300, 303-306

Viscous detonation, 191-199

Waves, forward and backward facing, 93
Weak final state, computed for ozone, 57-58
Weak point, 3, 19
Weak solution, 7-8, 133
Work done by explosive, 35-38

ZND model: definition, 4-5, 13-15; assumptions, 42; description, 42-51